D1074178

Molecular Similarity in Drug Design

Molecular Similarity in Drug Design

Edited by

P. M. Dean
Department of Pharmacology
University of Cambridge
Cambridge
UK

BLACKIE ACADEMIC & PROFESSIONAL
An Imprint of Chapman & Hall
London · Glasgow · Weinheim · New York · Tokyo · Melbourne · Madras

Published by
Blackie Academic and Professional, an imprint of Chapman & Hall,
Wester Cleddens Road, Bishopbriggs, Glasgow G64 2NZ

Chapman & Hall, 2–6 Boundary Row, London SE1 8HN, UK

Blackie Academic & Professional, Wester Cleddens Road, Bishopbriggs, Glasgow G64 2NZ, UK

Chapman & Hall GmbH, Pappelallee 3, 69469 Weinheim, Germany

Chapman & Hall USA, One Penn Plaza, 41st Floor, New York NY 10119, USA

Chapman & Hall Japan, ITP-Japan, Kyowa Building, 3F, 2-2-1 Hirakawacho, Chiyoda-ku, Tokyo 102, Japan

DA Book (Aust.) Pty Ltd, 648 Whitehorse Road, Mitcham 3132, Victoria, Australia

Chapman & Hall India, R. Seshadri, 32 Second Main Road, CIT East, Madras 600 035, India

---

First edition 1995

Typeset in 10/12pt Times by AFS Image Setters Ltd., Glasgow
Printed in Great Britain by St. Edmundsbury Press, Bury St. Edmunds, Suffolk

ISBN 0 7514 0221 4

A catalogue record for this book is available from the British Library
Library of Congress Catalog Card Number: 94-78528

∞ Printed on acid-free text paper, manufactured in accordance with ANSI/NISO Z39.48-1992 (Permanence of Paper)

Preface

The purpose of this book is to advance our methods of drug design by exploring novel concepts of molecular similarity. What appears to be similar to one mind may not necessarily be so to another. This lack of a conviction about the approximate nature of similarity, as opposed to identity, has plagued the whole field of similarity research. How similar are two different molecules? What methods can be used to quantify similarity? Once molecular structure can be precisely expressed, then vague and perhaps intuitive notions of similarity can be coded algorithmically; the validity of our ideas about molecular similarity may then be tested against sets of molecules. If similarity is to be used in drug design, then it has to be related to key features that are also associated with activity. Research into new methods for handling molecular similarity and molecular complementarity progress in parallel. The possibility of design from complementarity has become an exciting reality over the last decade and has focussed our attention on specific features related to ligand interaction with its binding site. That work has provided designers with a perfectly logical strategy for design. At the same time, lessons derived from complementarity-directed design can be applied as a check to the use of molecular similarity as a tool for drug design.

If molecular similarity is to have a serious impact on the way we go about drug design, then the theoretical methods have to be made accessible to the drug design profession. In this volume we are fortunate in having the participation of scientists from four pharmaceutical companies, Abbot Laboratories, Rhône-Poulenc Rorer, SmithKline Beecham Pharmaceuticals and Zeneca Pharmaceuticals; they outline their work at the cutting edge of molecular similarity usage. It is hoped that a mix from academe and industry will make this book readily accessible to a wide variety of readers.We have drawn together many strands from the molecular similarity field which hitherto have only been available in widely dispersed literature. It is hoped that this collection of material will help newcomers in the field to absorb what is known about molecular similarity, but at the same time there has to be an awareness of current problems in the application of these methods to drug design.

The first chapter is meant to be cautionary to prospective users of molecular similarity. What are the assumptions in assessing similarity between molecules binding to the same receptor site? What bearing have these assumptions on the design process? To what extent can similarity be used in automated *de novo* methods for design? The second chapter by Good, provides an extensive survey of all the molecular similarity indices currently in use and assesses their value. An illustration of the use of molecular similarity in QSAR is given to provide a feel for what can be achieved in a design problem by new statistical methods.

Chapters 3 and 4 by Leach and Perkins, respectively, probe, from different perspectives, the difficult problem of incorporating flexibility into similarity research. This is a combinatorial problem that can be treated in many ways. At the moment, considerable approximations have to be made for the problem to become tractable with current computing technology. Each of these avenues of exploration has to be carefully examined. Many different matches for similarity are possible within a set of molecules. Each conformation has to be assessed against all conformations of the other molecules; some measure of similarity within the set has to be derived before the processed data can be used for drug design.

Chapters 5 (Willett) and 6 (Mason) consider searching for molecular similarity in databases of chemical structures. Willett describes the algorithms for clustering databases of 2D structures and for similarity searching through databases of 3D structures. Mason gives an account of his experiences of similarity searching in conformationally flexible 3D databases. Emphasis is placed on the identification of new leads from pharmacophore searching. This gives rise to the idea of reducing a database to a number of scaffolds which may be used to span across binding sites. Thus it should be possible to optimize the search for new leads from a large database of known structures.

Masek in Chapter 7 examines the difficulty of determining molecular surface similarity by treating the surface as a skin with thickness. The problems here are to maximize the relative positional similarities of knobs and indentations in the surfaces. Matches between surface features are important in drug design to provide a tight fit to the receptor site. Chapter 8 by Livingstone and Salt explores the use of neural networks in the search for molecular similarity. This is an important new approach and is well described for newcomers to the field. Preliminary work on the application of neural networks to different similarity problems is given to enable the reader to gauge the value of this approach for drug design problems.

The ultimate method for assessing molecular similarity, or complementarity, is by comparing the electronic charge densities of the molecules. Three different approaches to this problem are outlined in Chapters 9–11. Popelier sketches out how charge density may be calculated and utilized by a topological description for similarity and complementarity. Mezey develops methods to describe the topology of molecular shape by shape-group methods and their expression numerically by shape codes. Whitley and Ford illustrate how these novel methods for topological depiction of molecular shape can be employed in drug design by shape-graph descriptions.

The final chapter, by Kim, reviews our current knowledge of comparative molecular field analysis (CoMFA) applied to drug design. This chapter is vital for anyone contemplating using the CoMFA methodology. Its approach is systematic and takes the reader through every step of the procedure and provides all the background material which would enable a user to chart a way through the minefield of potential difficulties that can be encountered in CoMFA. At each step of the procedure there are warnings about pitfalls that can trip the unwary. With proper usage, the CoMFA method can lead to the design of novel candidate structures with predicted activity.

Drug design is a multifaceted problem; molecular similarity is one important tool that can be used to provide new ideas for design. Much has already been achieved in this field but our knowledge is far from complete. It is anticipated that this book will provide a stimulus to further research in molecular similarity and at the same time will be of practical use in helping with day-to-day drug design problems.

Finally, I would like to acknowledge The Wellcome Trust for the encouragement and financial support given to me over many years.

P.M.D

Contributors

P.M. Dean	Drug Design Group, Department of Pharmacology, University of Cambridge, Tennis Court Road, Cambridge, CB2 1QJ, UK
M. Ford	School of Biological Sciences, University of Portsmouth, Portsmouth, PO3 3NP, UK
A.C. Good	Department of Pharmaceutical Chemistry, University of California, 513 Parnassus Avenue, San Francisco, CA 94143-0446, USA
K.H. Kim	Department of Structural Biology, Pharmaceutical Division, Abbot Laboratories, Abbot Park, IL 60064, USA
A.R. Leach	Department of Chemistry, University of Southampton, Southampton, SO9 5NH, UK
D.J. Livingstone	SmithKline Beecham Pharmaceuticals, The Frythe, Welwyn, Hertfordshire, AL6 9AR, UK
B.B. Masek	Department of Medicinal Chemistry, Zeneca Pharmaceuticals, 1800 Concord Pike, Willmington, DE 19897, USA
J.S. Mason	Computer-Aided Drug Design, Rhône-Poulenc Rorer, Collegevile, PA 19426, USA
P.G. Mezey	Mathematical Chemistry Research Unit, Department of Chemistry, University of Saskatchewan, Saskatoon, S7N 0W0, Canada
T.D.J. Perkins	Drug Design Group, Department of Pharmacology, University of Cambridge, Tennis Court Road, Cambridge, CB2 1QJ, UK
P.L.A. Popelier	University Chemical Laboratory, Lensfield Road, Cambridge, CB2 1EW, UK
D.W. Salt	School of Mathematical Studies, University of Portsmouth, Mercantile House, Hampshire Terrace, Portsmouth, PO1 2EG, UK
D. Whitley	School of Mathematical Studies, University of Portsmouth, Mercantile House, Hampshire Terrace, Portsmouth, PO1 2EG, UK
P. Willett	Department of Information Studies, University of Sheffield, Western Bank, Sheffield, S10 2TN, UK

Contents

1 Defining molecular similarity and complementarity for drug design

P.M. DEAN

1.1 Introduction

The ultimate objective in molecular similarity research applied to drug design is to take a set of flexible and dissimilar active molecules, extract the features of similarity between them and use that information to design novel dissimilar molecules with biological activity. This book attempts to draw together many different strands of research into molecular similarity which have a bearing on the goal in mind.

Molecular similarity is a complex notion that can only be described with reference to the immediate use for which it is intended. Similarity encompasses bonding patterns, atomic positions, conformation, shape and spatial disposition of molecular properties. The reverse of similarity is complementarity. In between lies molecular dissimilarity. Often a molecular designer may have ostensibly dissimilar molecules but may wish to extract features from them which show some similarity. The impetus for molecular similarity work comes primarily from molecular design; the pharmaceutical industry is probably the single greatest user of these new methods. The objective of this chapter is to define various aspects of similarity, and complementarity, and point towards their eventual usage in the exciting, newly evolving, methods of automated *de novo* drug design.

All notions of similarity involve perception of patterns and subsequent attempts to classify the patterns. What patterns we wish to search for, and how we classify the information within them, are the fundamental problems in molecular similarity research. Problems in molecular design dictate which pattern to focus on; methods for the quantification of that pattern are intricately linked to usage. In general, we can divide all similarity research methods into those which are based on discrete properties and those which are founded on continuous properties. For example, a pharmacophore may be defined by three or more atoms separated by a set of distances; there may be some tolerances on the distances but the general pattern is discrete. A searching algorithm then finds only those structures from a database that satisfy the pharmacophore query. In a continuous space problem, the user may wish to optimize the overlap of a continuous field such as the electrostatic potential projected onto defined molecular surfaces. The result can then be expressed by a similarity coefficient.

Complementarity descriptions are equivalent to trying to express the relationship of the fit between a cast and its mould. If similarity can be expressed by a correlation coefficient lying between 0 and 1, then complementarity can be related on the same scale but with coefficients lying between 0 and −1. Ideally, if we knew the structure of a pharmacological site at an atomic level, drug design would proceed rationally from notions of complementarity and chemical knowledge. Studies of complementarity can throw light on the value of similarity work as a tool for drug design. The intriguing question that arises is: to what extent can similarity research lead towards a description of its complement? Can approximate site maps be built up from similarity methods applied to a group of dissimilar molecules acting at the same site?

1.1.1 Molecular similarity: a basic tool for drug design

Although molecular similarity can be defined rigorously for each problem, there are many difficulties in the application of this to drug design. Firstly, the similarity between two structures has to be optimized with respect to the features being examined. If the features are to be optimized stepwise, then the optimum superposition may differ for different features. If multiple features are to be considered simultaneously, how are the features to be weighted? Conformational flexibility in the molecular structures adds another layer of complexity to the analysis of molecular similarity. Secondly, if molecular structures can be superposed satisfactorily, how are we to create novel structures with similarity optimized to a single representation of the set of original superposed structures? Structure generation *de novo*, within the bounds of some hypothetical shell, is a combinatoric problem involving the generation of 3D molecular graphs, atom placement on the graphs and conformational restriction.

1.1.2 Molecular shape similarity and complementarity

Molecular shape is determined by atom positions, but what elements of molecular structure do we use to compare shapes? Identity in atom positions, although easy to observe, is not necessarily the most important feature in molecular similarity usage. Dissimilar molecules may have quite remarkably similar overall shapes. Families of protein molecules may possess little homology in their amino acid sequence but nevertheless have approximately the same shape. Ligand binding sites can accommodate significant degrees of dissimilarity in ligand structure. The HIV protease peptide binding site is probably the best studied example to date. Here we have good X-ray crystallographic data for numerous inhibitors bound to the site (Appelt, 1993). The site is very large and made from 10 binding pockets and cleaves eight different sequences; each sequence contains 10 residues. There is a wide structural variation in the eight sequences. Small structural modifications to these sequences, particularly in the scissile bond region can produce effective inhibitors of the protease (Clare, 1993). Disregarding the medical importance of the HIV protease site, the complexity of its structure will provide a very rich, but challenging, test set for detailed studies of the link between molecular similarity and complementarity. Natural substrates for the HIV protease site are quite flexible structures; how much flexibility is needed by a ligand to enter the site? Can ligands with rigidified shapes, but with complementarity to the site be made with high affinity?

The degree of molecular shape change that is possible in a molecule is related to the number and position of the rotatable bonds. A measure of this change can be obtained from the distance matrix of the molecule: the complexity of the change C for j rotatable bonds is given by

$$C = \prod_{h=1}^{j} k_h n r_h \tag{1.1}$$

where the factor $k_h n$ is the product of the number of rows (k_h) and the number of columns (n) of the distance matrix which are changed by a single rotation step round the rotatable bond r_h. This expression can be expanded to accommodate ring flexion as well (Dean, 1993).

1.1.3 *Molecular electrostatic potential similarity*

Much effort has been expended on the development of methods to study molecular similarity between electrostatic potentials. Three-dimensional displays of potential have provided an enormous impetus to this work. How good are the comparisons in similarity and how relevant are they to drug design? Most methods, and consequently the software, to date, have been geared to electrostatic potential comparisons of conformationally frozen structures. Moreover, it appears historically that many studies in electrostatic potential similarity have evolved from comparisons between bio-isosteres. Gnomonic projection methods (Chau and Dean, 1987; Dean and Chau, 1987; Dean and Callow, 1987; Dean *et al.*, 1988) place the electrostatic potentials onto a sphere and rotate the structure until maximum similarity is obtained. This approach has focussed attention on rather special methods of comparison which are not necessarily applicable to the general case of molecular similarity between dissimilar molecules. These methods, which rely on rotating molecules against each other, are critically dependent on the choice of origin for the rotations. In bio-isosteres the use of common centroids may be valid for comparisons but rotation about the centroid will be a poor method for non-globular structures since the centroid is an arbitrary centre for rotation. These methods will undoubtedly fail when flexibility is introduced into the comparisons of similarity since the centroid will shift for each flexion. New methods need to be developed that are independent of comparisons of similarity based on molecular rotations. Exactly the same criticisms apply to identical methods used to compare similarity in molecular shape.

1.1.4 Molecular electrostatic potential complementarity

Molecular similarity in electrostatic potentials has been studied widely as a tool for drug design and is incorporated into many pieces of molecular design software. In contrast, molecular electrostatic complementarity has received scant attention. This is quite extraordinary since most rational methods of drug design implicitly assume some electrostatic complementarity. The assumption is based on the Coulombic term in the following equation for the molecular interaction (Clementi, 1980).

$$E = \sum_{i}^{\mathrm{I}} \sum_{j \neq i}^{\mathrm{J}} \left(-\frac{A_{ij}^{ab}}{r_{ij}^{6}} + \frac{B_{ij}^{ab}}{r_{ij}^{12}} + \frac{C_{ij}^{ab} q_i q_j}{r_{ij}} \right) \tag{1.2}$$

Molecule I, with atoms, i, may represent the site and molecule J with atoms, j, may represent the ligand. A, B and C are fitting constants for atom pairs i and j; the superscripts a and b represent the atom types and the class to which they belong; r_{ij} is the interatomic distance; q_i and q_j represent the charges on each atom. For the energy, E, to be minimized the pairwise summation of the Coulombic term should have a negative sign. The Coulombic term is also long range. This component can be analysed critically by comparing the electrostatic potentials generated by the ligand and the site projected onto one of their surfaces. In this case the complementarity can be measured by correlation; the coefficients have an opposite sign from similarity coefficients. Let the electrostatic potential $P_{\mathrm{A}k}$ be defined at point k from molecule A, $P_{\mathrm{B}k}$ is similarly defined, then the correlation coefficient r for the n points is given by

$$r = \frac{\left\{ \sum_{k=1}^{n} (P_{\mathrm{B}k} - \overline{P_{\mathrm{B}}})(P_{\mathrm{A}k} - \overline{P_{\mathrm{A}}}) \right\}}{\left\{ \sum_{k=1}^{n} (P_{\mathrm{B}k} - \overline{P_{\mathrm{B}}})^2 \right\}^{1/2} \left\{ \sum_{k=1}^{n} (P_{\mathrm{A}k} - \overline{P_{\mathrm{A}}})^2 \right\}^{1/2}} \tag{1.3}$$

Recently many ligand–protein co-crystal complexes have been studied (Dean *et al.*, 1992; Chau and Dean, 1994a,b,c). The first feature to be examined was the disposition of charge on the ligand and site atoms. In the 41 ligand-site examples studied, there was no correlation between the sign of the ligand atom charge and either a contact atom in the site or with atoms nearly in contact. The charges themselves did not appear to be complementary. However, when the potentials surrounding the ligand and the site were compared, most complexes showed electrostatic complementarity but values for individual complexes were found right through the range -0.9 to 0. The pattern of electrostatic complementarity was not changed significantly by using different models for the dielectric.

Complementarity cannot be easily broken down into an additive contribution by simple chemical moieties. This is an important finding for drug designers because it suggests that the addition of moieties, with precalculated electrostatic properties, to an evolving structure may not be an efficient way to build an electrostatically complementary molecule. Detailed scrutiny of these complexes reveals that the complementarity is frequently dominated by the relative positioning of the maximally charged groups. Complexes with complementary ionic charges can have poor electrostatic complementarity. This observation suggests that considerable caution should be used if similarity methods for molecular electrostatic potentials are to be employed directly in drug design; it may just happen that the archetype ligand shows poor complementarity with its site. There would be little point, in that case, in designing a structure to have electrostatic similarity.

1.1.5 *Molecular hydrophobic similarity*

In molecular interactions hydrophobic similarity is believed to be a driving force to enhance the energy of interaction. The hydrophobicity potential Φ_j, at the point j, (in $\mathrm{kJ\,mol^{-1}}$) can be computed from:

$$\Phi_j = -2.3RT\sum_{i}^{n} f_i e^{-r_{ij}} \tag{1.4}$$

where n is the number of fragments, f_i is the fragment hydrophobicity potential of the group i, r_{ij} is the distance between i and j, R is the gas constant and T is the temperature (Fauchère *et al.*, 1988). Hydrophobic maps can be constructed from this equation. In a *de novo* design project, it should be possible to project a hydrophobicity potential from the site onto the ligand skeleton surface and optimize the similarity in the construction of a novel molecule.

1.1.6 *Hydrogen bonding similarity and complementarity*

Hydrogen bonds are strongly localized and directional; they fall into two groups, acceptors and donors. Until recently, hydrogen bonds were thought only to occur between two strongly electronegative atoms, such as O–H...O, N–H...O or O–H...F-like structures. However, Desiraju (1991) has collected evidence from crystal structure analysis that C–H...O types of hydrogen bonds are widespread and may be important in ligand–site complexes. Hydrogen bonds may also be found with a donor group perpendicular to the π system of aryl rings.

The use of hydrogen-bond data in molecular similarity studies, or in pharmacophore generation, is problematical. The difficulties in analysis stem from the fact that although the bonds are localized, there is a significant directional tolerance. This difficulty is exemplified in hydrogen-bonding probability maps (Danziger and Dean, 1989a,b). Around each hydrogen bond acceptor atom there is a large arc where hydrogen-bond formation with a donor is probable; donor regions, on the other hand, are much more restricted spatially. Thus it is possible for different ligands, co-crystallized with a protein site, to show poor similarity in positions of hydrogen-bonding atoms but nevertheless make good hydrogen bonds and exhibit excellent complementarity. Clearly, attempts to match hydrogen-bonding atoms in a molecular similarity exercise is a poor strategy. What is needed is a method for matching the ligand hydrogen-bond probability shells. If these shells can be matched and optimally superposed, the maps may provide strong clues about the disposition of hydrogen-bonding site points, which in turn may then be used rigorously in a *de novo* drug design procedure.

1.2 Problems in using molecular similarity methods for drug design

Suppose we are given a set of two or more structures suspected of binding competitively to the same site; our objective is to design a novel molecule differing significantly from the test set, in terms of molecular structure, but having similar features that lead to activity. What are the problems in this exercise?

1. Structural bias in the test set needs to be minimized, otherwise the analysis will be distorted by weighting predominant features.
2. An appropriate measure of similarity needs to be used for the features. There are many measures to choose from; each is best suited to the problem type for which it was devised (see Chapter 2 for an extensive review of the methods). If more than one feature is to be used, how are the separate features to be weighted?
3. Most potential drug molecules have significant degrees of flexibility. There may be many regions of probable conformation space that each molecule can populate (see Chapter 4). The bound conformation may not be that with the global minimum conformational energy (see Chapter 3).
4. The molecules must be superposed before the similarity can be assessed. There may be numerous superpositions which show close similarity.
5. The binding face of the molecule is unknown. To what extent should the non-binding face be included in the measure of similarity?
6. With a set of molecules, how should the similarity of the set be expressed? Should similarity be expressed with respect to a defined member, or with each member taken in turn, or with some consensus measure of the set?

7. The molecules, when bound to the site, may show different binding modes. They may bind to different subsets of site points. Alternative binding modes need to be distinguished before any detailed 3D QSAR analysis is performed.
8. If the set is correctly superposed in the correct binding mode(s), a straightforward application of comparative molecular field analysis (CoMFA) should, providing that the biological activity has been properly determined, highlight spatially disposed features that need to be mimicked (see Chapter 12).
9. An envelope needs to be defined in which to create a novel molecule. This envelope should incorporate the key features, related to activity, generated from the CoMFA analysis.
10. Design a molecule with a shape defined by the envelope and exhibiting the spatially desired features.

Ideally the purpose of molecular similarity work, applied to drug design, is to build up a picture of the site by attempting to generate the complement of the similar superposed parts of the test molecules. This whole process is fraught with difficulties. It is illuminating to consider what would happen if we knew the structure of the site. Could we then foresee the difficulties posed to drug designers working only with limited data from molecular similarity?

1.2.1 Problems with combinatorics of site points

The strategy of dividing a target site into discrete regions and then using a number of these regions as a basis for designing a drug, is an important concept for site-directed drug design. It also has an important bearing on the validity of any strategy for drug design based on molecular similarity. These regions can be approximated, for argument's sake, to points called site points so that we can calculate the combinatorics easily. Let us suppose that there are n site points; if r site-point interactions are needed for a molecule to have measurable affinity/activity, the number of combinations of r site points, $C(n, r)$, is given by

$$C(n,r) = \frac{n!}{r!\,(n-r)!} \tag{1.5}$$

$C(n, r)$ is a maximum when $r \sim n/2$. Table 1.1 gives an example where $n = 10$ and r varies between 1 and 5. With $r = 5$ there are 252 different combinations of site points. Even with three site points, a minimal number needed to define a corresponding ligand pharmacophore, there are 120 triplet combinations. Each triplet would correspond, by complementarity, to a different pharmacophore. If drug design were to proceed from molecular similarity we must be aware that the test set may span different possible pharmacophores. Ideally, the test set would need to be partitioned into clusters of molecules either

Table 1.1 Combinatorics $C(n, r)$, for site points ($n = 10$) taken r at a time. S is the number of the subsets with m-tuplets in common

r	$C(n, r)$	S			
		$m = 2$	$m = 3$	$m = 4$	$m = 5$
1	10				
2	45	1			
3	120	8	1		
4	210	28	7	1	
5	252	56	21	6	1

containing particular pharmacophores, or interacting with specific sets of complementary site points.

1.2.2 Problems with common subsets of site points

The analysis given above warns about the number of possible combinations of site points that may complicate the use of molecular similarity for drug design. However, in practice, it is even more complex. Although the combinations are unique, there are commonalities between them. For example, if we label the site points 1, 2, 3, ... 10 there will be 252 combinations of five site points. 21 of these combinations will contain any specified triplet (e.g. 1, 2, 3), six will contain a specified quadruplet (e.g. 1, 2, 3, 4) and of course there is only one specified quintuplet (e.g. 1, 2, 3, 4, 5). The size, S, of the subset of commonalities from $C(n, r)$ which has m members in common is

$$S = \frac{(n - m)!}{(r - m)!\,(n - r)!} \tag{1.6}$$

Table 1.1 illustrates the magnitude of S for different values of m varying between 2 and 5 for values of r between 1 and 5 with n set to 10.

Therefore, if we were to make 252 unique molecules to fit the combinations of five site points, there would be expected to be some similarities, in terms of site points bound, between them. However, there would also be significant differences; this would depend on how many site points were not in common.

1.2.3 Problems with regional similarity and dissimilarity

If the test set is to be used in a paradigm for drug design, it needs to be partitioned into clusters of molecules showing related similarity. The molecules belonging to each cluster then have to be optimally superposed before being used in a CoMFA-like procedure to highlight field disposition related to activity. The crucial problem here is that similarity and superposition are intimately interdependent.

Consider the example of the subset where $n = 5$ and $m = 4$ in Table 1.1; six molecules would contain the specified quadruplet. How do we assess the similarity? If we measure the similarity between the whole of each molecule and try to optimize it, it is possible that the poorly matched features in the regions where there is no subset commonality would force a different superposition on the set compared with the binding modes found in the site. The only way to tackle this problem is to develop a procedure that maximizes regions of similarity between molecules and ignores regions of dissimilarity. The null correspondence method was developed to maximize regional similarity using a coordinate invariant procedure employing the minimization of the difference-distance matrix as discussed below (Barakat and Dean, 1991). Superposition is then based only on corresponding regions of maximized similarity.

1.2.4 Problems with different binding modes

Recent X-ray crystallographic studies of a number of ligand–site complexes show that there are, from time to time, different alignments of the ligand in the site (Mattos and Ringe, 1993). These are sometimes, but not always, derived from different conditions when seeding the crystal. There are a variety of circumstances that can lead to the existence of different binding modes.

1. The same ligand points may bind to a different set of site points.
2. Different ligand points may be involved in the binding of the ligand to the same set of site points.
3. Different ligand points may bind to a different set of site points.
4. A conformational change in the ligand may re-orient part of the ligand.

What possibility is there of detecting any of these four cases in the absence of crystal data for all complexes in a ligand similarity study? Consider case (1); similarity studies alone will not provide evidence for different binding modes of identical ligand points. With case (2), the key point is to determine how many closely equivalent, but different, superpositions of the molecules are possible for the feature classes being examined. Replicate runs of the simulated annealing procedure, outlined below, can give an indication of alternative superpositions. The number of alternatives increases if the number of null correspondences is also increased (Papadopoulos and Dean, 1991). Cluster analysis of the superposition transformation matrix indicates the alternatives quite clearly. Case (3) cannot be resolved by similarity studies alone. In principle, evidence for case (4) can be obtained from molecular similarity studies by a systematic application of annealing and cluster analysis on different conformations (Perkins and Dean, 1993).

The possibility of alternative binding modes in a CoMFA style of study needs urgent investigation since the critical problem in CoMFA lies in obtaining the molecular alignment for the set of molecules. If there are alternative binding modes, they would have to be investigated by considering clusters of similar alignments.

1.2.5 Problems with hydration at the site

The Brookhaven Protein Data Bank (PDB) (Bernstein *et al.*, 1977) reveals that in ligand–protein co-crystal complexes there are many examples where water molecules are intimately involved in ligand binding to the site, either directly through bridging waters or by supplementary stabilizing structures of networks of water molecules (Prive *et al.*, 1992; Lewendon and Shaw, 1993). Clearly, some of these waters are of mechanistic importance in hydrolysis reactions. The possibility exists that some of these waters may not be conserved in ligand binding but shift to other positions, thus they may create alternative site points within the maximum void of the site. Thus complementarity will vary between these possibilities and there will be concomitant changes in molecular similarity. This difference will significantly affect any design strategy based on molecular similarity.

1.3 Extraction of molecular similarity from two or more molecules

Three problems can be identified when an unbiased estimation of similarity is to be attempted: (1) what measure should be used for similarity? (2) how should the similarity be optimized? (3) how much of the molecular structures should be ignored in the similarity optimization? The choice of a measure for similarity is almost entirely dependent on the design problem to which similarity methods are put. Chapter 2 outlines a wide variety of methods for categorizing similarity. It should be borne in mind, however, that if the measure is to be calculated many times in the search for optimum similarity, it should be simple to compute. In this chapter we shall focus attention on only one method of searching for similarity, namely simulated annealing. Other chapters deal extensively with different methods.

Optimization of similarity can be performed by two general methods, firstly by transforming the coordinates of the molecule and reassessing the similarity each time, or secondly by a discrete method that changes the atom correspondences and only in the final stage is a superposition performed. If the entire molecular structure is used for similarity assessment, low values will be obtained for similarity between ostensibly dissimilar structures, which may mask out well-matched regions of molecular similarity. The omission of portions of structure to optimize regions of similarity is fraught with difficulties. Only a 'leave-one-out' procedure is satisfactory but the combinatorics of this can rapidly become intractable if sensible optimization procedures are not taken.

1.3.1 Simulated annealing as an optimization method

The great advantage of this method over downhill optimization procedures is that the technique is well understood from its analogy with physical systems and uphill excursions are catered for through the Metropolis condition. The method can be applied to either optimization approach by coordinate transformation or by atom correspondences. A coordinate transformation procedure will be described in Chapter 4. Here we concentrate on optimizing atom correspondences by minimizing the difference-distance matrix. Similarity is performed pairwise onto a base molecule, A. Molecule B is to be fitted to molecule A. The atoms in both molecules are numbered and an order in B is to be found, by random perturbations, which maximizes the similarity with the specified order of A. Molecule B is then transformed onto A using the corresponding orders.

Let there be N_A and N_B atoms respectively in molecules A and B; they should be numbered so that $N_B \geqslant N_A$. Let Z denote the configuration space for all possible orderings, π; $E_{\pi i}$ is the value of the objective function at the configuration π_i, we have to find the optimal configuration π_{opt}

$$E_{\pi_{opt}} = \min\{E_{\pi_i} \mid \pi_i \in Z\} \tag{1.7}$$

The size of the configuration space Z is given by

$$Z = N_B!/(N_B - N_A)! \tag{1.8}$$

The objective function E is analogous to a statistical mechanical quantity of energy whose behaviour can be monitored by Monte Carlo methods. Consider an objective function to have a value E_s at a state s and at a temperature t. Apply a random perturbation to shift the system to a new state s' with an objective function of value $E_{s'}$. The difference between the values of the objective function ΔE is calculated. If ΔE is negative, the perturbation is accepted unconditionally and the new state s' is maintained. If ΔE is positive, a probability value $P(s'|s)$ is calculated from

$$P(s'|s) = e^{-\Delta E/kt};\ kt = T \tag{1.9}$$

where k is the Boltzmann constant, t is a temperature in kelvin; kt is replaced by an annealing temperature, T, having the same units as E. The value of $P(s' \mid s)$ is compared with a random number, R, taken from the distribution $[0,1)$; if $R < P(s'|s)$ then new state s' is accepted. The process is repeated at the same temperature until equilibrium is reached; the temperature is then diminished by a small amount and the cycle repeated until the change in E is deemed to be insignificant. The decrease in temperature progressively restricts the uphill transitions between successive Markov chains. In theory the optimization is ergodic; the optimum can be reached from any starting position, although, in practice, this may not always be the case due to the finite time allowed for the optimization.

Let d_{ij}^{A} and d_{ij}^{B} be elements in the distance matrices of A and B; the corresponding element in the difference-distance matrix is

$$\Delta d_{ij}^{AB} = |d_{ij}^{A} - d_{ij}^{B}| \tag{1.10}$$

The sum of these elements for the subdiagonal difference-distance matrix is

$$E = \sum_{i=2}^{N_A} \sum_{j=1}^{N_B} \Delta d_{ij}^{AB} \tag{1.11}$$

E is the value of the objective function for the proposed transition. At each step in the minimization, a pair of atom numbers in B is swapped to create a new ordering π_i, the ordering in A is never changed; thus a current state s is perturbed to s' and ΔE can be computed.

1.3.2 Null correspondences

The null correspondence method enables a certain number of correspondences to be omitted from the objective function. These null correspondences are examined randomly. If there are to be k null correspondences then there are $N_A - k$ atoms in A to be matched by a similar number in B. The k nulls are randomly assigned to A; coresponding points in B are also considered in the objective function to be nulls simply by virtue of their current correspondences to nulls in A. In the case where $N_B > N_A$, some points in B will have no initial correspondence with A; these are identified as an outer subset of B (Figure 1.1). Points in A and B that are denoted as nulls do not enter into the calculation of the objective function. Two types of change are separately allowed in the perturbation of s to s'. The nulls in A can be swapped for non-nulls in A; the order of A is always maintained, only the nulls are moved. In contrast, the order of B is changed, but only certain changes are permitted. For convenience, B is divided into two subsets—an inner subset where correspondences exit with A and an outer subset where there are no correspondences with A. Excluded exchanges in B are two points from the outer subset, two null correspondences from the inner subset, a null correspondence from the inner subset and a point from the outer subset. The algorithm is given in Figure 1.2. The decision whether to perturb either the combination in A, or the order in B, is determined by the parameter V obtained from the formula

$$V = k/(N_A + N_B) \tag{1.12}$$

If a random number generated from the uniform distribution $[0, 1)$ is less than V then a perturbation is made to A; otherwise the order of B is changed.

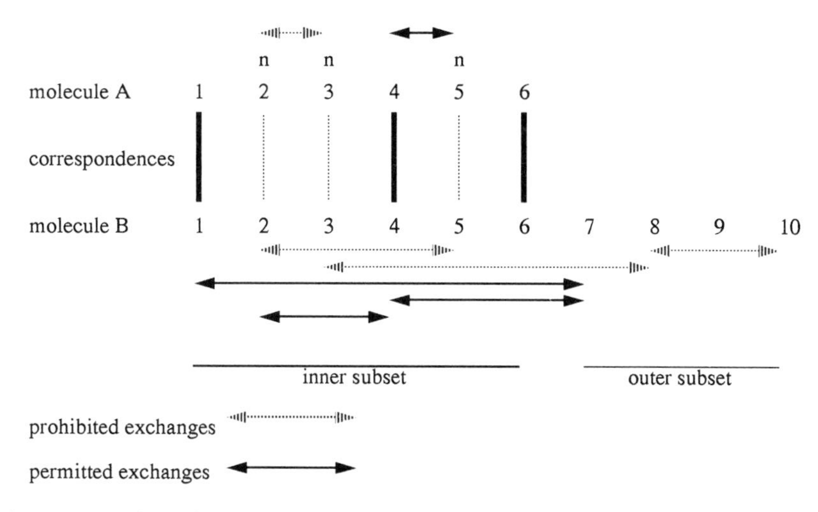

Figure 1.1 A scheme for incorporating null correspondence (n) into the simulated annealing algorithm for optimizing the search for molecular similarity.

```
read coordinates of A and B
read number of Markov chains, chain length, annealing parameters
read number of null corrspondences to be used
rescale input data
calculate distance matrices
allocate nulls randomly
compute current value of objective function
    do i = 1, number of Markov chains
        do j = 1, length of Markov chains
            randomly decide to perturb null correspondence or order of B
            if (a null correspondence is to be perturbed) then
                get new configuration for A
                compute new value of objective function
            else
                change order of B
                compute new value of objective function
            endif
            compare proposed new value of objective function with current value
            if (Metropolis test successful) then
                update current value of objective function
                update either null correspondence or order of B
            endif
            if (stop criterion satisfied) stop
            decrement temperature
        end do
    end do
```

Figure 1.2 An algorithm in pseudo code for optimizing the search for molecular similarity using null correspondences.

Thus, with small numbers of nulls, changes are made predominantly to B. In practice, the difficulty lies in specifying the number of nulls to include since it is completely problem-dependent. The only safe way to proceed is to start with no nulls and gradually increase the number. At the same time the objective function can be monitored for a sharp improvement, such as a better rms of the fitted portions of the molecules.

1.3.3 Annealing schedule

A critical feature of all annealing schedules is that the initial temperature should be set to give a ratio of (accepted exchanges/proposed exchanges) close to 1. Normally this will vary with individual problems unless steps are taken to standardize the problems by scaling the objective function. Thus if the actual difference in the objective function is ΔE, this can be scaled by a preliminary assessment of $\sigma_{\Delta E}$ from the statistical distribution of ΔE, thus

$$\Delta E_{\text{scaled}} = \Delta E c/(3\sigma_{\Delta E}) \tag{1.13}$$

c is a constant which controls the initial frequency of acceptance and consequently the initial temperature.

The length of the Markov chains, l_{m}, needs to be set to

$$l_{\text{m}} = \min\{2N_{\text{A}}(N_{\text{B}} - 1)\ln C(N_{\text{A}}, N_{\text{A}} - k), 50\,000\} \tag{1.14}$$

where $C(N_{\text{A}}, N_{\text{A}} - k)$ is the binomial coefficient. A large Markov chain is needed to take into account changes to A which take some time to equilibrate.

The temperature decrement is achieved dynamically to make the cooling efficient (Barakat and Dean, 1990a). Although simulated annealing is said to be ergodic, in practice it is not really so. Two problems occur: (1), minor mismatches are found between the atom corespondences; (2), mismatches are found due to a pseudo-symmetry problem in the order of correspondences. These are due to the algorithm cooling too quickly or equilibrium not being reached in the Markov chains. One way to circumvent the problems is to perform two re-annealing steps. The first re-anneal is performed on the worst 25% of atom correspondences and re-annealing starts at T set to 75% of its initial value. This strategy is frequently successful for rectifying minor mismatches. A second strategy is to reverse the order for B and re-anneal with T set to 75% of the initial value; a further re-anneal of the worst 25% is carried out as before.

The objective function for molecular similarity is a difference-distance matrix method. The great advantage is that the procedure is rotationally and translationally invariant. Once atom correspondences between the molecules have been found, the molecules have to be superposed. Only corresponding atoms in the two molecules are used to obtain the superposition transformation matrix by McLachlan's algorithms (McLachlan, 1982). The matrix is then used to transform the whole of the molecules to get the final superposition.

This type of annealing procedure works well for atom positional similarity research and has been used numerous times (Barakat and Dean, 1990a,b, 1991; Papadopoulos and Dean, 1991; Perkins and Dean, 1993). A different strategy is needed for finding the similarity directly between molecular surfaces (see Chapter 7). It is possible to use the fact that the surface is continuous and has a specified thickness, in a routine to employ a null correspondence implicitly between the surface points. Strict criteria for surface overlap can lead to a maximization of molecular surface similarity.

1.3.4 Ergodicity problems with annealing

In theory, simulated annealing is ergodic but in practice, because we allow the algorithm a restricted amount of time to converge, it is not. Widely separated equivalent local minima can trap the algorithm in a non-global position. This configurational landscape problem has been extensively investigated (Barakat and Dean, 1990b). However, for many applications of molecular similarity the non-ergodicity turns out to be of significant value for drug designers. There may be a variety of different matches for superposition with equivalent similarity values. If replicate runs of the algorithm are performed, it is possible to identify equivalent minima by cluster analysis on the elements of the transformation matrix for the eventual superposition (Papadopoulos and Dean, 1991). Suppose we have k replicates for the superposition of two molecules A and B; each superposition is performed through the transformation matrix of 12 elements, nine for rotation and three for translation. The transformation elements can be used as a metric in 12-space. Thus we can cluster on a data matrix with $k \times 12$ elements. Ward's method of clustering, as opposed to other cluster methods, is ideal because it maximizes cluster isolation and does not violate the ultrametric inequality (Murtagh and Hecht, 1987). The resulting dendrogram gives a visual illustration of how the superpositions are clustered. The number of significantly different clusters can be ascertained from Mojena's (Mojena, 1977) stopping rules (see Chapter 4 for a more detailed explanation). This procedure enables a user to distinguish between minor and major differences in the superpositions obtained through the replicate runs. If a set of molecules, rather than just a pair, now has to be examined for similarity and superposition, then two procedures are available for this. The base reference method chooses one molecule and superposes the best possible superpositions of all the other molecules, including nulls (Papadopoulos and Dean, 1991). An alternative procedure is the consensus method where we seek the best similarity possible in the set of molecules, from their replicate superpositions (Perkins and Dean, 1993). The value of this method in a CoMFA alignment has yet to be determined.

1.3.5 Similarity between molecules with multiple features

Molecular recognition between a ligand and its site reaches an optimum when the summation of all the contributions from the individual forces is minimized (see equation 1.2). At the same time, the relative positions between the two interacting molecules change to optimize that interaction. In the absence of information about the site, it is impossible to predict the relative alignments of a set of molecules, docked into the site, based on the similarity of their molecular features alone. However, bearing that caveat in mind, it is tantalizing, and may be expedient, to develop methods for determining the similarity between a set of molecules with multiple features. The overriding difficulty encountered is, how should each of the features be weighted in the similarity exercise?

In an ideal world we would have a very large number of different binding sites expressing a range of features, such as electrostatic potential, hydrophobic potential, hydrogen-bonding potential, and steric potential information which could be calculated from the crystal structure of the site. There would also need to be an ideal test set of molecules bound to the site along with their binding energies. It might then be possible to derive, by statistical analysis or by using artifical neural networks (see Chapter 8), some general weighting scheme for the contributions of each feature that might be considered in a similarity study. However, even with that information, the problem of multiple binding modes derived from attachment to different site points, would still leave a strong element of uncertainty in the interpretation of the similarity from multiple features.

Attempts to grapple with this problem are legion. Klebe (1993) has provided an excellent critical summary. Kearsley and Smith (1990) in their SEAL program use an alignment function that computes an electrostatic similarity score. The SUPER program of Hermann and Herron (1991) takes the van der Waals surfaces mapped by a grid with the electrostatic potential calculated at each grid point; matching proceeds by minimizing the differences of both parameters. Other methods consider two separate assessments; firstly a fit is obtained by matching a steric or accessible surface and secondly the surface features are matched separately. This method has the advantage that windowed regions can be assessed for the similarity parameters (Borea *et al.* 1992); the disadvantage is that the alignment is performed only on one parameter.

A very different approach is to use all the parameters as dimensions in a higher dimensional space and construct the distance matrix for each molecule. Matching can then be performed using an objective function based on minimizing the difference-distance matrix (Dean, 1993). Consider five independent variables and the three variables of Cartesian space to give an eight-dimensional space. The five independent variables need to be autoscaled independently; the three Cartesian variables need to be autoscaled together. The distance between any two points, i and j, is given by

$$d_{ij} = \left[\sum_{k=1}^{8} (x_{ik} - x_{jk})^2 \right]^{1/2} \tag{1.15}$$

Now if the variables are to be given weights, w, the Cartesian coordinates should have the same weight, thus the equation can be expanded

$$d_{ij} = \left\{ \left[\sum_{k=1}^{3} w_1 (x_{ik} - x_{jk})^2 \right] + \left[\sum_{k=1}^{5} w_k (x_{ik} - x_{jk})^2 \right] \right\}^{1/2} \tag{1.16}$$

Clearly the weights should be adjusted to

$$3w_1 + \sum_{k=1}^{5} w_k = 1 \tag{1.17}$$

The difference–distance matrix can then be constructed from the distance matrix elements by equation (1.10). This equation can then be used in equation (1.11) to create an objective function for minimization by simulated annealing.

1.4 Molecular design based on molecular similarity

The type of similarity assessment used will determine the subsequent design strategy. For example, if the properties have been made discrete in some way to create a similarity map of discrete regions, it may be possible to build novel structures spanning the space to exhibit these properties. The simplest possible procedure is to make use of molecular databases and search for structures having similar property dispositions. An alternative would be to attempt an automated *de novo* design within the constraints of some supersurface of the set of similar molecules.

1.4.1 Molecular design using similarity and molecular databases

This procedure is equivalent to pharmacophore searching through the databases. Fast searching techniques have been pioneered by Willett (see Chapter 5). Mason (see Chapter 6) describes a number of test cases where searches have revealed different structures from a company database; subsequent testing showed biological activity. This method is not strictly a design procedure initially; it is simply a searching process to find a novel lead with specified similarities. Once a lead has been found, however, it will be possible to modify the structure to improve the similarity and perform further optimizations to change the structure from a ligand into a putative drug molecule. The overwhelming advantage of this approach is that the lead structure, because it is known to exist, can usually be readily synthesized. *De novo* methods for drug design, on the other hand, may create structures that are difficult to make.

Similarity studies generate molecular superpositions which reveal either atom correspondences or functional group similarities. These similarities are expressed as points in space with associated properties. A distance matrix is constructed between the points; this serves as a search query for interrogating the database. Molecules in the database can be placed into sets of sensible conformations so that a variety of conformational states of each structure can be searched. Hits, from the database that satisfy the query, can then be generated for biological assessment. The last step is to examine the similarity of the active hits with the original similarity test set to see if the similarity of the new structures can be improved by small conformational changes and optimal superpositioning.

1.4.2 Conformational changes to maximize similarity

Hits from databases are found to lie within a certain region of tolerance from the template screen. Tests need to be performed on the hit to ascertain whether adjustments to the conformation of the hit can match the similarity with the template more closely. Template forcing methods need to be applied using a combination of similarity and conformational energy within the objective function.

1.4.3 Molecular design using similarity and automated de novo *methods*

In principle, this approach offers exquisitely sensitive methods for optimizing molecular similarity of a novel structure with the test set. It is convenient to break down the design problem into two components: the construction of novel structures with similar shapes through a space-filling network of bonds, and the incorporation of atoms into the network to create a molecule with the desired similar properties.

1.4.4 The design surface envelope

If molecular similarity studies have been performed on two or more molecules and a superposition obtained, the first step in design is to create the supersurface, that is the outermost surface, of the set of superposed molecules (Figure 1.3). This supersurface functions as an envelope in which the design must take place. The specification of the 'active' surface is critical for the superposition. It is this surface which lies in contact with that of the site and its 3D nature will be critical for the interactions through the short-range forces in equation (1.2). The non-interaction surface is less critical and only defines an artificial boundary to constrain the structure generation. Thus

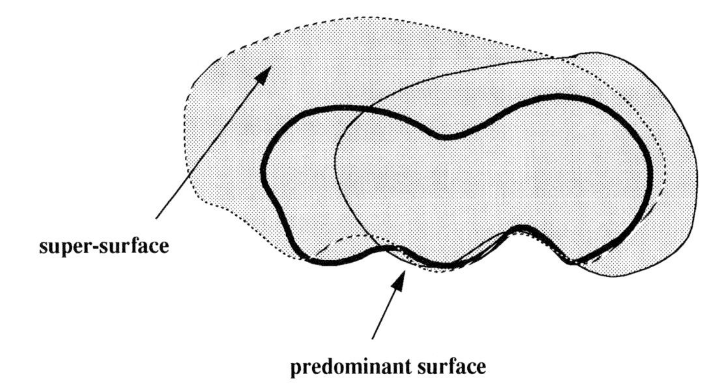

Figure 1.3 Construction of the molecular supersurface. Three molecules are superposed so that one face predominates in the superposition. The supersurface is the outermost surface of the three structures (filled with shading) and acts as a constraining boundary. Drug design should be optimized towards the predominant surface.

some form of null correspondence procedure, or an equivalent in terms of surface points, is needed to assign which surfaces are important. The supersurface must be constructed very carefully within that region. An important problem is to ensure that the supersurface is a continuous manifold without holes in it. This can be achieved in most cases by expanding the surface a little way from the van der Waals surface.

1.4.5 The design surface molecular similarity parameters

Molecular surface parameters can be projected onto this precisely defined envelope. Currently, there is little experience of how the projection should be performed. A variety of schemes can be advocated: (1), using the projection of the parameters from one molecule in the set onto the supersurface; (2), using the parameters from different molecules in turn as a separate design paradigm; (3), using a projection of averaged parameters; (4), using a projection of maximum or minimum parameters generated by the molecular set. *De novo* structure design within the envelope is a combinatoric procedure which is difficult to optimize; molecular similarity between the projected parameters is only one of the combinatoric difficulties. Schemes (1) and (2) at least offer the possibility of using actual parameter values at each surface point. Schemes (3) and (4) are artificial and may constrain the design program to search for less meaningful structures.

1.4.6 Creation of novel molecular skeletons with a similar supersurface

If we neglect to consider the database searching methods, such as CAVEAT (Laurie *et al.*, 1989; Laurie and Bartlett, 1994), on the grounds of not being true *de novo* techniques, we are left with fragment-based methods for novel generation of molecular structures. All these methods have in common an assembly procedure based on selection of fragments from a library of small pieces. At the heart of this approach is the combinatoric problem called the 'knapsack problem'. Stated simply: given a set of rigid small pieces, how can a selection of pieces be arranged optimally within a prescribed closed surface? The problem is compounded by the strong possibility that there may be many significantly different arrangements that give closely similar, near optimal solutions. In other words, many molecular frameworks may have a surface similar to the target supersurface under consideration.

De novo drug design methods have been geared to a molecular complementarity problem where the structure of the site is known and molecules are created to fit the site directly. In the absence of site structure, methods using molecular similarity are the only way forward. To the author's knowledge, little has yet been published on the problem; it may be helpful to describe our early experience in adapting *de novo* design methods to similarity-based design.

Filling the void can be achieved either by using molecular fragments as in the DOCK algorithm of Kuntz *et al.* (1994), or by using small pieces of bonding networks with appropriate angles and bond lengths to create molecular skeletons without atomic identities. Both alternatives have their advantages and drawbacks. Actual molecular fragments can rapidly lock the evolving structure into a particular molecule; bonding networks dominate the formation of a putative molecule. The advantage is that the procedure is rapid; the disadvantage is that the configuration landscape problem is dire and optimization is likely to be weak. Small portions of molecular networks give good spatial fitting of skeletons but atoms need to be added during a separate optimization step (Dean *et al.*, 1995); the weakness in the method is that optimization takes place in two stages and convergence is slow.

Conformational restriction to mould the ligand into the supersurface can be achieved by ring-bracing methods. Dangers arise when rings are fused to have common bonds or common vertices. The structures generated may rapidly become too complex to synthesize.

1.4.7 Creation of structures with parameter surfaces similar to the supersurface

If design proceeds via the molecular skeleton approach, then atoms must be added to the skeletons so that the parameters projected onto the surface match the target parameters on the supersurface. An optimization method for molecular similarity is essential. Barakat (1993) suggests that the optimization can be achieved by annealing an atom placement procedure using transferable properties from a small library of acyclic and aromatic pieces of combinations of atoms from the set $\{H, C, O, N, F, S, Cl\}$ (Chau and Dean, 1992a,b,c). The objective function for optimization is Pearson's product moment correlation for the molecular similarity between the evolving solution and the parameters of the target supersurface. The method converges well to optimize the similarity between the two surfaces.

1.5 Conclusions

Molecular similarity is a fast-emerging tool for drug design. Most research efforts so far have concentrated on the development of methods for expressing molecular similarity. Each method has evolved with a particular problem in mind. That is why this book examines so many different approaches to molecular similarity. There is no generally agreed algebraic expression of similarity, or even what we mean by molecular similarity; this is not a criticism but simply illustrates the fertile nature of this research field. Similarity is the converse of complementarity. Therefore, because molecular interactions take place by complementarity between the interacting species, the problems that arise in assessing complementarity should not be forgotten when we attempt to use molecular similarity for drug design. Atom arrangements in sites can give rise to different combinations of site points which provide alternative binding points for a ligand; these in turn can give rise to alternative binding modes. It is perilous to ignore the lessons of molecular complementarity if we are to use molecular similarity in a sensible way for drug design.

Not all the molecular surface of the ligand is involved in an interaction with the site. Thus an overall expression for similarity may weaken its use in drug design. What is needed are optimization methods that focus attention on the strongly matched regions of molecular surfaces. Null correspondences provide one method to help with this problem but there may be more efficient ways of handling these unmatched portions of surface. When multiple features are involved in molecular similarity, the optimization is non-trivial; unfortunately, our experience with multiple features is very limited.

Molecular design strategies derived from molecular similarity studies are only sparsely documented. Database searching methods have provided some good examples of finding structurally dissimilar molecules, but at the same time containing some similar property disposition, that are biologically active (see Chapter 6). However, *de novo* drug design strategies are only just emerging for site-directed design. It is too early to assess whether they will be of value for similarity-based design procedures. These novel design methods are certainly ripe for testing now that combinatoric optimization techniques are being tailored to many parts of the drug design problem.

Acknowledgements

I would like to thank the Wellcome Trust for continuous personal support over many years through the Principal Research Fellowship scheme, and

Rhône-Poulenc Rorer for generous research funding. Thanks go to all my colleagues in the Drug Design Group and in the Cambridge Centre for Molecular Recognition for their skills and dedication in investigating many of the problems considered in this chapter.

References

Appelt, K. (1993) Crystal structures of HIV-1 protease–inhibitor complexes. *Perspectives in Drug Design and Discovery*, **1**, 23–48.

Barakat, M.T. (1993) Optimization of the Atom Assignment Problem in Computer-Aided Drug Design, PhD Thesis, University of Cambridge, Cambridge.

Barakat, M.T. and Dean, P.M. (1990a) Molecular structure matching by simulated annealing. 1. A comparison between different cooling schedules. *Journal of Computer-Aided Molecular Design*, **4**, 295–316.

Barakat, M.T. and Dean, P.M. (1990b) Molecular structure matching by simulated annealing. 2. An exploration of the evolution of configuration landscape problems. *Journal of Computer-Aided Molecular Design*, **4**, 317–330.

Barakat, M.T. and Dean, P.M. (1991) Molecular structure matching by simulated annealing. 3. The incorporation of null correspondences into the matching problem. *Journal of Computer-Aided Molecular Design*, **5**, 107–117.

Bernstein, F.C., Koetzle, T.F., Williams, G.J.B., Meyer, E.F., Brice, M.D., Rogers, J.R., Kennard, O., Shimanouchi, T. and Tasumi, M. (1977) The Protein Data Bank: a computer-based archival file for macromolecular structures. *Journal of Molecular Biology*, **112**, 535–542.

Borea, P.A., Dean, P.M., Martin, I.L. and Perkins, T.D.J. (1992) The structural comparison of two benzodiazepine receptor antagonists: a computational approach. *Molecular Neuropharmacology*, **2**, 261–268.

Chau, P.-L. and Dean, P.M. (1987) Molecular recognition: 3D structure comparison by gnomonic projection. *Journal of Molecular Graphics*, **5**, 97–100.

Chau, P.-L. and Dean, P.M. (1992a) Automated site-directed drug design: the generation of a basic set of fragments to be used for automated structure assembly. *Journal of Computer-Aided Molecular Design*, **6**, 385–396.

Chau, P.-L. and Dean, P.M. (1992b) Automated site-directed drug design: searches of the Cambridge Structural Database for bond lengths in molecular fragments to be used for automated structure assembly. *Journal of Computer-Aided Molecular Design*, **6**, 397–406.

Chau, P.-L. and Dean, P.M. (1992c) Automated site-directed drug design: an assessment of the transferability of atomic residual charges (CNDO) for molecular fragments. *Journal of Computer-Aided Molecular Design*, **6**, 407–426.

Chau, P.-L. and Dean, P.M. (1994a) Electrostatic complementarity between proteins and ligands. 1. Charge disposition, dielectric and interface effects. *Journal of Computer-Aided Molecular Design*. In press.

Chau, P.-L. and Dean, P.M. (1994b) Electrostatic complementarity between proteins and ligands. 2. Ligand moieties. *Journal of Computer-Aided Molecular Design*. In press.

Chau, P.-L. and Dean, P.M. (1994c) Electrostatic complementarity between proteins and ligands. 3. Structural basis. *Journal of Computer-Aided Molecular Design*. In press.

Clare, M. (1993) HIV protease: structure-based design. *Perspective in Drug Design and Discovery*, **1**, 49–58.

Clementi, E. (1980) Computational aspects of large chemical systems. *Lecture Notes in Chemistry*, **19**, Springer, Berlin.

Danziger, D.J. and Dean, P.M. (1989a) Automated site-directed drug design: a general algorithm for knowledge acquisition about hydrogen-bonding regions at protein surfaces. *Proceedings of the Royal Society of London*, B, **236**, 101–113.

Danziger, D.J. and Dean, P.M. (1989b) Automated site-directed drug design: the prediction and observation of ligand point positions at hydrogen-bonding regions on protein surfaces. *Proceedings of the Royal Society of London*, B, **236**, 115–124.

Dean, P.M. (1993) Molecular similarity. In *3D QSAR in Drug Design: Theory, Methods and Applications* (ed. H. Kubinyi), pp. 150–172. ESCOM, Leiden.
Dean, P.M. and Callow, P. (1987) Molecular recognition: identification of local minima for matching in rotational 3-space by cluster analysis. *Journal of Molecular Graphics*, **5**, 159–164.
Dean, P.M. and Chau, P.-L. (1987) Molecular recognition: optimized searching through rotational 3-space for pattern matches on molecular surfaces. *Journal of Molecular Graphics*, **5**, 152–158.
Dean, P.M., Callow, P. and Chau, P.-L. (1988) Molecular recognition: blind-searching for regions of strong structural match on the surface of two dissimilar molecules. *Journal of Molecular Graphics*, **6**, 28–34.
Dean, P.M., Chau, P.-L. and Barakat, M.T. (1992) Development of quantitative methods for studying electrostatic complementarity in molecular recognition and drug design. *Journal of Molecular Structure (Theochem.)*, **256**, 75–89.
Dean, P.M., Barakat, M.T. and Todorov, N.P. (1995) Optimization of combinatoric problems in structure generation for drug design. In *New Perspectives in Drug Design* (eds P.M. Dean, G. Jolles and C.G. Newton). Academic Press, London. In press.
Desiraju, G.D. (1991) The C–H... O hydrogen bond in crystals: what is it? *Accounts in Chemical Research*, **24**, 290–296.
Fauchère, J-L., Quarendon, P. and Kaetterer, L. (1988) Estimating and representing hydrophobicity potential. *Journal of Molecular Graphics*, **6**, 202–206.
Hermann, R.B. and Herron, D.K. (1991) OVID and SUPER: two overlap programs for drug design. *Journal of Computer-Aided Molecular Design*, **5**, 511–524.
Kearsley, S.K. and Smith, G.M. (1990) An alternative method for the alignment of molecular structures: maximizing electrostatic and steric overlaps. *Tetrahedron Computer Methodology*, **3**, 615–633.
Klebe, G. (1993) Structural alignment of molecules. In *3D QSAR in Drug Design: Theory, Methods and Applications* (ed. H. Kubinyi), pp. 173–199. ESCOM, Leiden.
Kuntz, I.D., Meng, E.C. and Shoichet, B.K. (1994) Computer assisted drug design: results and new directions. In *New Perspectives in Drug Design* (eds P.M. Dean, G. Jolles and C.G. Newton). Academic Press, London. In press.
Lauri, G. and Bartlett, P.A. (1994) CAVEAT—a program to facilitate the design of molecules. *Journal of Computer-Aided Molecular Design*, **8**, 51–66.
Lauri, G., Shea, G.T., Waterman, S., Telfer, S.J. and Bartlett, P.A. (1989) *CAVEAT: A Program to Facilitate the Design of Organic Molecules. Version 2.0.* University of California, Berkeley.
Lewendon, A. and Shaw, W.V. (1993) Transition state stabilization by chloramphenicol acetyltransferase. *Journal of Biological Chemistry*, **268**, 20997–21001.
Mattos, C. and Ringe, D. (1993) Multiple binding modes. In *3D QSAR in Drug Design: Theory, Methods and Applications* (ed. H. Kubinyi), pp. 226–254. ESCOM, Leiden.
McLachlan, A.D. (1982). Rapid comparison of protein structures. *Acta Crystallographica*, **A38**, 871–873.
Mojena, R. (1977) Hierarchical grouping methods and stopping rules: an evaluation. *Computer Journal*, **20**, 359–363.
Murtagh, F. and Hecht, A. (1987) *Multivariate Data Analysis*, Reidel, Dordrecht.
Papadopoulos, M.C. and Dean, P.M. (1991) Molecular structure matching by simulated annealing. 4. Classification of atom correspondences in sets of dissimilar molecules. *Journal of Computer-Aided Molecular Design*, **5**, 119–133.
Perkins, T.D.J. and Dean, P.M. (1993) An exploration of a novel strategy for superposing several flexible molecules. *Journal of Computer-Aided Molecular Design*, **7**, 155–172.
Prive, GG., Milburn, M.V., Tong, L., de Vos, A.M., Yamaizumi, Z., Nishimura, S. and Kim, S-H. (1992) X-ray crystal structures of transforming *p21 ras* mutants suggest a transition-state stabilization mechanism for GTP hydrolysis. *Proceedings of the National Academy of Sciences USA*, **89**, 3649–3653.

2 3D molecular similarity indices and their application in QSAR studies

A.C. GOOD

2.1 Introduction

In order to undertake 3D QSAR calculations or database searches for a biomolecular system of interest, the creation of some form of binding mode model is essential. It is often the case that while a number of ligands for a given receptor are known, the receptor structure itself is not. In this instance, one must infer the critical interactions from ligand structure. While the use of atom–atom matches between ligands are frequently used to this end, such a correspondence is not always obvious when bonding topologies show significant divergence. The two molecules rolipram and denbufylline (Figure 2.1) provide a good illustration of this. While both compounds are inhibitors of phosphodiesterase, their structures show no simple atom–atom correspondence. In these circumstances the search for complementarity in molecular properties such as molecular electrostatic potential (MEP) and shape provides an attractive alternative for determining potential pharmacophores. 3D molecular similarity indices provide an efficient way by which these comparisons may be undertaken. A 3D molecular similarity index is essentially a mathematical function of the molecular properties it is used to compare. By overlaying two molecules and applying the index over property space, a

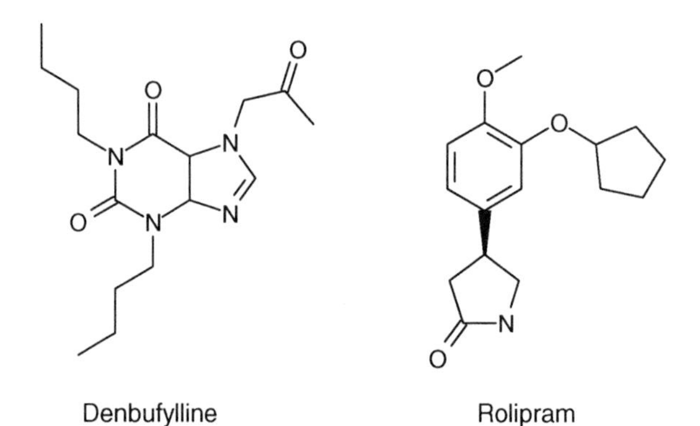

Figure 2.1 The phosphodiesterase inhibitors denbufylline and rolipram. Adapted from Blaney *et al.* (1993a).

numerical value for the molecular similarity is determined. The numbers obtained from molecular similarity calculations contain much information regarding the quantitative relationships between ligand properties. By applying this similarity value as a maximisable function in molecular alignment optimisations, the index can be used to help determine the binding models required for 3D QSAR studies and database searching. For rapidly evaluable properties, molecular similarity indices can be used as the central scoring functions within 3D database searches. The quantitative nature of molecular similarity calculations also means that the index values can be used directly as data within QSAR studies. The indices, the techniques used in their evaluation, and their resulting applications will all be considered in this chapter.

2.2 An index of indices

Many different formalisms have been proposed for the calculation of overall molecular similarity between two overlaid molecules. These indices can be broken down into two broad families. The first and largest family is best described as comprising 'cumulative' similarity indices. They obtain this name from the fact that overall similarity is determined through the accumulation of property overlap or difference values over all space. The second family of indices can be defined as being 'discrete' in nature. These formulae determine similarity at discrete points in space, with the overall similarity determined from the average of the point values. Within these families two further subtypes can be found. The first subtype undertakes some form of overlap or difference measurement in order to describe molecular complementarity, and uses the result in an unmodified form. The second type is constructed so that the resulting similarity value is bounded, for example through normalisation. Indices of all types will be considered.

2.2.1 *Cumulative indices*

The most widely used form of 'cumulative' index applied to the calculation of 3D molecular similarity was pioneered by Carbo *et al.* (1980):

$$C_{AB} = \frac{\int P_A P_B \, dv}{\left(\int P_A^2 \, dv\right)^{1/2} \left(\int P_B^2 \, dv\right)^{1/2}} \tag{2.1}$$

Using the Carbo index, molecular similarity C_{AB} is determined from the structural properties P_A and P_B of molecules A and B being compared over all space.[1] The numerator measures property overlap while the denominator normalises the similarity result. The resultant equation is a modified version of the Pearson correlation coefficient, which measures the deviation of two data sets from linearity.[2] As a consequence of this, the Carbo index measures the deviation of two molecular properties from proportionality, and is thus sensitive to the shape of the property distribution rather than to its magnitude. This is highlighted by the fact that when the measured properties of two molecules correlate, the similarity index tends towards unity. Thus if $P_A = nP_B$, then R_{AB} equals unity.

To increase the sensitivity of the formula to property magnitude, Hodgkin and Richards (1987, 1988) proposed a modification of the Carbo index:

$$H_{AB} = \frac{2\int P_A P_B \, dv}{\int P_A^2 \, dv + \int P_B^2 \, dv} \tag{2.2}$$

This index has the effect that if $P_A = nP_B$, then

$$H_{AB} = \frac{2n}{1 + n^2}$$

Another variant in formulation has recently been presented by Petke (1993):[3]

$$J_{AB} = \frac{\sum_k P_{A_k} P_{B_k}}{\max\left(\sum_k P_{A_k}^2, \sum_k P_{B_k}^2\right)} \tag{2.3}$$

By using only the largest of the two normalising denominator terms, the sensitivity of the index to differences in property magnitude is further enhanced. The resulting effect is that when $P_A = nP_B$

$$J_{AB} = 1/n \text{ if } -\infty < n < -1 \text{ or } 1 < n < \infty, \; J_{AB} = n \text{ if } -1 < n < 1$$

[1] P will be used to denote a generalised molecular property where applicable throughout this section.

[2] Pearson correlation coefficient is

$$P = \frac{\sum_k (u_k - \bar{u})(v_k - \bar{v})}{\left[\sum_k (u_k - \bar{u})^2\right]^{1/2} \left[\sum_k (v_k - \bar{v})^2\right]^{1/2}}$$

[3] Summation and integral symbols may be used interchangeably with most of these formulae. Which is relevant depends on whether molecular property summation was undertaken numerically or analytically. The techniques used for comparing properties through space will be considered further in section 2.4.

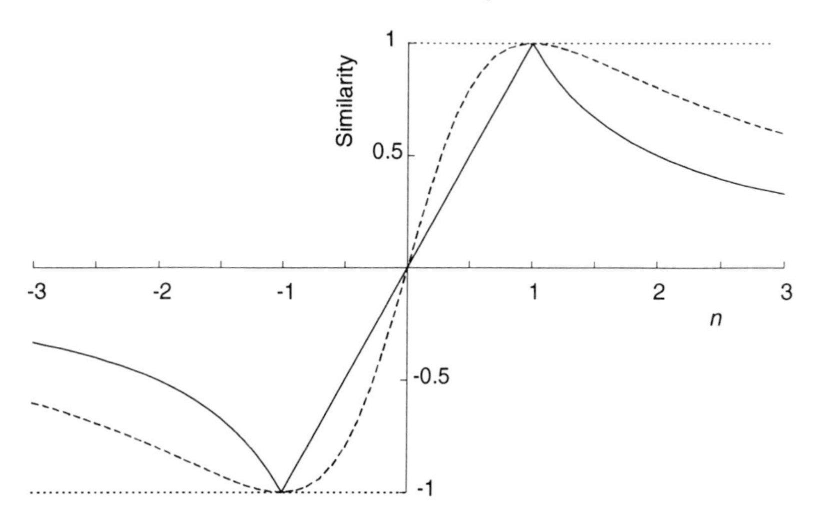

Figure 2.2 Variation of equations (2.1)–(2.3) with the constant of proportionality n between molecular property values under comparison. ———, J; ····, C; – – –, H. Adapted from Petke (1993).

Figure 2.2 shows the variation of C, H and J with differing proportionality constants. Note that other variants are possible, for example using only the template molecule in the denominator term in order to measure substructure similarity.

C and H have been widely applied in the measurement of molecular similarity. Originally C was used to measure the similarity of quantum mechanically derived electron density (Carbo *et al.*, 1980; Bowen-Jenkins and Richards, 1985, 1986a,b; Hodgkin and Richards, 1986; Carbo and Domingo, 1987). This has the virtue of being analytical and firmly grounded in quantum chemistry. Electron density, however, is not a very discriminating property and its calculation is extremely central-processor unit (cpu) intensive. In order to increase evaluation speed, Cioslowski and Fleischmann (1991) and Amovilli and McWeeny (1991) created variants of C, which used the natural spin orbitals derived from *ab initio* quantum mechanical calculations, in order to compare electron distributions. Cooper and Allen (1991) used the momentum space wavefunction property in order to improve discrimination. For the purposes of medicinal chemistry, however, MEPs and shape properties tend to be preferred for comparison due to their case of calculation and improved discrimination. To this end, calculations have been extended to allow the evaluation of MEP/molecular electrostatic field (MEF) (Hodgkin and Richards, 1987; Burt *et al.*, 1990; Burt and Richards, 1990; Richard, 1991; Good, 1992; Good *et al.*, 1992) and shape (Hermann and Herron, 1991; Meyer and Richards, 1991; van Geerestein *et al.*, 1992; Moon and Howe, 1992; Good and Richards, 1993; Nilakantan *et al.*, 1993). J has so far been used purely in the measurement of MEP and MEF similarity (Petke, 1993), though there is no reason why this formula could not be applied in the same way as C and H. The Spearman rank correlation coefficient, a distant relative of the Carbo index, has also been used to measure molecular similarity:

$$S_{AB} = 1 - \frac{6 \sum_{i=1}^{n} d_i^2}{n^3 - n} \tag{2.4}$$

Rather than using the actual numerical values of the properties being measured, this index measures the Pearson correlation coefficient of the relative ranks of the data. The numerator term d_i is the difference in the property ranking at point i of two structures, and n is the total number of points over which the property is measured. This index has been used in a number of studies (Namasivayam and Dean, 1986; Dean and Chau, 1987; Chau and Dean, 1987; Manaut *et al.*, 1991; Sanz *et al.*, 1993) for molecular similarity calculations involving MEP and accessible surface. Hopfinger (1980) proposed a number of indices for measuring the common volume of steric overlap. One example is shown here:

$$V_{O_{ij}} = \sum_{i=1}^{n_A} \sum_{j=1}^{n_B} (V_{Ai} \cap V_{Bj}) \tag{2.5}$$

Each atom is described by a sphere of van der Waals radius. The atomic overlap volume for all intermolecular atom pair combinations is calculated. Summation of these molecular overlap data provides a measure of shape similarity. Hermann and Herron (1991) and Masek *et al.* (1993) have both proposed variants of this approach.

Another Hopfinger (1983) index comes in the form of the ISDFP (integrated spatial difference in field potential) index. This equation provides a measure of dissimilarity in non-bonded forces outside the van der Waals surface:

$$M_P(A, B) = \frac{1}{\Phi}\left[\int_{\Phi} [P_A(R, \otimes, \varphi) - P_B(R, \otimes, \varphi)]^2 \, d\Phi\right] \tag{2.6}$$

$P_A(R, \otimes, \varphi)$ is the potential due to compound A as determined by a probe, P, at spherical co-ordinate position $(R, \otimes, \varphi)$. Φ is the integration volume over which $M_P(A, B)$ is determined.

A number of variants of equation (2.6) have been used in molecular dissimilarity calculations. Perry and van Geerestein (1992) used the root mean squared difference for optimising molecular surface shape similarity:

$$\mathrm{RMSD}_{AB} = \left[\sum_{i=1,n} \{(P_{A,i} - P_{B,i})^2\}/n\right]^{1/2} \tag{2.7}$$

Other variants including the sum squared error and mean squared difference have also been used as a dissimilarity measure (Dean and Chau, 1987; Badel *et al.*, 1992; van Geerestein *et al.*, 1992; Blaney *et al.*, 1993a,b). Blaney *et al.*

(1993a) have also used mean squared error in a discrete manner to analyse graphically molecular similarity.

The SYBYL molecular modelling program uses another variant of this form of equation for the optimisation of Lennard-Jones and/or Coulombic fields:

$$E_{\text{pen}} = \frac{Sff}{\sum\limits_{i=1,n} W_i} \sum_{i=1,n} [W_i(P_{\text{A},i} - P_{\text{B},i})^2] \tag{2.8}$$

For each point in space considered, the squared difference in field values is determined. Sff and W_i are user supplied weights for the overall function and individual lattice point i, respectively. The resultant function provides a measure of molecular dissimilarity in terms of an energy penalty.

2.2.2 *Discrete indices*

While the statistical dissimilarity formulae described above can be considered as discrete indices, only indices specifically designed to provide discrete similarity data will be described in this section. Indices which measure similarity at discrete points in space have recently been developed in order to permit more control over the nature of molecular similarity calculations. The use of discrete similarities allows both graphical (Good, 1992) and statistical (Petke, 1993) analysis to be undertaken on the resultant similarity grid. Overall molecular similarity is obtained by summing the similarities over all points in space and then dividing by the total number of points considered. The resultant value is the average similarity over measured space. The first specifically designed discrete similarity index was developed by Reynolds *et al.* (1992), who described a 'linear' similarity index:

$$l_{\text{AB}} = 1 - \frac{|P_{\text{A}} - P_{\text{B}}|}{\max(|P_{\text{A}}|, |P_{\text{B}}|)} \tag{2.9}$$

$\max(|P_{\text{A}}|, |P_{\text{B}}|)$ equals the larger molecular property magnitude between P_{A} and P_{B} at the grid point where the similarity is being calculated.

An exponential variant of this index was also proposed by Good (1992):

$$e_{\text{AB}} = \exp^{\frac{|P_{\text{A}} - P_{\text{B}}|}{\max(|P_{\text{A}}|, |P_{\text{B}}|)}} \tag{2.10}$$

Petke (1993) created two further discrete indices based on H and J.

$$h_{\text{AB}} = \frac{2P_{\text{A}}P_{\text{B}}}{P_{\text{A}}^2 + P_{\text{B}}^2} \tag{2.11}$$

$$j_{\text{AB}} = \frac{P_{\text{A}}P_{\text{B}}}{\max(P_{\text{A}}^2, P_{\text{B}}^2)} \tag{2.12}$$

The discrete indices shown have been applied to the measurement of MEP and MEF similarity (Good, 1992; Petke, 1993).[4]

2.3 The ultimate index

There are clearly distinct patterns of form in the similarity indices shown above. They suggest that the basic requirement for a successful (dis)similarity index is a function able to describe the product or difference of a property for two overlaid molecules over all (relevant) space. The question that probably springs to mind while leafing through all these equations is, which of these myriad indices is best? Unfortunately, there is no simple answer. Figure 2.2 shows how different indices exhibit significant differences in sensitivity. Petke (1993) explored issues of sensitivity in some detail. Similarity index values were calculated for MEPs and MEFs obtained across selected surface regions of guanine. MEPs and MEFs determined using the Huzinaga/Dunning 'double zeta' basis sets (Dunning, 1970) were compared with those derived from 4-31G, 3-21G and STO-3G basis sets (Hehre *et al.*, 1969; Ditchfield *et al.*, 1971; Binkley *et al.*, 1980). Selected results from these studies are shown in Figure 2.3. Similar studies were also carried out by Good (1992), in which the behaviour C, H, $\bar{l}$ and $\bar{e}$ were compared at a single discrete point in space for different MEP values. Both these studies suggested that discrete indices tend to be more sensitive than their cumulative counterparts, and that J is more sensitive than C, H and S. These systems are, however, rather artificial in nature. The results shown in Figures 2.2 and 2.3 are essentially describing index behaviour for properties that are similar and correlate strongly across the whole region of calculation. Similarity calculations involving actual ligand series are likely to yield more complicated relationships. This point is highlighted in another study undertaken by Good (1992). Nine dopamine D_2-receptor agonist structures were superimposed onto *np*-apomorphine according to the receptor model of Manallack and Beart (1988), and their similarity to *np*-apomorphine was compared. Selected results from these studies are illustrated in Figure 2.4. The data show that while C, H, $\bar{l}$ and $\bar{e}$ behave similarly, C and H are in fact more sensitive in terms of similarity value variation. The reasons for this are twofold. Firstly, the low sensitivity of C and H to correlating properties will emphasise significant regions of dissimilarity as well as similarity (Figure 2.2 illustrates this in the negative proportionality area of the graph). It is also the case that in many regions of comparison, properties will not in fact correlate. The overall index sensitivity on property summation is thus difficult to predict and use as a quality measure.

[4] Averaged overall similarity values for the discrete indices will henceforth be known as $\bar{l}$, $\bar{e}$, $\bar{h}$ and $\bar{j}$.

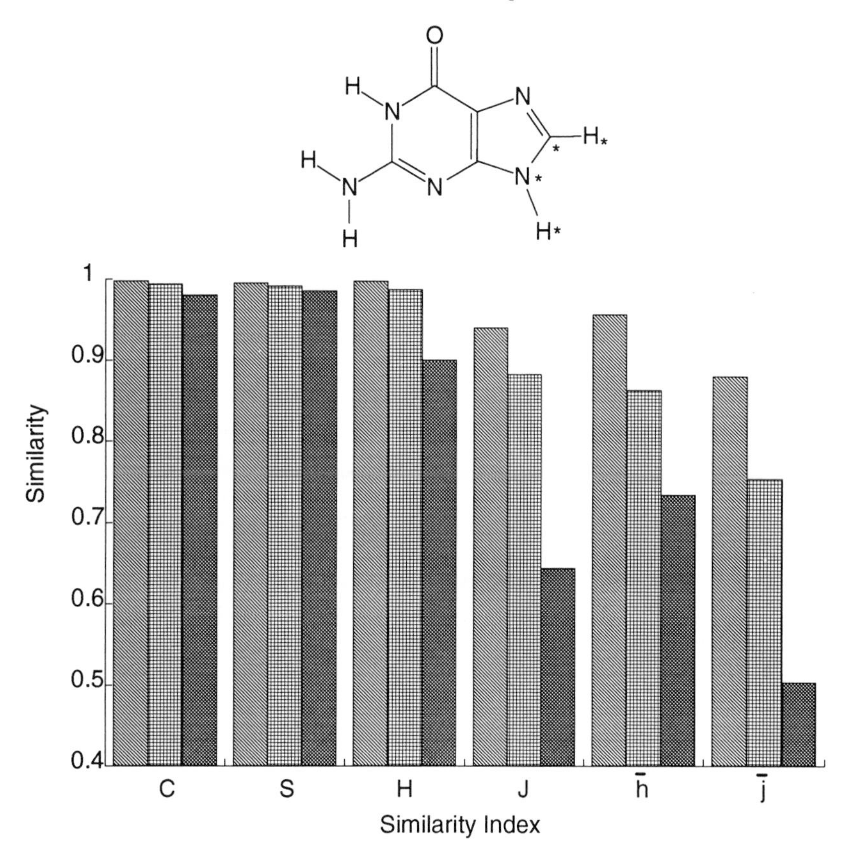

Figure 2.3 Similarity index comparisons for equations (2.1)–(2.4), 2.11 and 2.12. Results derived from comparisons of MEPs on the van der Waals surface of the starred atoms of guanine, computed from SCF wavefunctions using different basis sets. ▧, DZ:4-31G; ▦, DZ:3-21G; ▩, DZ:STO-3G. Adapted from Petke (1993).

Another way in which the relative quality of an index can be measured is in terms of its ability to describe molecular properties such as biological activity. Good *et al.* (1993a) undertook QSAR calculations using similarity matrices on nine different molecular series applying a number of different indices and index calculation conditions. Selected results from these investigations are shown in Figure 2.5. They show that there is some variation in the ability of an index to predict molecular properties. For instance, the discrete indices $\bar{I}$ and $\bar{e}$ appear more predictive than *C*, *H* and *S* given the same calculation conditions. However, the structure of *C*, *H* (and *J*) are such that analytical Gaussian function approximations can be applied to MEP and electron density similarity calculations (Good *et al.*, 1992; Good and Richards, 1993). MEP similarity calculations using *C* in conjunction with these Gaussian functions are actually more predictive than all the numerically evaluated

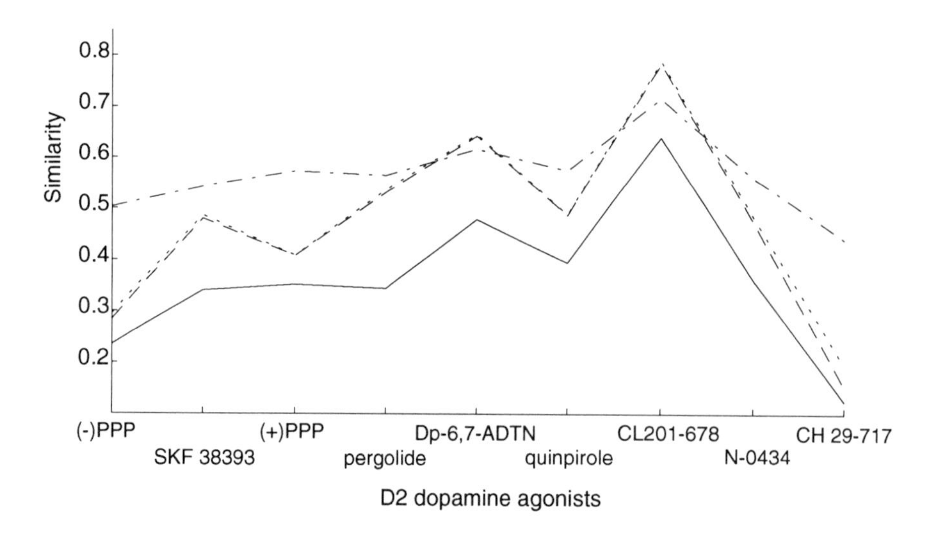

Figure 2.4 Variation of MEP similarity index calculations from equations (2.1), (2.2), (2.9) and (2.10) for *np*-apomorphine versus nine other dopamine agonists. ———, $\bar{I}$; - - - -, $\bar{e}$; - - - - - -, C; – – –, H. Adapted from Good (1992).

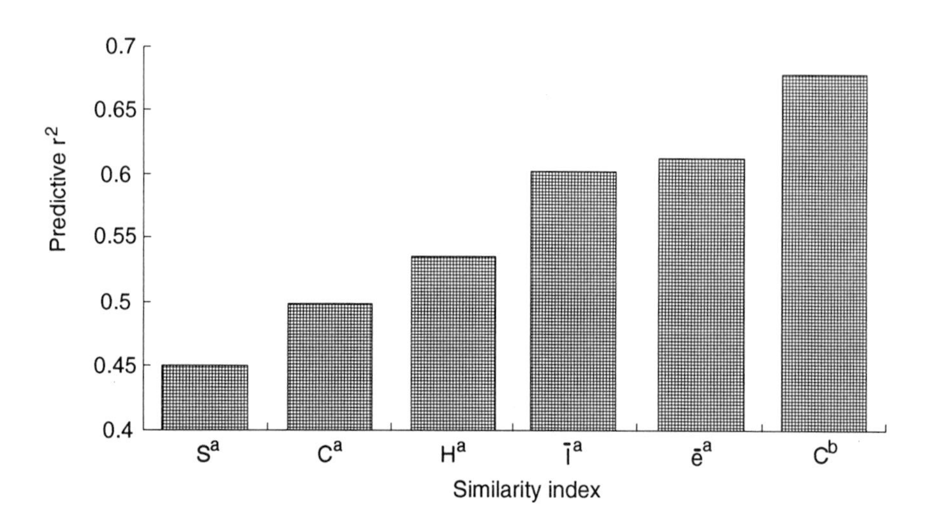

Figure 2.5 Average predictive r^2 values derived from QSAR studies of nine different molecular series, using similarity matrices calculated employing equations (2.1), (2.2), (2.4), (2.9) and (2.10). Predictive r^2 provides a quality test of the QSAR model (see p. 46). a, MEP similarities calculated numerically using a rectilinear grid with an increment of 1.0 Å and an extent of 4.0 Å. See p. 33 for further details. b, MEP similarities calculated analytically using the three-Gaussian function approximation for $1/r$ distance dependence of point charge derived MEP (Good *et al.* 1992). See p. 37 for further details. Adapted from Good *et al.* (1993a).

similarity matrices. It is clear from these, and other results of this study, that it is not just the index but the method in which properties are calculated that will determine the quality of the resulting similarity data. The above observations illustrate the difficulty in defining which is the best index. Each formula has its own strengths and weaknesses. For example, discrete indices such as $\bar{l}$ and $\bar{j}$ allow greater control and graphical/statistical analysis of similarity evaluations, while C, H and J have structures that allow rapid analytical evaluation using Gaussian functions. Also, while S performs poorly in the studies detailed above, its structure is such that it is scale invariant. This can be useful when considering systems with a net charge, as it prevents the charged region from dominating all other system features. As we shall see in sections 2.4 and 2.5, many of the indices have been applied with a good deal of success to a wide variety of problems. The best advice is that the choice of index and its method of evaluation should be made with careful consideration to the molecular design problem at hand. The information provided in section 2.4 should make the nature of these considerations clearer.

2.4 Molecular superposition and property evaluation

The search for potential pharmacophores is one of the central roles to which similarity indices may be applied. When a similarity index was first applied to a QSAR molecular design problem (Hopfinger, 1980), single point evaluations of molecules already superimposed in the postulated binding mode were considered. The electron density comparisons undertaken parallel to this work by Carbo *et al.* (1980) also restricted similarity calculations to single point calculations. It was not until the work of Namasivayam and Dean (1986) and Hodgkin and Richards (1987) that indices were considered as optimisable functions and applied to molecular superposition calculations. Since then, a host of similarity index-based approaches have been devised and applied to the problem of pharmacophore determination.

In many ways, it is not so much the similarity index but the methodology applied to property evaluation that determines the success of a molecular similarity calculation. Detailed consideration will now be given to evaluation tchniques used by the Richards group, together with the indices used and their application in molecular superposition calculations. Related techniques applied by other groups will also be mentioned.

2.4.1 A grid-based view of space

Rectilinear grids are commonly used to approximate continuous molecular properties in computer-aided design techniques, and this is no less the case in molecular similarity calculations.

MEP and MEF evaluation. When originally applied to the problem of measuring molecular similarity, C was used to measure molecular similarity based on quantum mechanically derived electron density (Carbo *et al.*, 1980; Bowen-Jenkins and Richards, 1985, 1986a,b; Hodgkin and Richards, 1986; Carbo and Domingo, 1987). Cpu hungry calculations and lack of results discrimination led Hodgkin and Richards (1987) to apply C and H in numerical evaluations of MEP and MEF in order to determine molecular similarity. For example, MEP was calculated using the standard point charge approach, where the charges (q_i) assigned to each atom (i), create an electrostatic potential at a point r for a molecule of n atoms according to the following equation:

$$P_r = \sum_{i=1}^{n} \frac{q_i}{(r - R_i)} \qquad (2.13)$$

where R_i is the nuclear coordinate position of atom i. MEP and MEF values were determined at the intersections of a rectilinear grid constructed around the two molecules. In order to avoid singularities at the atomic nuclei (where $1/(r - R_i)$ tends to infinity), electrostatic properties were only determined outside the van der Waals volumes of the molecules involved in the calculation. C and H index evaluations were then undertaken numerically. This procedure is summarised in Figure 2.6.

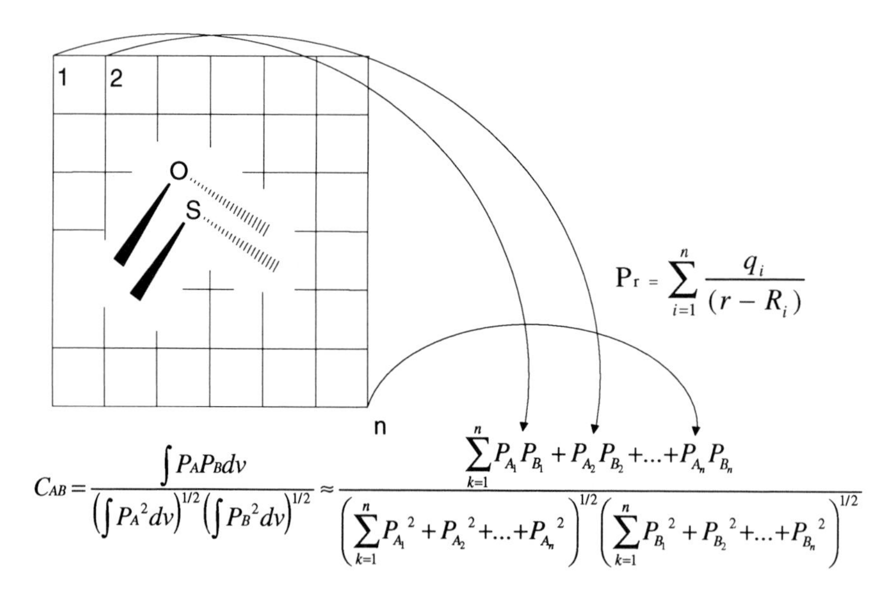

Figure 2.6 Schematic 2D representation of the numerical grid-based evaluation of equation (2.1) using MEP data.

In order to execute molecular similarity optimisations, the following steps were undertaken. First one molecule was defined as fixed and the other as mobile. An initial choice of overlap was made, and then the mobile molecule was allowed to rotate and translate as a rigid body in order to maximise the similarity index (C or H in this case) using the Simplex method (Nelder and Mead, 1965).

The structures chosen for comparison in the above studies were Me_2CH_2, Me_2S and Me_2O. These molecules were chosen primarily to allow direct comparisons with the calculations undertaken previously using electron density. This should give the reader a good indication of the kind of compounds used in conjunction with *ab initio* electron-density molecular similarity calculations. The increased speed of MEP and MEF similarity calculations allows systems of greater biological significance to be chosen for study. With this in mind, the methodology was further developed by Burt *et al.* (1990) using a series of nitromethylene insecticides. Various grid increments, extents and charge schemes were tested in order to determine the optimum compromise between speed and accuracy of calculation. The conclusions drawn from these studies were that a 1.0 Å grid increment with a 4 Å extent was sufficient to produce consistent results. The functionality was developed further by Burt and Richards (1990) who added the ability to include flexible fitting during similarity optimisation. Using a series of hypoglycaemic/hypolipidaemic active ligands, they found that by allowing torsional flexibility, structural orientations with significantly higher similarity could be realised. It was noted, however, that energetically unfavourable conformations were sometimes being obtained in order to achieve high similarity. In order to alleviate this problem, the similarity index was weighted with a Boltzmann factor in order to penalise large increases in internal energy:

$$C_{AB} = \left(\frac{\int P_A P_B \, dv}{\left(\int P_A^2 \, dv \right)^{1/2} \left(\int P_B^2 \, dv \right)^{1/2}} \right) e^{-(c\Delta E/RT)} \tag{2.14}$$

where c = weighting factor, ΔE = (energy of rotated conformation − energy of initial conformation), R = gas constant and T = temperature. A weighting factor of 0.1 was found to produce a good compromise between internal energy increases and optimum similarity. All the software functions described above have been integrated into a single software package known as ASP.

Similar functionality to that provided above has also been developed by Manaut *et al.* (1991). They have developed MEPCOMP and related programs which use S to determine and optimise MEP similarity from *ab initio* derived MEP grids, as well as those calculated from point charges. They employ a novel method for determining which grid points are included in the optimisation. Rather than simply excluding the van der Waals volume of the molecules from the calculation, inner and out exclusion volumes are defined proportional to the van der Waals radii of the constituent atoms. The internal shell is usually set to a small value so that the high MEP values can attempt to simulate steric interactions. Sanz *et al.* (1993) also created MEPSIM in order to integrate MEPCOMP and other similarity calculation programs into a common design framework.

One of the problems of a rectilinear grid with uniform point density is that the system gives equal importance to points close to and far from the molecules being compared. Richard (1991) proposed an alternative in the form of a molecular atom centred radial (MACRA) grid. As its name suggests, a MACRA grid produces grid points emanating from the atom centres of a molecule. A template sphere with a fixed (2 Å) or van der Waals radius (R1) with approximately uniformly distributed points is centred on each atom. This forms the first layer of the grid. The second layer is created by scaling the sphere radius through the addition of the average distance between grid points on the lower layer (for a sphere with 56 points this translates to a distance of ca. $(4\pi R1^2/56)^{1/2}$). Points clashing with lower shells of other atoms are removed. As the layers are built up, a radial grid is formed which requires around 1/50th of the points to cover the same volume as a 1.0 Å increment rectilinear grid around a ligand of drug-like proportions. Figure 2.7 shows a schematic representation of a MACRA grid. The methodology has certain problems, not least of which is its inability to deal adequately with concave molecular surfaces. Nevertheless, the technique is an interesting one.

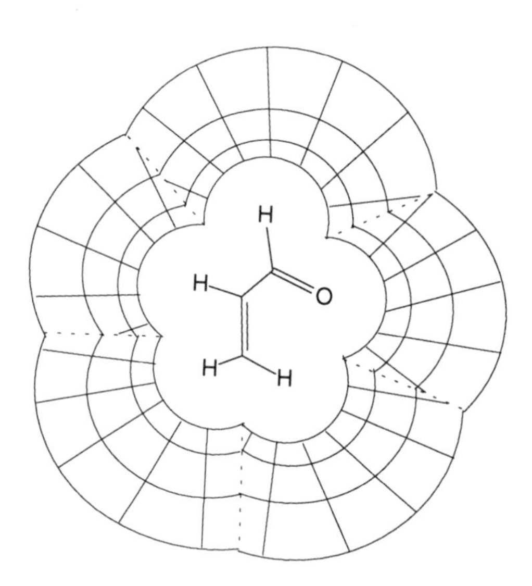

Figure 2.7 Schematic 2D representation of a MACRA grid. Adapted from Richard (1991).

Shape similarity calculations. Molecular shape complementarity between ligand and receptor is a fundamental aspect of molecular recognition. In order to allow the calculation of molecular shape similarity using a rectilinear grid, Meyer and Richards (1991) turned C and H into point counting functions through modifications of the ASP program. The calculation mechanics are similar to those applied in numerical MEP and MEF evaluations. For shape, every grid point is tested to see whether it falls inside the van der Waals surface of each molecule. The results are then applied to modified versions of C and H. The modified version of C is shown below.

$$C_{AB} = \frac{B}{(T_A T_B)^{1/2}} \tag{2.15}$$

B is the number of grid points falling inside both molecules, while T_A and T_B are the total number of grid points falling inside each individual molecule. Fine grids of 0.2–0.5 Å are required to produce consistent results using this methodology.

Calculation times using these techniques allow for a large number of single point evaluations, for example in the pursuit of QSAR data. Similarity optimisation is also possible, though somewhat time consuming when molecules of significant size are considered. Optimisations using these numerical techniques also have the tendency to suffer from premature convergence. The analytical techniques discussed in section 2.4.2 answer many of these problems.

2.4.2 *Gaussian function evalution*

While the use of grid-based similarity evaluation techniques is popular, its numerical foundations contain certain weaknesses. The biggest of these is that in order to gain computation speed, the grids used are normally coarse, with the unfortunate consequence that the numerical evaluations of spatial properties is somewhat rough. In particular, the optimisation of similarity through the adjustment of relative molecular positions is coarse and crude. It is, for example, very difficult for a grid-based similarity optimisation to superimpose a molecule on top of itself, since the program tends to converge prematurely at some discrete point.

The mathematical structures of indices C, H (and J), are such that an analytical alternative may be applied to their evaluation. This alternative comes in the form of molecular property Gaussian function approximations. These functions can be rapidly evaluated and their analytical nature ensures robust similarity optimisation.

MEP evaluation. It is general procedure to measure the MEP of a molecule using the point charge approximation shown in equation (2.13). It is possible to substitute the inverse distance dependence term of equation (2.13) with a Gaussian function approximation:

$$P_r = \sum_{i=1}^{k} q_i(\gamma_1 e^{-\alpha_1 r^2} + \gamma_2 e^{-\alpha_2 r^2} + \ldots \gamma_k e^{-\alpha_k r^2}) \tag{2.16}$$

When this potential function is placed into equation (2.1) (C), the following function is obtained:

$$R_{AB} = \frac{\sum_{i=1}^{n}\sum_{j=1}^{m} q_i q_j \int (G_1^i + G_2^i + \ldots G_k^i)(G_1^j + G_2^j + \ldots G_k^j)\,\mathrm{d}v}{\left(\sum_{i=1}^{n}\sum_{i=1}^{n} q_i q_i \left(\int (G_1^i + G_2^i + \ldots G_k^i)^2\,\mathrm{d}v\right)^{1/2}\right)\left(\sum_{j=1}^{m}\sum_{j=1}^{m} q_j q_j \left(\int (G_1^j + G_2^j + \ldots G_k^j)^2\,\mathrm{d}v\right)^{1/2}\right)} \tag{2.17}$$

where $G_k^i = \gamma e^{-\alpha_k (r-R_i)^2}$.

The integral terms shown in equation (2.17) expand into a series of two-centre Gaussian overlap integrals. This can be seen by considering a two Gaussian function term expansion:

$$\begin{aligned}
&\int (\gamma_1 e^{-\alpha_1 (r-R_i)^2} + \gamma_2 e^{-\alpha_2 (r-R_i)^2})(\gamma_1 e^{-\alpha_1 (r-R_j^2)} + \gamma_2 e^{-\alpha_2 (r-R_j)^2})\,\mathrm{d}v \\
&= \gamma_1^2 \int e^{-\alpha_1 (r-R_i^2)} e^{-\alpha_1 (r-R_j)^2}\,\mathrm{d}v + \gamma_2^2 \int e^{-\alpha_2 (r-R_i)^2} e^{-\alpha_2 (r-R_j)^2}\,\mathrm{d}v \\
&\quad + \gamma_1\gamma_2 \int e^{-\alpha_1 (r-R_i)^2} e^{-\alpha_2 (r-R_j)^2}\,\mathrm{d}v + \gamma_1\gamma_2 \int e^{-\alpha_2 (r-R_i)^2} e^{-\alpha_1 (r-R_j)^2}\,\mathrm{d}v
\end{aligned} \tag{2.18}$$

A two-centre Gaussian overlap integral has a simple solution based on the exponent values and distances between atom centres (Szabo and Ostland, 1982):

$$\int e^{-\alpha_1 (r-R_i)^2} e^{-\alpha_2 (r-R_j)^2}\,\mathrm{d}v = \left(\frac{\pi}{\alpha_1 + \alpha_2}\right)^{3/2} \exp\left(\frac{\alpha_1\alpha_2}{\alpha_1 + \alpha_2}|R_i - R_j|^2\right) \tag{2.19}$$

The similarity calculation can thus be broken down into a succession of readily calculable exponent terms. As a result of this, it is possible to evaluate MEP similarity rapidly and analytically, and as no singularity exists when the potential aproaches an atomic nucleus (Figure 2.8), the calculation need not be restricted to regions outside the atomic van der Waals radii.

Good *et al.* (1992) used these mathematical properties to determine one-, two- and three-term Gaussian functions to approximate the inverse distance dependence term of equation (2.13):

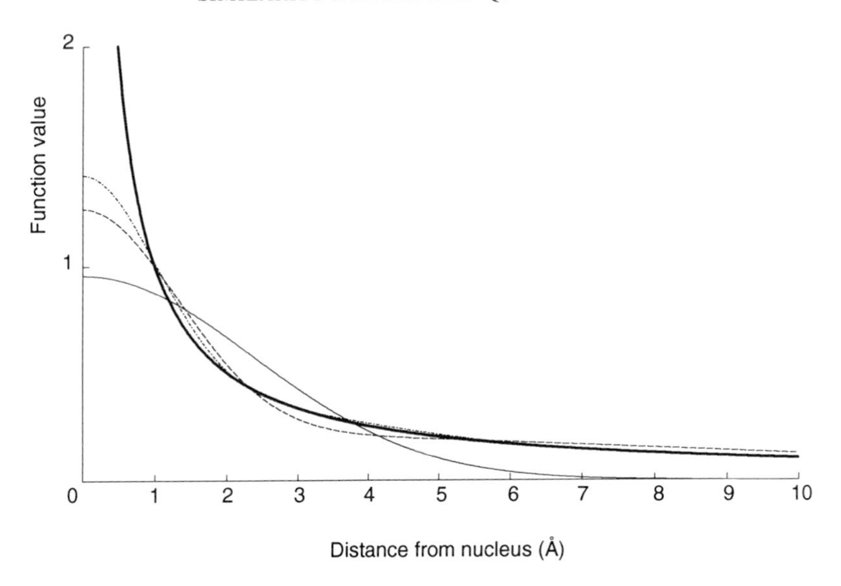

Figure 2.8 Inverse distance and Gaussian functions as functions of distance. ▬▬, $1/r$; ——, 1 Gaussian; - - - - -, 2 Gaussians; –––·, 3 Gaussians. Adapted from Good *et al.* (1992).

$$\text{One Gaussian} \quad 0.9437e^{-0.0890r^2} \tag{2.20}$$

$$\text{Two Gaussians} \quad 0.2181e^{-0.0058r^2} + 1.0315e^{-0.2889r^2} \tag{2.21}$$

$$\text{Three Gaussians} \quad 0.3001e^{-0.0499r^2} + 0.9716e^{-0.5026r^2} + 0.1268e^{-0.0026r^2} \tag{2.22}$$

The fit of these terms to the inverse distance dependence term is shown in Figure 2.8. These Gaussian functions were incorporated into the ASP program, and MEP similarity optimisation calculations were undertaken on identical molecules in different spatial orientations. The results confirmed that the analytical functions were able to superimpose the two molecules back on top of each other (assuming both orientations were in the same local similarity well) while the numerical evaluations would always converge prematurely. The analytical evaluations were also found to be two orders of magnitude faster than equivalent numerical calculations on a per simplex iteration basis. In the example cited by Good *et al.* (1992), one apomorphine molecule was optimised onto another apomorphine molecule rotated by 30° and translated 1 Å in the X-, Y- and Z-axes through its amine nitrogen atom. The time for the three-Gaussian function-based calculation to optimise to unity was approximately 4 min.[5] In contrast, a numerical calculation using a 1.0 Å

[5] The machine used for this calculation was a VAXstation 3520. Its central processing unit speed is comparable to an Iris 4D/20.

increment grid and 4.0 Å extent required almost 2 hours and converged prematurely with a similarity of 0.724. Good *et al.* (1993a) also showed that when used as parameters in QSAR calculations, analytically derived MEP similarity matrices were more predictive than their numerical counterparts (Figure 2.5). It would therefore seem that the use of these Gaussian functions can actually improve data quality, as well as increase the speed and robustness of the optimisation calculations.

Electron density evaluation. The incorporation of Gaussian functions into C, H (and J) need not be limited to MEP calculations. In principle, any property which can be approximated to a set of Gaussians could be compared in this way. Electron density is one such property. Perhaps the first attempt to apply Gaussians in such a way was undertaken by Hodgkin and Richards (1986). In this study, an attempt was made to parameterise Gaussian functions centred at electron density maxima, so as to reproduce the *ab initio* electron density distributions of molecular fragments. The technique was found to work well for the small fragments investigated, but was not pursued because of the difficulty of breaking a larger drug molecule down into a suitable set of fragments.

More recently, Good and Richards (1993) proposed the use of atom-centred Gaussian functions to approximate electron density. The approach taken is somewhat more elementary than that of Hodgkin and Richards (1986), since electron density is simply approximated to the square of the STO-3G atomic orbital wavefunctions used in the GAUSSIAN 88 *ab initio* quantum mechanical program (Frisch *et al.*, 1988). While this is a rough approximation to molecular electron density, the results produced using this technique were found to be akin to those achieved using numerical shape analysis (Meyer and Richards, 1991). As before a large increase in evaluation speed was obtained: two to three orders of magnitude depending on the grid increment used in numerical analysis (0.2–0.5 Å is usual). Employing Gaussian shape data in QSAR calculations produced similar, but slightly inferior, results when compared to the numerical approach (Good *et al.*, 1993a).

Overall, it is clear that Gaussian function-based calculations offer much potential. Their analytical nature and ease of calculation make for robust and rapid similarity optimisations, while their empirical nature seems to impact little on their efficacy. They are thus excellent tools for use in pharmacophore searches and as QSAR parameter generators.

2.4.3 *Other evaluation techniques*

Many innovative methods have been developed to measure spatial properties for the purposes of quantifying overall molecular similarity. Some will be considered in detail in other chapters but deserve a second mention.

Gnomonic projection. This technique was first applied by Chau and Dean (1987) in order to measure MEP similarity using *S*. The method projects the properties of a molecule onto the surface of a sphere. The sphere is approximated by a tessellated icosahedron (Chau and Dean, 1987; Perry and van Geerestein, 1992; Blaney *et al.*, 1993a), or an icosahedron and dodecahedron oriented so that the vertices of the dodecahedron lie on the vectors from the centre of the sphere through the mid-points of the icosahedral faces (Bladen, 1989; van Geerestein *et al.*, 1992). This technique has many useful features including the dramatic simplification of how to map two irregular surfaces to one another. Calculation times are rapid, especially when elements of symmetry inherent to the system are applied to re-orient the projections (Bladen, 1989; van Geerestein *et al.*, 1992; Perry and van Geerestein, 1992). As a result of these features, gnomonic projection techniques have been widely applied to quantitative similarity measurement, from exhaustive similarity comparisons between molecular pairs (Chau and Dean, 1987; Dean and Chau, 1987; Dean *et al.*, 1988), through interactive graphical and numerical analysis (Blaney *et al.*, 1993a,b), to full-scale database searching (van Geerestein *et al.*, 1992; Perry and van Geerestein, 1992). One problematic feature of gnomonic projections is that the technique is essentially restricted to comparing rotational space, and thus the centres of projections used must be chosen with care depending on the choice of application.

Volume measurements. The measurement of molecular shape through the use of volume overlap measurements was one of the first ways in which quantitative similarity measures were made. Hopfinger (1980) first used such measures to determine sets of QSAR parameters for a set of Baker triazines. Hopfinger simply measured the sum of the overlaps between all pairs of atoms in the molecules being compared, using the hard sphere approximation for each atom. Hermann and Herron (1991) developed the program OVID, which measures and optimises the overlap volume between ligand atoms prespecified as being key to activity. Masek *et al.* (1993) use the more accurate analytical volume calculations presented by Connolly (1985) in order to determine optimum molecular shape comparisons (MSC) (see Chapter 7). Extra functions such as only including overlap between atoms with matching chemistry may also be included in the evaluation. OVID is fast but requires the user to define the atoms considered important in the binding mode. The technique is thus essentially measuring fit quality to a postulated pharamacophore. MSC searches are more exhaustive and as a consequence have a higher central processor unit requirement.

Atom distributions. A number of techniques have been developed to allow shape-similarity measurements based on matching atom distributions. Moon and Howe (1992) developed a 3D database search system in which queries are built up from single or multiple overlapped active ligands. Atoms from the query are defined as required or optional: database molecules must have an atom match with required atoms in order to be retrieved as a hit, while optional atom matches can further augment the shape match score. Database molecules are fitted to the query in multiple orientations using the clique detection algorithms of Kuntz *et al.* (1982). Atoms are matched when their centres are found within a predetermined distance of each other. The score applied to a match depends on the degree of chemical complementarity between matched atoms. The atom match scores are used as the molecular property measure in the similarity calculations that follow. Nilakantan *et al.* (1993) have developed another 3D database search system based on the distribution of atom triplet distances within a molecule. Atom triplet distances are defined for every combination of three atoms. The distances are sorted and scaled by length, integerised and packed into a 2048-bit signature in order to save space.[6] The signatures of the template molecules and database molecules are compared during the first stage of a database search, and the triplets of those molecules deemed similar enough are regenerated on the fly. Triplets matches are used as the molecular shape property to quantify similarity.

Bemis and Kuntz (1992) have applied a simplified version of triplet matching in order to measure shape similarity from the perspective of clustering a 3D database. Their technique is to measure the perimeter of each triplet in a molecule, and use the resultant distance to augment the appropriate bin of a molecular shape histogram. These histograms are then compared in order to quantify shape similarity.

Dot surface evaluation. Hermann and Herron (1991) developed an alternative to the OVID program for measuring surface shape complementarity. The software, named SUPER, is used to undertake more exhaustive pharmacophore determination searches. The program basically works by comparing molecular dot surfaces of overlaid molecules, with surface points considered as matching when they are within a predefined distance of each other. MEP data can also be taken into account by discarding matches for points with an MEP difference greater than a predefined cut-off.

Badel *et al.* (1992) use what they define as bidimensional surface profiles to measure surface complementarity. A 2D slice is made through the molecule, and points of a Connolly surface (Connolly, 1983) considered to lie on this plane are used to create a closed polygon line with fixed distance between points. The angular profile of the resulting surface contour is then determined by moving along the line calculating the angles formed by three successive points. Sections of this resulting profile are then compared in order to measure shape complementarity.

[6] A molecule with n atoms generates $n(n-1)(n-2)/6$ triples (5456 triples for a methotrexate-size molecule—heavy atoms only), which is too much data, in the context of a database, to store on hard disk.

2.5 Molecular similarity data in QSAR

The ability of similarity indices to quantify the complementarity of molecular properties between compounds suggests their potential utility within QSAR calculations. A number of techniques have been applied in order to correlate index and activity data. The relative merits of these methodologies are discussed in this section. Materials and methods detail is also given, in order to provide further insight regarding similarity data acquisition and application.

2.5.1 Regression correlation of similarity results with biological data

The simplest way in which to correlate molecular similarity results with biological data is through simple regression analysis. A good example of this comes from the work of Seri-Levy and Richards (1993), who used similarity calculations in order to determine a QSAR for the potency ratio of two ligand enantiomers (eudismic ratio). 3D structures of the ligand enantiomers were built using the CHEM-X modelling program, and then optimised using the semi-empirical quantum mechanical MOPAC program (Stewart, 1990). MEP point charge data were obtained using the program RATTLER (Ferenzcy and Reynolds, 1991). Each enantiomeric pair was then superimposed by a least squares fit so as to maximise the overlap of their stereogenic centres, and the ASP program was used to measure shape and MEP dissimilarity. Shape dissimilarity (1-similarity) calculations were undertaken using the C variant devised by Meyer and Richards (1991). MEP dissimilarity was measured using C in conjunction with the three-Gaussian function approximation of Good *et al.* (1992). The eudismic ratios were correlated against either the shape or MEP dissimilarities, or an average of the two. Selected results from this study are shown in Figure 2.9. Calculations of this nature have also been carried out by Dughan *et al.* (1991), who used molecular similarity calculations in order to quantify the complementarity of the peptide bond to a series of isosteres. Burt *et al.* (1990) correlated the activity of a group of nitromethylene insecticides with their MEP similarity to the most active molecule in the series. Similar studies were also undertaken by Hopfinger (1980, 1983), who used M_P and V_O (equations 2.5 and 2.6) in the construction of QSAR models for Baker triazine and 2,4-diamino-5-benzylpyrimidine dihydrofolate reductase (DHFR) inhibitors (Figure 2.10). A slightly different approach was used in deriving the QSAR compared to our first two examples. Rather than undertaking similarity calculations against a single lead molecule, Hopfinger evaluated similarity using each data set molecule in turn as the template. Each set of similarity values was then correlated with activity, and the group yielding the best result retained for the QSAR model. The final model was derived using multiple linear regression (MLR), with the substituent hydrophobicity parameter π used in conjunction with the shape similarity data. The best model determined for

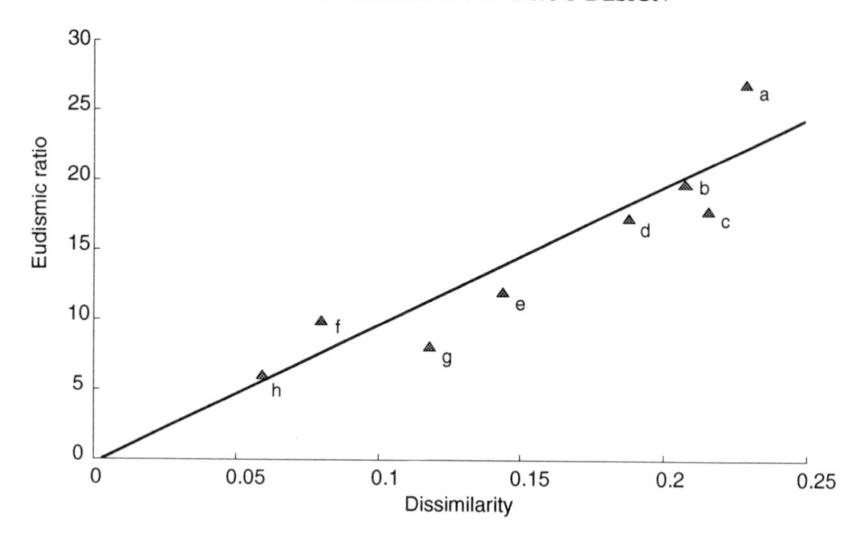

Figure 2.9 Correlation between eudismic ratio and averaged MEP/shape dissimilarity for eight potent chiral molecules: (a) norepinephrine, (b) atropine, (c) epinephrine, (d) scopolamine, (e) amphetamine, (f) dromoran, (g) metamphetamine, (h) methadone. $y = -0.4 + 97x$, $r = 0.930$. Adapted from Seri-Levy and Richards (1993).

the Baker triazine inhibitors was

$$\text{Log}\,(1/C) = -21.31 V_O + 2.39 V_O^2 + 0.44(\pi_3 + \pi_4) + 52.23 \quad (r = 0.931) \tag{2.23}$$

It is clear from the above results that molecular similarity-based regression equations can produce good QSARs. There are, however, significant constraints placed on the use of similarity data when applied in conjunction with standard multiple regression calculations. For example, it should be noted that in the studies of Hopfinger (1980, 1983) and Burt *et al.* (1990), QSARs were constructed using similarity data determined from calculations undertaken with respect to a single template molecule. It is unlikely that a single molecule will contain all, or most, of the structural information inherent to a given ligand data set. It would be preferable, therefore, if one could use the full N

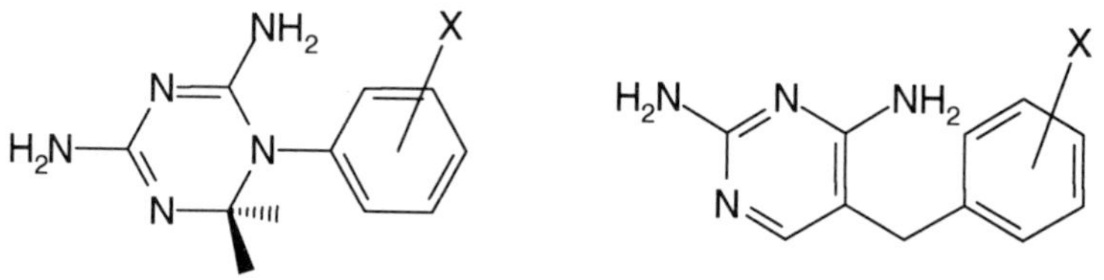

Figure 2.10 Generalised form of the Baker triazines (left) and 2,4-diamino-5-benzylpyrimidines (right) used in the QSAR studies of Hopfinger (1980, 1983).

by N (similarity of each ligand in data set calculated against all other ligands) molecular similarity results matrix in the QSAR calculation. This would make all the SAR data embedded within the molecular similarity values available for model generation. If a full data matrix is used, however, the resultant QSAR parameter set is highly underdetermined (high variable (similarity data) to observation (ligand activity data) ratio). Topliss and Edwards (1979) showed how the use of underdetermined parameter sets in conjunction with MLR can lead to a high probability of chance correlation. For example, in a system with 27 observations and 35 variables, the probability of obtaining an $r^2 \geqslant 0.9$ using purely random variables was found to be approximately 45%. If one wished to use full N by N similarity matrices for both shape and MEP on a system with 27 ligands, a parameter set containing 54 variables would result. It is thus clear that if we are to make full use of molecular similarity results, alternative data handling techniques are required.

2.5.2 *Similarity matrices and multivariate statistical techniques*

Classical MLR suffers from the assumption that the physicochemical parameters used in deriving a QSAR model for activity are independently and normally distributed. For parameters such as similarity matrices, this is unlikely to be the case, as many of the ligand structures in a given series tend to be closely related, and as a consequence their similarities tend to correlate. This feature of many physicochemical parameters can produce significant chance correlation within QSAR models generated using MLR (Cramer *et al.*, 1988a). To combat this problem and allow the analysis of large data sets, techniques able to construct a small number of orthogonal component indices from large numbers of intercorrelated parameters (dimension reduction) must be applied. These methodologies must be used in conjunction with advanced model validation procedures in order to test the predictivity of the resulting statistical models. The use of such techniques is essential to exploit more fully the large amount of physicochemical data available to the computational chemist.

Principal component analysis. First described by Pearson (1901), principal component analysis (PCA) has been a widely used statistical tool for variable dimension reduction for some time. The basic object of PCA is to convert the variables present in a given parameter set into a series of orthogonal components. Each component is made up of a linear combination of data variables, for example, the ith component of a system with n variables (V)

would be,[7]

$$Z_i = a_{i1}V_1 + a_{i2}V_2 + \cdots + a_{in}V_n \tag{2.24}$$

The principal components are sorted by the amount of data variance they explain (greatest first). Usually only the first two or three components are required to describe the majority of the variance. The resulting principal component data have a number of uses. For example, plotting the principal component data for a ligand series can provide a useful graphical SAR model. Carbo and Calabuig (1992) used PCA and other SAR plots to analyse the molecular similarity matrices of a number of systems.[8] The technique has been applied in conjunction with factorial design for choosing the QSAR training set compounds which best span property space (Norinder, 1991a,b, 1992).[9] The major components of PCA calculations can also be used within multiple linear regression calculations in order to reduce the probability of chance correlation (principal component regression, PCR).

Partial least-squares regression. While PCR offers significant advantages over MLR, there is still room for improvement. The PCA derived components are created with data variance description rather than correlation to activity in mind. As a consequence, it is possible to have a high noise-to-signal ratio in the resulting PCA components. Partial least-squares regression (PLS) provides the answer to this problem. PLS essentially works by constructing orthogonal components which model data variance as in PCA, but the components are also constructed in order to maximise correlation with the observed property. In this way the resulting components signal to noise ratio is enhanced, thus improving the resultant correlation.[10]

Cross-validation. It is possible to create as many PLS components as there are variables in a given data set. In general, only the first few components (five or less) will enhance the quality of the QSAR model, with the rest just adding noise. Cross-validation provides a useful method by which to test the model predictivity. The technique works by randomly choosing a subset of ligands in order to create a QSAR model. This model is used to predict the

[7] For a detailed exposition on the matrix algebra involved in the derivation of principal components consult any good book on multivariate statistics. A particularly well-written example with much relevant subject matter is authored by Manly (1986).

[8] Graphical plots of data derived from PCA provide an attractive method for molecular property analysis. A number of alternative pattern recognition techniques have been presented for creating such plots, for example non-linear mapping (Hudson *et al.*, 1988) and neural networks (Livingstone *et al.*, 1991). See Chapter 8 for further details.

[9] Plummer (1990) provides further details regarding problems of experimental design and many other helpful insights on QSAR construction.

[10] For PLS algorithm details consult Wold *et al.* (1984), Dunn *et al.* (1984) or Geladi and Kowalski (1986). The third reference provides a detailed description and comparison of MLR, PCA and PLS algorithms.

activity of the remaining ligands. The process is repeated many times until every compound activity has been predicted using a model from which it was excluded. A cross-validated r^2 can then be generated which provides a measure of model predictivity by comparing the predictive sum of squares (press) against the sum of the squared deviations of each observed biological activity to its mean

$$\text{cross-validated } r^2 = 1 - \frac{\sum (y_{\text{obs}} - y_{\text{pred}})^2}{\sum (\bar{y}_{\text{obs}} - y_{\text{obs}})^2} \tag{2.25}$$

Use of this technique allows multiple models containing differing numbers of components to be tested, in order to determine which is the most predictive. Cramer *et al.* (1988a) have undertaken extensive studies to confirm the value of PLS and cross-validation compared with MLR[11] when dealing with underdetermined QSAR data matrices.

Practical application. With the enormous battery of physicochemical data available to computational chemists, PLS (with cross-validation) is becoming an increasingly popular method for QSAR generation. These techniques were brought to the attention of the computational chemists with the publication of the 3D QSAR technique CoMFA (Cramer *et al.*, 1998b; see Chapter 12 for more details). CoMFA requires techniques other than MLR to deal with the hugely underdetermined systems it uses in deriving QSAR models. The methodology behind CoMFA uses the basic premise that non-covalent forces dominate drug–receptor interactions, and that these forces can be described in terms of steric and electrostatic effects. CoMFA attempts to describe these forces in terms of electrostatic, steric (and other related property) grids calculated around a given series of overlaid ligands. The properties at each grid point are then used to generate a QSAR for the system. The use of N by N similarity matrices contains many parallels to this approach. When an N by N similarity calculation is used to describe a given property, for example MEP, the structural information within the resulting matrix implicitly contains much of the location dependence of the MEP data inherent in CoMFA. The difference between the systems is that while CoMFA will describe a region around a group of molecules using a large number of grid points, the similarity matrix will attempt to describe the same region using just a few numbers. As we have already intimated, the problem of underdetermination exists for similarity matrices in the same way as for CoMFA, and as a consequence PLS likewise becomes the technique of choice in QSAR generation. QSARs have been generated using N by N similarity

[11] Cramer *et al.* (1988a) also considered an alternative QSAR model test called bootstrapping, which they used to determine the confidence interval of r^2 in order to measure model stability. This technique has yet to gain a major foot hold in the QSAR community in comparison to cross-validation, but has been widely applied elsewhere (Leger *et al.*, 1992).

matrices in conjunction with PLS for a number of biological systems (Good *et al.*, 1993a,b). We will now illustrate how these matrices can be generated and applied for QSAR model generation with an example.

Similarity matrix generated QSAR for a series of benzodiazepine receptor inverse agonist site ligands. Allen *et al.* (1990) were the first to study the benzodiazepine system using 3D QSAR techniques, employing CoMFA in order to characterize a pharmacophore for the site. This ligand set provides a good example of a system where 3D QSAR has significant advantages over more traditional methods. The ligand structures (Table 2.1) of the data set show significant variation in both the central skeleton and the peripheral functional groups. The system would thus not lend itself to approaches such as the one applied by Hopfinger to the DHFR inhibitors of Figure 2.10, since it is virtually impossible to break the ligands down into well-defined substituents. The spread of functionality of the aligned ligands (Figure 2.11) further emphasises this point.

Model building. 3D ligand structures were constructed from scratch within CHEM-X using standard bond lengths and angles, ensuring an identical starting geometry for any given substituent. Once built, each structure was minimised within CHEM-X using the default force field. AM1 MOPAC (Stewart, 1990) calculations were then undertaken from which point charges were back calculated to fit the consequent MEPs using the RATTLER software. The structural and activity data for the 37 benzodiazepine receptor inverse agonist site ligands are shown in Table 2.1. The structures were superimposed by least-squares fitting, using the starred atoms shown in Table 2.1. The most active molecule (**37**) in the series was used as the fitting template. Figure 2.11 shows the resultant molecular alignment.

Evaluation of molecular similarity[12]. Molecular complementarity was determined using the ASP program. Similarity calculations were run using MEF in conjunction with H. A 1.0 Å increment and 4.0 Å extent was used for the calculation grid. Shape similarity was evaluated using the grid based approach of Meyer, applying a grid increment of 0.2 Å to evaluate C. For both calculations each ligand was compared to every other ligand, leading to a parameter set of two 37×37 matrices (diagonal values of ligand similarity to itself set to unity).

Statistical analysis. The MEF and shape matrices were merged to form a single 37×74 matrix. After being read into the GOLPE PLS program (Baroni *et al.*, 1993), each matrix column was scaled to have zero mean and unit

[12] Data derived from a previous study (Good *et al.*, 1993a).

Table 2.1 Benzodiazepine receptor inverse agonist site ligand structure and affinity data. Adapted from Good *et al.* (1993a)

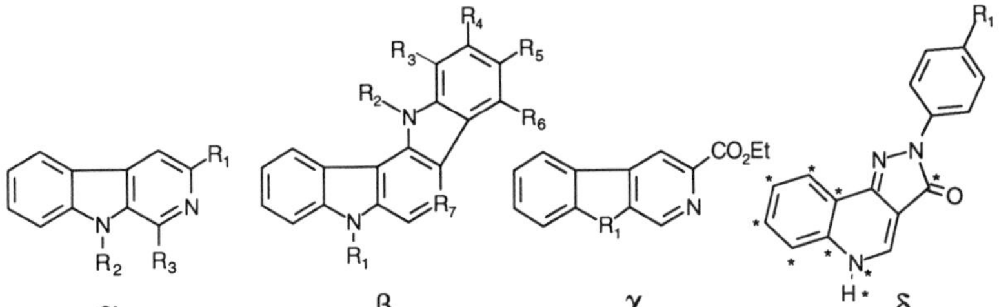

No.	Parent	R_1	R_2	R_3	R_4	R_5	R_6	R_7	log $1/IC_{50}$
1	α	CO_2CH_3	H	H	–	–	–	–	–0.699
2	α	$CO_2CH_2CH_3$	H	H	–	–	–	–	–0.699
3	α	OCH_2CH_3	H	H	–	–	–	–	–1.380
4	α	$OCH(CH_3)_2$	H	H	–	–	–	–	–2.699
5	α	$OCH_2CH_2CH_2CH_3$	H	H	–	–	–	–	–1.991
6	α	OCH_3	H	H	–	–	–	–	–2.093
7	α	$OCH_2CH_2CH_3$	H	H	–	–	–	–	–1.042
8	α	$COCH_2CH_2CH_3$	H	H	–	–	–	–	–0.447
9	α	$CH_2CH_2CH_2CH_3$	H	H	–	–	–	–	–2.389
10	α	H	H	H	–	–	–	–	–3.210
11	α	$CO_2C(CH_3)_3$	H	H	–	–	–	–	–1.000
12	α	Cl	H	H	–	–	–	–	–1.653
13	α	NO_2	H	H	–	–	–	–	–2.097
14	α	$CO_2CH_2C(CH_3)_3$	H	H	–	–	–	–	–2.875
15	α	CO_2CH_3	H	CH_2CH_3	–	–	–	–	–3.877
16	α	H	H	CH_2CH_3	–	–	–	–	–5.398
17	α	H	H	CH_3	–	–	–	–	4.093
18	β	H	H	H	H	H	H	N	–0.602
19	β	H	H	CH_3	H	H	H	N	–1.919
20	β	H	H	H	CH_3	H	H	N	–1.000
21	β	H	H	H	H	CH_3	H	N	–2.350
22	β	H	H	H	H	H	CH_3	N	–3.836
23	β	CH_3	H	H	H	H	H	N	–3.066
24	β	H	CH_3	H	H	H	H	N	–2.196
25	β	CH_3	CH_3	H	H	H	H	N	–3.283
26	β	H	H	H	H	H	H	CH	–3.295
27	β	H	H	H	H	H	OCH_3	N	–2.398
28	β	H	H	H	H	H	Cl	N	–2.854
29	χ	C(=O)	–	–	–	–	–	–	–4.415
30	χ	C(=NOH)	–	–	–	–	–	–	–3.699
31	χ	O	–	–	–	–	–	–	–3.964
32	χ	CH_2	–	–	–	–	–	–	–2.833
33	χ	C(=O)N(H)	–	–	–	–	–	–	–3.380
34	χ	S	–	–	–	–	–	–	–3.230
35	δ	H	–	–	–	–	–	–	0.398
36	δ	Cl	–	–	–	–	–	–	0.222
37	δ	OCH_3	–	–	–	–	–	–	1.000

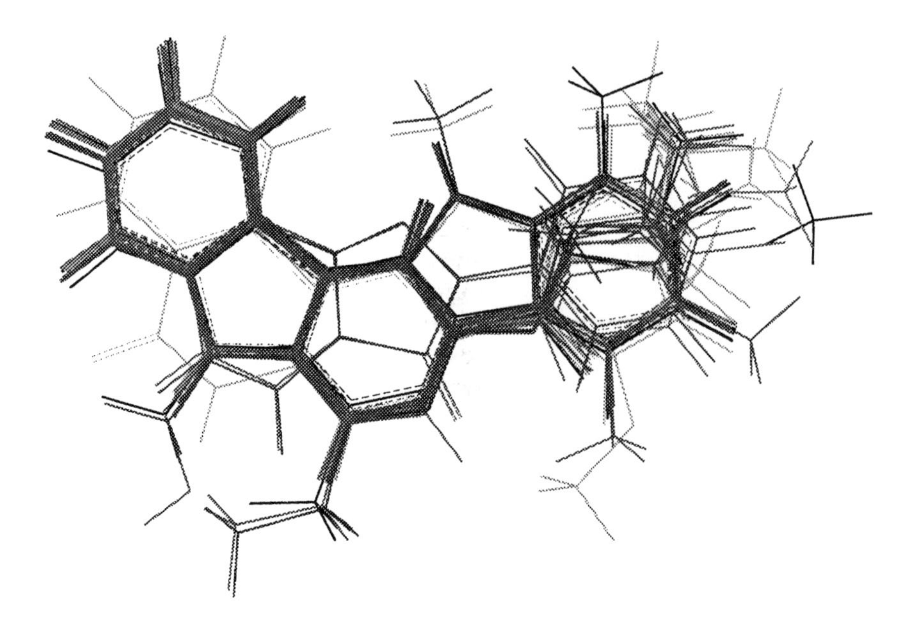

Figure 2.11 Benzodiazepine receptor inverse agonist site ligands aligned for 3D QSAR generation.

variance. This ensures that all matrix data columns are given equal weighting during statistical analysis, which is important for similarity matrices, since MEF/MEP similarity tends to vary more than that of shape. PLS calculations were then undertaken on the resulting matrix using leave-one-out (LOO) cross-validation. LOO means that a series of PLS calculations are undertaken, leaving out each ligand in turn. The activity of the excluded ligand is predicted using the model generated without it, and the resulting predictions are used to generate a cross-validated r^2 value. The best PLS model generated from this study has a cross-validated r^2 of 0.655 in three components.

Many PLS calculations would end at this point. However, while PLS is a significant improvement compared to other statistical methods for 3D QSAR, the technique is still prone to adding noise in with signal when constructing component descriptions. To counter this, a number of variable selection techniques are available to sort the wheat from the chaff. The simplest technique is simply to remove variables which show little standard deviation. This prevents the introduction of unimportant data, which when scaled produces much unwanted noise. This technique is not suitable for similarity calculations, since variables tend to exhibit similar deviations within their property subset. Another technique is to analyse the variables for cross-correlation, keeping only the most orthogonal parameters in order to reduce the dimensionality of the data matrix while losing the minimum of information. Such a method is available within the GOLPE package in the form of the D-optimal design (Mitchell, 1974). The procedure probes the distribution of variables in the GAMMA or LOADING space of the PLS analysis, and attempts to keep the *n* most orthogonal parameters in that space for a given number of components (*n* is user defined). For this example, the PLS model which yielded the best cross-validated r^2 (three components) was chosen for analysis, and 50% of the variables were retained. A LOO PLS run on the resulting 37×37 matrix yielded a model with a cross-validated r^2 of 0.745 in four components.

A second GOLPE variable selection procedure known as progressive fixing exclusion was then utilised (Baroni *et al.*, 1993) to further prune the data matrix. This process attempts to determine which variables aid correlation through the application of design matrices processed using fractional factorial design (Box *et al.*, 1978). Large numbers of different data matrices are created in which variables are included or excluded according to the current design cycle. The resultant matrices are used to form a new cross-validated PLS model, and the predictivity of the models used to determine which variables improve predictive correlation. All variables which are determined to be noise in the system are excluded. GOLPE retains all variables determined to be signal together with those of uncertain property. It is recommended, however, that data of unknown effect are also excluded to maximise the robustness of the model (Baroni *et al.*, 1993). These data were thus also removed manually, leaving only variables proven to aid predictive correlation. The procedure was executed on the reduced matrix using the four-component model. Cross-validation was undertaken using 50 random groups of 10 molecules from the test set. This form of cross-validation is considered a more robust test than the LOO procedure, especially during variable selection. Multiple random selections help avoid biasing variable selection towards predicting activity for ligand types which predominate within the data set (deviations from a normal distribution of data). The ratio of variables to dummies was set to 1 with a design combinations to variables ratio of 1.0 (these numbers control the nature of the design matrices applied to variable selection). More dummies and design combinations increase technique utility and cpu requirements (not always in equal amounts). For this experiment, the most exhaustive run possible was used. The final LOO PLS model had a cross-validated r^2 of 0.804 in four components. The final data matrix contained 18 variables, 8 MEP and 10 shape.

At this point, an analysis of the model is required. A graph of predicted versus observed activity is shown in Figure 2.12. The equation for the PLS model rotated back into regression space is

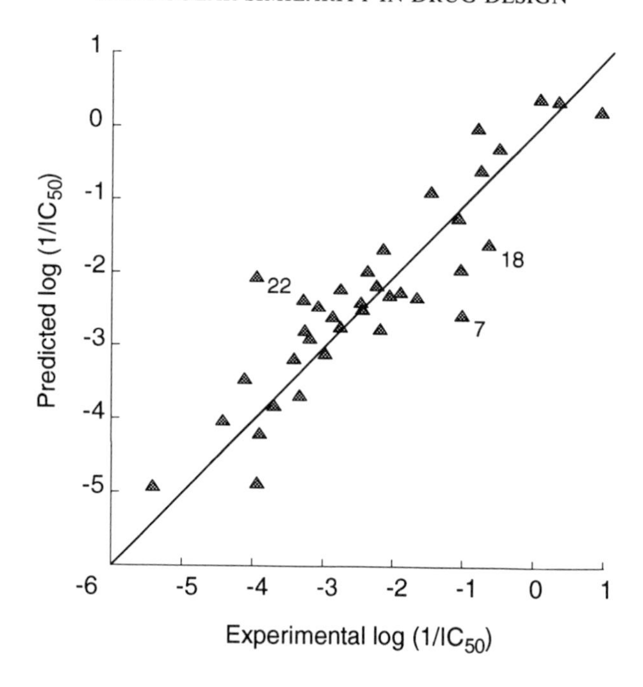

Figure 2.12 Graph of predicted versus experimental activity, for the benzodiazepine receptor inverse agonist site ligands detailed in Table 2.1.

$$\begin{aligned} y = -18.096 &+ 0.277S_{\mathrm{MEP19}} + 0.307S_{\mathrm{MEP20}} + 2.082S_{\mathrm{MEP35}} + 1.869S_{\mathrm{MEP36}} \\ &+ 2.378S_{\mathrm{MEP37}} + 2.542S_{\mathrm{MEP13}} - 1.813S_{\mathrm{MEP16}} - 1.227S_{\mathrm{MEP17}} \\ &+ 1.312S_{\mathrm{MEP33}} + 6.811S_{\mathrm{SHA18}} + 6.759S_{\mathrm{SHA8}} + 0.358S_{\mathrm{SHA31}} \\ &+ 0.443S_{\mathrm{SHA32}} + 0.456S_{\mathrm{SHA24}} + 0.362S_{\mathrm{SHA37}} + 0.471S_{\mathrm{SHA27}} \\ &+ 0.532S_{\mathrm{SHA17}} - 0.356S_{\mathrm{SHA18}} \end{aligned} \tag{2.26}$$

The equation illustrates a major weakness of QSARs derived from similarity matrices. While CoMFA is able to display the coefficients of its QSAR equations as maps of favourable and unfavourable structural interactions, no such elegant method exists for extracting chemical meaning from similarity matrix equations. Nevertheless, overall activity is well modelled as is generally the case (Good *et al.*, 1993a). Only three molecules have an error greater than one order of magnitude (**7**, **18** and **22**). At first it would appear that the model is unable to make out some of the structural subtleties of the type β molecules. However, closer inspection of the activity data reveals that methyl group addition (**22**) has more impact on activity than the addition of a methoxy group (**27**). Ligand **27** appears as a positive influence in the QSAR (according to the equation 2.26) and has its methoxy group directly overlapping the methyl group of **22**. The direction of the methoxy group in the alignment may thus need refinement. Similarly, the R_1 group of **7** is aligned in the same way as all other type α structures. It may be, however, that R_1 groups containing other oxygens need to be realigned to a position closer to a postulated pharmacophore hydrogen bond (Allen *et al.*, 1990). One way to alter alignments, of course, would be to apply similarity optimisation to the system, since at the moment only a simple least-squares fit of the central skeletons has been undertaken. Since the alignment problem is the single largest obstacle to successful 3D QSAR models (iterative alignment alterations for honing a 3D QSAR are not uncommon), this would deserve further consideration in a more extensive study. The strong correlation achieved nonetheless portrays the abundance of structural information stored within the similarity matrices, and illustrates their potential as tools in QSAR research.

2.6 Conclusions

In this chapter we have considered many of the techniques used to quantify ligand similarity through the application of molecular similarity indices. An array of molecular properties have been utilised in conjunction with this methodology, from the detailed comparison of electrostatic fields and molecular volumes, through to the rapid analysis of triangle perimeters formed between atom triplets. The consequent adaptability allows the indices to be evaluated rapidly enough for database searching, yet provide a sufficiently accurate reflection of structural equivalence to be applicable in QSAR studies. As a result, molecular similarity index calculations have been used in a wide variety of molecular design problems, including pharmacophore elucidation, 3D QSAR calculations and shape-based database searches. This versatility makes such quantitative molecular similarity calculations important tools in the hands of the computational chemist.

References

Allen, M.S., Tan, Y., Trudell, M.L., Narayanan, K., Schindler, L.R., Martin, M.J., Schultz, C., Hagen, T.J., Koehlar, K.F., Codding, P.W., Skolmich, P. and Cook, J.M. (1990) Synthetic and computer assisted analysis of the pharmacophore for the benzodiazepine receptor inverse agonist site. *Journal of Medicinal Chemistry*, **33**, 2343–2357.

Amovilli, C. and McWeeny, R. (1991) Shape and similarity: two aspects of molecular recognition. *Journal of Molecular Structure*, **227**, 1–9.

ASP, Oxford Molecular Ltd, The Magdalen Centre, Oxford Science Park, Sandford on Thames, Oxford OX4 4GA, UK.

Badel, A., Mornon, J.P. and Hazout, S. (1992) Searching for geometric molecular shape complementarity using bi-dimensional surface profiles. *Journal of Molecular Graphics*, **10**, 205–211.

Baroni, M., Constantino, G., Cruciani, G., Riganello, D., Valigi, R. and Clementi, S. (1993) GOLPE: an advanced chemometric tool for 3D QSAR problems. *Quantitative Structure–Activity Relationships*, **12**, 9–20.

Bemis, G.W. and Kuntz, I.D. (1992) A fast efficient method for 2D and 3D molecular shape description. *Journal of Computer-Aided Molecular Design*, **6**, 607–628.

Binkley, J.S., Pople, J.A. and Hehre, W.J. (1980) Self consistent molecular orbital methods: small split valence basis sets for first row elements. *Journal of the American Chemical Society*, **102**, 939–946.

Bladen, P. (1989) A rapid method for comparing and matching the spherical parameter surfaces of molecules and other irregular objects. *Journal of Molecular Graphics*, **7**, 130–137.

Blaney, F.E., Finn, P., Phippen, R. and Wyatt, M. (1993a) Molecular surface comparison: application to molecular design. *Journal of Molecular Graphics*, **11**, 98–105.

Blaney, F.E., Naylor, D. and Woods, J. (1993b) MAMBAS: a realtime graphics environment for QSAR. *Journal of Molecular Graphics*, **11**, 157–165.

Bowen-Jenkins, P.E., Cooper, D.L. and Richards, W.G. (1985) *Ab initio* computation of molecular similarity. *Journal of Physical Chemistry*, **89**, 2195–2197.

Bowen-Jenkins, P.E. and Richards, W.G. (1986a) Molecular similarity in terms of valence electron density. *Journal of the Chemical Society. Chemical Communications*, 133–135.

Bowen-Jenkins, P.E. and Richards, WG. (1986b) Quantitative measures of similarity between pharmacologically active compounds. *International Journal of Quantum Chemistry*, **30**, 763–768.

Box, G.E.P., Hunter, W.G. and Hunter, J.S. (1978) *Statistics for Experimenters*. Chapter 12. John Wiley, New York.

Burt, C. and Richards, W.G. (1990) Molecular similarity: the introduction of flexible fitting. *Journal of Computer-Aided Molecular Design*, **4**, 231–238.

Burt, C., Richards, W.G. and Huxley, P. (1990) The application of molecular similarity calculations. *Journal of Computational Chemistry*, **11**, 1139–1146.

Carbo, R. and Domingo, L. (1987) LCAO-MO similarity measures and taxonomy. *International Journal of Quantum Chemistry*, **32**, 517–545.

Carbo, R. and Calabuig, B. (1992) Molecular quantum chemistry measures and *N*-dimensional representation of quantum objects. II. Practical applications. *International Journal of Quantum Chemistry*, **42**, 1695–1709.

Carbo, R., Leyda, L. and Arnau, M. (1980) An electron density measure of the similarity between two compounds. *International Journal of Quantum Chemistry*, **17**, 1185–1189.

Chau, P.-L. and Dean, P.M. (1987) Molecular recognition: 3D surface structure comparison by gnomonic projection. *Journal of Molecular Graphics*, **5**, 97–100.

CHEM-X, Chemical Design Ltd, Roundway House, Cromwell Business Park, Chipping Norton, Oxon, OX7 5SR, UK.

Cioslowski, J. and Fleischmann, E.D. (1991) Assessing molecular similarity from results of *ab initio* electronic structure calculations. *Journal of the American Chemical Society*, **113**, 64–67.

Connolly, M.L. (1983) Analytical molecular surface calculation. *Journal of Applied Crystallography*, **16**, 548–558.

Connolly, M.L. (1985) Computation of molecular volume. *Journal of the American Chemical Society*, **107**, 1118–1124.

Cooper, D.L. and Allen, N.L. (1991) A novel approach to molecular similarity. *Journal of Computer-Aided Molecular Design*, **3**, 253–259.

Cramer, R.D. III, Bunce, J.D. and Patterson, D.E. (1988a) Cross-validation, bootstrapping, and PLS compared with multiple linear regression in conventional QSAR studies. *Quantitative Structure–Activity Relationships*, **7**, 18–25.

Cramer, R.D. III, Patterson, D.E. and Bunce, J.D. (1988b) Comparative molecular field analysis (CoMFA). Effect of shape on binding of steroids to carrier proteins. *Journal of the American Chemical Society*, **110**, 5959–5967.

Dean, P.M. and Chau, P.-L. (1987) Molecular recognition: optimised searching through molecular 3-space for pattern matches on molecular surfaces. *Journal of Molecular Graphics*, **5**, 152–158.

Dean, P.M., Callow, P. and Chau, P.-L. (1988) Molecular recognition: blind searching for regions of strong structural match on the surfaces of two dissimilar molecules. *Journal of Molecular Graphics*, **6**, 28–34.

Ditchfield, R., Hehre, W.J. and Pople, J.A. (1971) Self consistent molecular orbital methods: an extended Gaussian type basis set for molecular orbital calculations of organic molecules. *Journal of Chemical Physics*, **54**, 724–728.

Dughan, L., Burt, C. and Richards, W.G. (1991) The study of peptide bond isosteres using molecular similarity. *Journal of Molecular Structure*, **235**, 481–488.

Dunn, W.J. III, Wold, S., Edlund, U. *et al.* (1984) Multivariate structure–activity relationships between data from biological tests and an ensemble of structure descriptors. The PLS method. *Quantitative Structure–Activity Relationships*, **3**, 131–137.

Dunning, T.H. (1970) Gaussian basis functions for use in molecular calculations. I. Contraction of (9*s*5*p*) atomic basis sets for the first row atoms. *Journal of Chemical Physics*, **3**, 2823–2833.

Ferenzcy, G., Reynolds, C.A. and Richards, W.G. (1991) Semi-empirical AM1 electrostatic potential and AM1 electrostatic potential derived charges, a comparison with *ab initio* values. *Journal of Computational Chemistry*, **11**, 159–166.

Frisch, M.J., Head, G.M., Schlegel, H.B., Raghavachari, K. *et al.* (1988) *Gaussian 88*. Gaussian Inc., Pittsburgh, Pennsylvania.

Geladi, P. and Kowalski, B.R. (1986) PLS regression: a tutorial. *Analytica Chimica Acta*, **185**, 1–17.

Good, A.C. (1992) The calculation of molecular similarity: alternative formulas, data manipulation and graphical display. *Journal of Molecular Graphics*, **10**, 144–151.

Good, A.C. and Richards, W.G. (1993) Rapid evaluation of shape similarity using Gaussian functions. *Journal of Chemical Information and Computer Sciences*, **33**, 112–116.

Good, A.C., Hodgkin, E.E. and Richards, W.G. (1992) The utilisation of Gaussian functions for the rapid evaluation of molecular similarity. *Journal of Chemical Information and Computer Sciences*, **32**, 188–191.

Good, A.C., Peterson, S.J. and Richards, W.G. (1993a) QSARs from similarity matrices. Technique validation and application in the comparison of different similarity evaluation methods. *Journal of Medicinal Chemistry*, **36**, 2929–2937.

Good, A.C., So, S. and Richards, W.G. (1993b) Structure–activity relationships from similarity matrices. *Journal of Medicinal Chemistry*, **36**, 433–438.

Hehre, W.J., Stewart, R.F. and Pople, J.A. (1969) Self consistent molecular orbital methods: use of Gaussian expansions for Slater type orbitals. *Journal of Chemical Physics*, **51**, 2657–2664.

Hermann, R.B. and Herron, D.K. (1991) OVID and SUPER: two overlap programs for drug design. *Journal of Computer-Aided Molecular Design*, **5**, 511–524.

Hodgkin, E.E. and Richards, W.G. (1986) A semi-empirical method for calculating molecular similarity. *Journal of the Chemical Society. Chemical Communications*, 1342–1344.

Hodgkin, E.E. and Richards, W.G. (1987) Molecular similarity based on electrostatic potential and electric field. *International Journal of Quantum Chemistry. Quantum Biology Symposia*, **14**, 105–110.

Hodgkin, E.E. and Richards, W.G. (1988) Molecular similarity. *Chemistry in Britain*, 1141–1144.

Hopfinger, A.J. (1980) A QSAR investigation of DHFR inhibition by Baker triazines based upon molecular shape analysis. *Journal of the American Chemical Society*, **102**, 7196–7206.

Hopfinger, A.J. (1983) Theory and analysis of molecular potential energy fields in molecular shape analysis: a QSAR study of 2,4-diamino5-benzylpyrimidines as DHFR inhibitors. *Journal of Medicinal Chemistry*, **26**, 990–996.

Hudson, B., Livingstone, D. and Rahr, E. (1988) Pattern recognition display methods for the analysis of computed molecular properties. *Journal of Computer-Aided Design*, **3**, 55–65.

Kuntz, I.D., Blaney, J.M., Oatley, S.J., Langridge, R. and Ferrin, T.E. (1982) A geometric approach to macromolecule–ligand interactions. *Journal of Molecular Biology*, **161**, 269–288.

Leger, C., Prolitis, D.N. and Romano, J.P. (1992) Bootstrap technology and applications. *Technometrics*, **34**, 378–395.

Livingstone, D.J., Hesketh, G. and Clayworth, D. (1991) Novel method for the display of multivariate data using neural networks. *Journal of Molecular Graphics*, **9**, 115–118.

Manallack, D.T. and Beart, P.M. (1988) A three dimensional receptor model of the dopamine D_2 receptor from computer graphic analyses of D_2 agonists. *Journal of Pharmacy and Pharmacology*, **40**, 422–428.

Maly, B.F.J. (1986) *Multivariate statistical methods: A Primer*. Chapman and Hall, London.

Manaut, M., Sanz, F., Jose, J. and Milesi, M. (1991) Automatic search for maximum similarity between MEP distributions. *Journal of Computer-Aided Molecular Design*, **5**, 371–380.

Masek, B.B., Merchant, A. and Matthew, J.B. (1993) Molecular shape comparisons of angiotensin II receptor antagonists. *Journal of Medicinal Chemistry*, **36**, 1230–1238.

Meyer, A.M. and Richards, W.G. (1991) Similarity of molecular shape. *Journal of Computer-Aided Molecular Design*, **5**, 426–439.

Mitchell, T.J. (1974) An algorithm for the construction of D-optimal experimental designs. *Technometrics*, **16**, 203–210.

Moon, J.B. and Howe, W.J. (1992) 3D database searching and *de novo* design construction methods in molecular design. *Tetrahedron Computer Methodology*, **3**, 697–711.

Namasivayam, S. and Dean, P.M. (1986) Statistical method for surface pattern matching between dissimilar molecules: electrostatic potentials and accessible surfaces. *Journal of Molecular Graphics*, **4**, 46–50.

Nelder, J.A. and Mead, R. (1965) Simplex method for function minimization, *Computer Journal*, **7**, 308–313.

Nilakantan, R., Bauman, N. and Venkataraghavan, R. (1993) New method for rapid characterization of molecular shapes: applications in drug design. *Journal of Chemical Information and Computer Sciences*, **33**, 79–85.

Norinder, U. (1991a) Experimental design based 3D QSAR analysis of steroid–protein interactions; application to human CBG complexes. *Journal of Computer-Aided Molecular Design*, **4**, 381–389.

Norinder, U. (1991b) An experimental design based QSAR study on beta-adrenergic blocking agents using PLS. *Drug Design and Discovery*, **8**, 127–136.

Norinder, U. (1992) Experimental design-based QSTR of some local anaesthetics using the PLS method. *Journal of Applied Toxicology*, **12**, 143–147.

Pearson, K. (1901) On lines and planes of closest fit to a system of points in space. *Philosophical Magazine*, **2**, 559–572.

Perry, N.C. and van Geerestein, V.J. (1992) Database searching on the basis of 3D molecular similarity using the SPERM program. *Journal of Chemical Information and Computer Sciences*, **32**, 607–616.

Petke, J.D. (1993) Cumulative and discrete similarity analysis of electrostatic potentials and fields. *Journal of Computational Chemistry*, **14**, 928–933.

Plummer, E.L. (1990) The application of quantitative design strategies in pesticide discovery. In *Reviews in Computational Chemistry* (eds K.B. Lipkowitz and D.B. Boyd), pp. 119–169. VCH, New York.

Reynolds, C.A., Burt, C. and Richards, W.G. (1992) A linear molecular similarity index. *Quantitative Structure–Activity Relationships*, **11**, 34–35.

Richard, A.M. (1991) Quantitative comparison of MEPs for structure–activity studies. *Journal of Computational Chemistry*, **12**, 959–969.

Sanz, F., Manaut, F., Rodriguez, J., Lozoya, E. and Loprezdebrinas, E. (1993) MEPSIM: a computational package for analysis and comparison of MEPs. *Journal of Computer-Aided Molecular Design*, **7**, 337–347.

Seri-Levy, A. and Richards, W.G. (1993) Chiral drug potency: Pfeiffer's rule and computed chirality coefficients. *Tetrahedron Asymmetry*, **4**, 1917–1921.

Shoichet, B.K., Bodian, D.L. and Kuntz, I.D. (1992) Molecular docking using shape descriptors. *Journal of Computational Chemistry*, **13**, 380–397.

Stewart, J.J. (1990) MOPAC: a semiempirical molecular orbital program. *Journal of Computer-Aided Molecular Design*, **4**, 1–105.

SYBYL, Tripos Associates Inc., 1699 S. Hanley Rd., Suite 303, St. Louis, Missouri 63144.

Szabo, A. and Ostland, N.S. (1982) *Modern Quantum Chemistry*, pp. 410–412. Macmillan, London.

Topliss, J.G. and Edwards, R.P. (1979) Chance factors in studies of quantitative structure–activity relationships. *Journal of Medicinal Chemistry*, **22**, 1238–1244.

van Geerestein, V.J., Perry, N.J., Grootenhuis, P.D.I. *et al.* (1992) 3D database searching on the basis of shape using the SPERM prototype method. *Tetrahedron Computer Methodology*, **3**, 595–613.

Wold, Ambano, Dunn *et al.* (1984) Multivariate data analysis in chemistry. In *Chemometrics—Mathematics and Statistics in Chemistry* (ed. B.R. Kowalski), pp. 17–95. Reidel, Dordrecht.

3 The treatment of conformationally flexible molecules in similarity and complementarity searching

A.R. LEACH

3.1 Introduction

The properties of a molecule are intimately linked to the 3D structures, or *conformations* that it can adopt. Consideration of the conformational properties of a molecule is therefore essential in any approach to rational drug design. Conformational analysis is the study of the conformations of a molecule and the relationships between them. This chapter considers the importance of conformational flexibility in similarity and complementarity searching and describes some of the approaches that have been used to tackle these problems. First, however, we must establish some of the fundamental concepts involved.

Crucial to any approach to quantitative conformational analysis is a means of calculating the energy of a given conformation. Methods for calculating conformational energies fall into two broad categories: approaches based on the use of quantum mechanics (encompassing semi-empirical approaches and *ab initio* methods) and empirical force-field approaches. Neither of these will be considered here, as there are many excellent reviews (Hehre *et al.*, 1986; Pople and Beveridge, 1970; Clark, 1985; Burkert and Allinger, 1982). The energy is often a complicated function of the Cartesian or internal coordinates which is difficult to visualise except for the very simplest molecules. The energy function is said to define a *potential energy surface*, a simple (1D) example of which is shown in Figure 3.1. There will usually be certain combinations of coordinates where the energy function passes through a minimum value; changing any of the coordinates of a conformation at an energy minimum will lead to an increase in the energy. These stable, minimum energy conformations are sometimes called *conformers*. The very lowest energy minimum is usually referred to as the global energy minimum. Vibrations contribute to the statistical weight of a given conformation and as a consequence, the global minimum energy conformation of an isolated molecule does not necessarily have the highest statistical weight. The narrow potential well in Figure 3.1 is of lower energy, but the broader well may have a larger statistical weight due to the proportionately larger contribution from the conformational entropy, because more vibrational states are accessible (the vibrational energy levels are closer together). It is important to remember

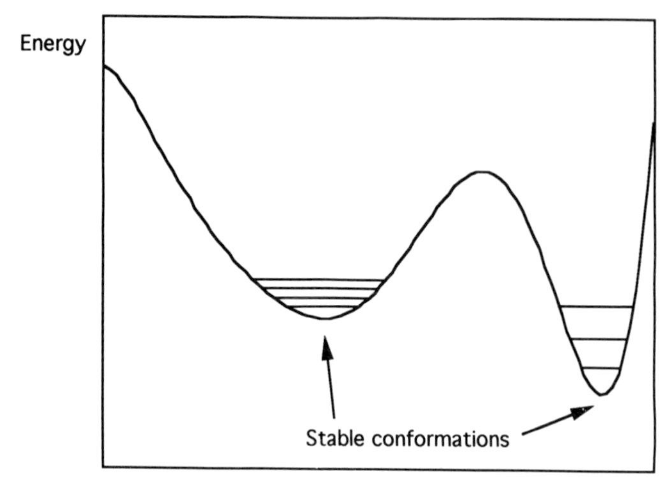

Figure 3.1 Schematic illustration of a simple one-dimensional potential energy hypersurface. The x-axis represents a conformational parameter; this might be a torsion angle, for example. Although the narrow well is of lower energy, it may have a smaller statistical weight due to entropic effects. Reproduced from *Reviews in Computational Chemistry* (eds K.B. Lipkowitz and D.B. Boyde), VCH Publishers Inc., 1991.

that the energies and relative populations calculated with one energy function may not necessarily correspond to those calculated with another. Moreover, it is crucial to recognise that the relative populations will also depend on the environment: a calculation performed on an isolated molecule, 'in the gas phase' may give a very misleading picture of the situation in solution, let alone in a binding site. In addition, the energy functions currently in use only provide the internal energy, not the free energy. Statistical perturbation methods can be used to determine relative free energies (Beveridge and DiCapua, 1989) (for example, the relative free energy of binding of two similar inhibitors), but such calculations require enormous amounts of computer time and are still in the early stages of development.

Our discussion of conformational analysis in similarity and complementarity searching will focus on the problem of finding compounds that interact with a biological macromolecule such as an enzyme, gene, receptor or polysaccharide. We will henceforth use the term *receptor* to denote all such macromolecules and will assume that the interaction occurs at a specific binding site. The obvious reason for the interest in this problem is that if the receptor is implicated in a disease then such compounds may act as the starting point for the development of new drugs. We will use the term *similarity searching* to refer to the situation when an active compound has been identified and we wish to find other molecules that would also be expected to be active. A complementarity search is used when a 3D model of the binding site is available and we wish to find molecules that would interact with it. The model may contain full coordinate information about all the atoms (e.g. an X-ray structure) or it may be much simpler (a pharmacophore), containing just the functional groups required for activity and their relative positions. The relative positions between groups in a pharmacophore are invariably expressed in terms of the distances (or distance ranges) between them. For all except rigid molecules, consideration of the conformational space is important when determining the similarity between two molecules or the complementarity of a molecule to a binding site. However, it is not sufficient simply to find the conformation in which the molecule is 'most similar' or 'most complementary' to the target, difficult though that task often is. Rather, we should also take into account the molecule's accessible conformational space. A molecule may achieve optimal similarity in a conformation whose energy is so high that it is extremely unlikely to be populated. The question we must therefore try to answer is, 'in which of its thermally accessible conformations is the molecule most similar/complementary to the target?' A common application of similarity and complementarity calculations is in searching databases of compounds to find those which are most similar to the target molecule, binding site or pharmacophore. The expectation is that some of the 'hits' obtained from such a search will provide new leads. A typical corporate database might contain several hundreds of thousands of diverse compounds, and so it is obviously important that the method used to perform the search is fast, able to operate without user intervention and able to deal with many different kinds of molecule. These then are the major requirements for any conformational procedure to be used as part of a similarity/complementarity calculation. In the remainder of this chapter we will discuss the methods currently available for exploring the conformational space of molecules and then consider which of them are applicable in similarity and complementarity calculations.

3.2 Methods for exploring conformational space: a summary

A variety of methods have been described for exploring the conformational space of molecules to identify minimum energy conformations (Howard and Kollman, 1988; Leach, 1991). As indicated above, the proper objective of any conformational search should be to identify all thermally accessible conformations. However, in some cases the number of energy minima is so huge that this objective is changed to finding just a single low-energy conformation, often the global energy minimum. In some cases experimental information may be available (for example from NMR experiments) which allows constraints to be placed on the conformational search, restricting it to those structures that satisfy the constraints. Minimisation algorithms can only locate the nearest minimum energy conformation from a given starting structure, and so it is necessary to provide some other method of generating the initial starting structures for subsequent minimisation. This is the purpose of a conformational search algorithm. Although conformational analysis is properly concerned with locating minimum energy conformations, in applications where large numbers of molecules have to be considered it is often not possible (due to time considerations) to minimise each structure. Moreover, as most energy functions are appropriate to the isolated, gas-phase molecule, it may in any case be misleading to base predictions of binding upon minimum energy structures. The methods to be considered will be divided into the following categories: systematic search, random search, distance geometry, genetic algorithms and molecular dynamics. As we shall see, not all methods for performing a conformational search are directly applicable to the similarity/complementarity search problem.

3.3 Systematic search methods

The systematic search is perhaps the most easily understood method for performing a conformational search because it is closely related to the way in which we might perform a conformational search using a set of molecular models. Most conformational changes arise from rotations about single bonds, although in many cases a small amount of angle opening is involved. A common approach, which has the effect of eliminating many degrees of freedom, is to assume that the bond lengths and bond angles are fixed and to permit rotation solely about single bonds. For each bond, we choose a series of torsional values which it will be permitted to adopt. The values are usually chosen to cover the full range of torsion angles from 0° to 360°. Alternatively, we can specify a dihedral increment (θ), in which case the values adopted by the angle would be $0, \theta, 2\theta, 3\theta, \ldots 360 - \theta$. The number of values for the ith rotatable bond is thus $360/\theta_i$. We then systematically generate the conformations which correspond to all possible combinations of torsional values, minimising each structure if desired.

If the molecule has N rotatable bonds then the total number of conformations that would be generated by this approach is:

$$\prod_{i=1}^{N} \left(\frac{360}{\theta_i} \right)$$

The number of possible solutions thus increases exponentially, a situation known as a *combinatorial explosion.* We can illustrate the implications of the combinatorial explosion as follows. If there are five rotatable bonds and the angular increment is 60° then 6^5 (7776) structures must be considered. If each structure takes 10 seconds to minimise, this search would require 22 hours. Seven rotatable bonds gives 279 936 structures and would require more than a month.

It is possible to make considerable improvements to the basic systematic search algorithm described above. A *tree representation* of the problem is useful here. Trees are often used to represent the interrelationships between the states a system can adopt. They consist of *nodes* connected by *edges*. There is often a single *root node* which represents the initial situation where no dihedral angle values have been assigned. From the root node there are one or more daughter (or leaf) nodes; these correspond to the options available for the first 'move'. In the systematic search these relate to the values that the first dihedral angle can adopt, and so there will be $360/\theta_1$ such nodes. From each of the nodes at this first level there will be an appropriate number of nodes which correspond to the values that the second dihedral angle is allowed, and so on. The tree in Figure 3.2 represents a molecule in which there are three values for the first torsion angle, two for the second and three for the third. A maximum of 18 ($= 3 \times 2 \times 3$) conformations would therefore be generated in this simple example.

Setting the first bond to its first value corresponds to moving from the root node of the tree to the first of its daughter nodes (numbered 2 in Figure 3.2). When the second bond is assigned its first value this corresponds to a move from node 2 to node 5, similarly for the third bond. As values have now been assigned to all of the variable bonds in the molecule, the conformation is fully defined and ready to be minimised. Having generated this first conformation, there are a variety of choices for the next move. A commonly used algorithm for searching trees is the *depth-first search*, which uses a *backtracking* method; here the nodes would be expanded in the order 1, 2, 5, 11, 12, 13, 6, 14, 15, 16, 3, ...

The efficiency of a systematic search can be improved by discarding structures which violate some form of energetic or geometric criterion (for example, close interatomic contacts) before the time-consuming energy minimisation stage. The efficiency may be improved further by checking the partially constructed conformations for such problems. If a violation is detected, then all conformations which lie below the current node in the search tree will also contain the problem and can be eliminated from further consideration. For example, if the first value of torsion angle 1 when combined with the second value of torsion 2 gives rise to some problem (for example, two atoms very close in space) then all conformations which contain this combination of torsion angle values would be invalid and can be immediately rejected. These conformations are represented by nodes 14, 15 and 16 in Figure 3.2. The portion of the tree that lies below node 6 is said to have been *pruned* from the search tree.

It is important to note that only those portions of the molecule whose relative orientations will not be changed later can be considered in such

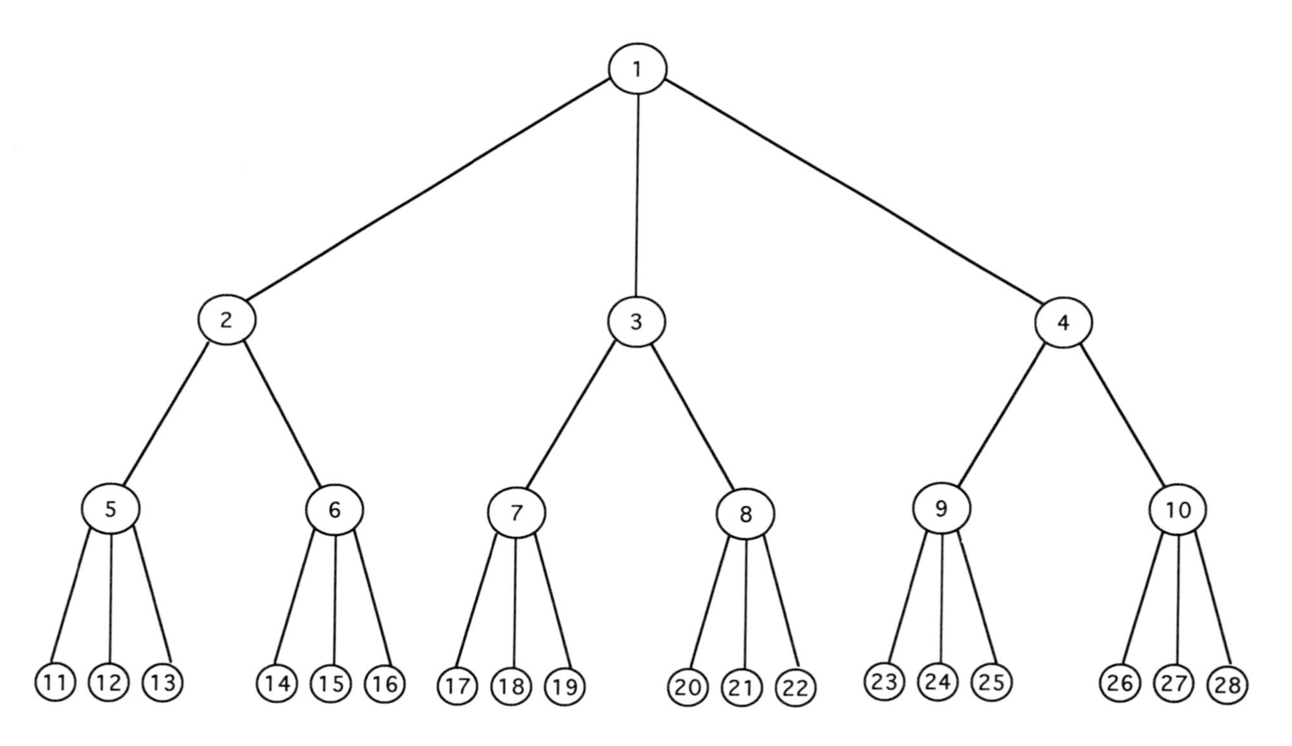

Figure 3.2 A search tree which represents the conformational space available to a molecule containing three rotatable bonds, with the first angle having three possible values, the second torsion angle having two values and the third rotatable bond having two values. The total number of possible solutions is thus 18. Reproduced from *Reviews in Computational Chemistry* (eds K.B. Lipkowitz and D.B. Boyd), VCH Publishers Inc., 1991.

checks and so the order in which the dihedral angles are altered will be crucial to ensure optimal efficiency.

Rings are particularly difficult for systematic search algorithms. The usual procedure is to open the ring to give a 'pseudo acyclic' molecule. This is then processed as above. It is necessary to check that the geometry of the ring is satisfactory. For example, six ring-closure constraints are used in the MULTIC (multi-conformer) program (Lipton and Still, 1988): the distance between the two atoms forming the ring, the two angles at the two atoms in the ring closure bond, and the three internal dihedral angles. One problem is that these ring-closure constraints can only be applied quite late in the process when most of the ring system has been completed and so, if the ring is rejected, a significant amount of computer processing will have been wasted.

3.4 Model-building approaches

Some of the combinatorial problems with the systematic search can be partially alleviated by using pre-formed *conformational fragments.* Chemists have for many years used molecular models to construct three-dimensional representations of molecules. The fragment- or model-building approach to conformational analysis can be regarded as a successor to that tradition.

In the approach, the program accesses a library of fragments, each of which may be able to exist in more than one conformation. For example, cyclohexane might have the chair, twist-boat and boat conformations. To search the conformational space of a molecule the program first determines which fragments are present in the molecule and then examines all possible combinations of fragment conformations. The tree-search algorithms described above can be used to do this. There are three important points to note about a fragment-based approach:

1. It is assumed that a given fragment shows the same conformational behaviour in a small molecule as in a large one, i.e. short-range forces are dominant in determining conformation.
2. The methods are restricted to those regions of conformational space implicitly defined by the fragment conformations (usually local energy minima). This is why the method can be more efficient than the bond-based systematic approaches (due to the smaller number of combinations to investigate). However, it can result in minima being missed.
3. Only molecules for which the required fragments are present can be analysed.

The WIZARD-II and COBRA programs (Dolata *et al.*, 1987; Leach *et al.*, 1990; Leach and Prout, 1990) use a fragment-joining approach to generate conformations of a molecule. These programs are designed to perform rapid and automatic conformational analysis. In the first stage of the analysis the programs decide which fragments (called 'conformational units') are to be

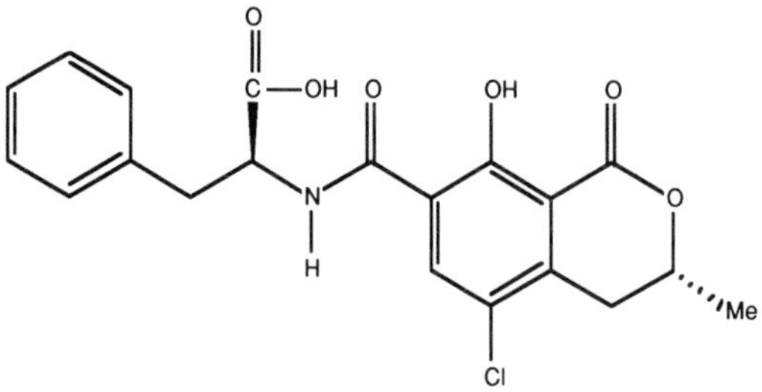

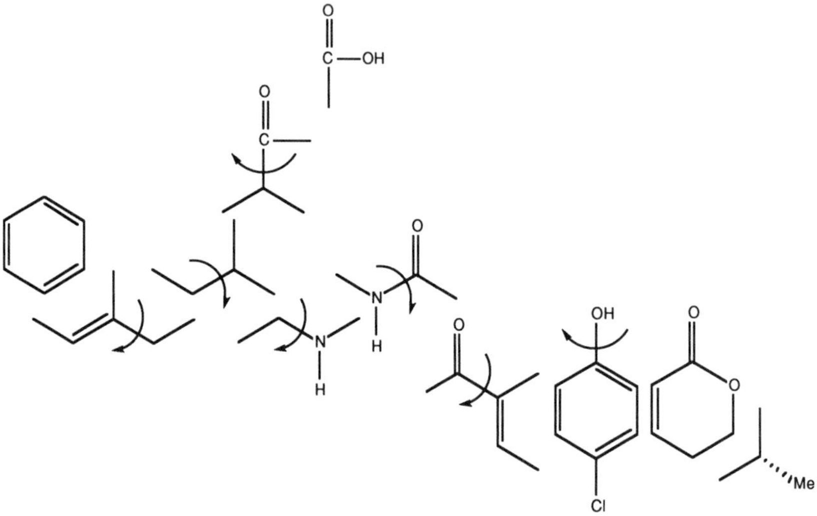

Figure 3.3 The conformational units used by the COBRA program to construct ochratoxin A. Reproduced from A.R. Leach (1994a).

used to construct the molecule. A substructure search algorithm matches the fragments from the programs' knowledge base onto the molecule. To illustrate the types of fragments used, Figure 3.3 shows the units used to construct ochratoxin A. Each unit is permitted to adopt one or more discrete conformations which are stored as separate template files. The templates are obtained by analysing experimental data (e.g. crystal structures) and from theoretical studies, and represent the conformations that the unit is observed to adopt in minimum energy conformations of molecules. A conformation of the molecule is constructed by selecting one template for each unit, and then joining the templates together in a stepwise fashion. Different conformations of the molecule are obtained by joining different combinations of templates. Combinations in which the templates do not fit together properly or in which there are high-energy steric interactions are rejected as soon as they are identified.

The conformational space explored by the program consists of all possible combinations of templates and is conveniently represented as a search tree. If each unit i has S_i templates (taking account of the fact that there may be more than one unique orientation possible for some of the templates) then the total number of potential conformations of the molecule is

$$\prod_{i=1}^{N_u} S_i$$

where N_u is the number of units in the molecule. As with the systematic search, not all terminal nodes in the search tree necessarily correspond to an acceptable conformation of the molecule; some combinations of templates may not fit together adequately, or may give rise to high-energy steric interactions. We have investigated the application of various algorithms in exploring the search tree. The basic algorithm is the depth-first search, with backtracking to find all solutions. A more sophisticated search method is the A* search (Hart *et al.*, 1968), which is an algorithm for finding the optimal or *least-cost path* to a goal node. This requires that each edge in the search tree has an associated cost. For example, in the classic travelling salesman problem an appropriate cost function would be the distance travelled. An obvious cost function to use in conformational analysis is the internal energy of the conformation, as might be calculated using an empirical force field. The least-cost goal node then corresponds to the global minimum energy conformation. The A* search algorithm enables the search to be directed towards those combinations of templates that correspond to the lowest energy structures (Leach and Prout, 1990). The A* algorithm has also been used in a combined clustering/conformational search procedure which directly generates a molecule's 'most different' conformations (Leach, 1994a): this provides an efficient alternative to a two-stage procedure in which the conformational search is performed and then the potentially large number of conformations is clustered.

3.5 Random search methods

Random search methods differ in philosophy from the systematic approaches. These are iterative methods, performing the same set of operations many times. At each stage, the 'current' structure is randomly changed to give a new structure (we will see later how such random changes are made). The new structure may then be further processed (for example, by minimising it). The cycle is completed by selecting a new 'current' structure from those previously generated. This process is repeated a large number of times, at the end of which it is hoped that the entire conformational space will have been covered and all the energy minima located.

The simplest form of random search is one which adds a random amount to the Cartesian coordinates (i.e. it makes a random change in the *x*-, *y*- and *z*-coordinates of all the atoms in the molecule) (Saunders, 1987; Ferguson and Raber, 1989; Saunders, 1989). The resulting structure is then minimised and compared with those already generated. If it has not been previously found it is saved. The 'randomisation' process is then performed again, and the cycle is repeated.

A more efficient random searching method varies the torsion angles, keeping bond lengths and angles fixed (Chang *et al.*, 1990). Thus, whereas the Cartesian random search changes all $3N$ variables the number of variables in a dihedral search is much smaller. Rings in the molecule are first broken to give a 'pseudo acyclic' structure. At each iteration some or all of the torsion angles are changed by a random amount between 0° and 180°. The resulting structure is checked for ring closure and, if satisfactory, it is minimised. A variety of different options are possible: for example, only a subset of the bonds may be changed in any one iteration.

3.6 Choosing the structure to be 'randomised'

At each iteration of the random search it is necessary to decide which structure should be chosen as the starting point for the next stage of 'randomisation'. The simplest method is to use the structure generated in the previous step. However, this might require much time to be spent exploring a high-energy region of conformational space. We could alternatively perturb the structure with the lowest energy found so far. Again, this might lead to much time being spent in a region of conformational space which does not in fact correspond to the lowest energy structures.

The *Metropolis method* provides an alternative. This tends to favour the lowest energy regions of the conformational space, but it also permits 'jumps' to different regions. It operates as follows. The energy of each new conformation is compared with the energy of its predecessor. If the energy is lower, then the new structure becomes the 'current' conformation. If it is higher in energy, however, then the value of the Boltzmann factor

$$e^{-[E_{new}-E_{old}]/kT}$$

is compared with a random number between 0 and 1. If the Boltzmann factor is greater than the random number then the new structure becomes the current conformation. By this means the algorithm favours the lowest energy states of the system, but also allows random fluctuations into higher-energy states (which may then lead to even lower energy states).

Random search methods have the advantage of being easy to implement and have so far proved to be very successful in finding minimum-energy conformations which other techniques have failed to locate. However, they do suffer from the problem that the same conformation may be generated many times, which represents a significant waste of computational resources.

3.7 Simulated annealing

Annealing is the process by which a substance is first heated until it melts and then the temperature is slowly lowered until it crystallizes into a perfect lattice. It is a widely used technique in many areas of manufacture, such as the production of large crystals of silicon for integrated circuits. A characteristic feature of annealing is the careful temperature control, with a long time being spent at the phase transition from liquid to solid. This is required to produce a crystal free of any imperfections. In simulated annealing (Kirkpatrick *et al.*, 1983) the computer is used to perform the analogous task. Simulated annealing is a powerful method for finding the optimal solution from among a large number of possible solutions. Finding a global minimum energy conformation constitutes such a problem.

The perfect lattice corresponds to the minimum of the global free energy. Imperfections arise when the substance becomes trapped in some locally optimal structure from which it is unable to escape because the temperature is reduced too rapidly. The simulated annealing method uses a control parameter to play the role of the temperature whose value is steadily decreased. At each 'temperature' it is necessary that the system achieves the equivalent of thermal equilibrium. This can be done using the Metropolis algorithm. As the temperature is reduced the lower energy states become more probable until at absolute zero the system occupies the lowest possible energy state.

Wilson and Cui have described an application of the simulated annealing algorithm to searching conformational space (Wilson *et al.*, 1988; Wilson and Cui, 1990). They have reported the results of using their program to locate the global energy minimum of a number of molecules, some of which have considerable degrees of torsional freedom (for example, arachidonic acid which has 15 rotatable bonds, polyalanines from Ala-20 to Ala-80, and Met-enkephalin). However, for arachidonic acid there are in fact *two* conformations of identical energy (Leach and Prout, 1990). This provides a clear illustration of the limitations of methods which only find a single conformation. One of the COBRA conformations is the same as that found by simulated annealing. The other is its 'mirror image' (Figure 3.4).

3.8 Distance geometry

Distance geometry is the study of geometric problems based on the distances between points. It has been investigated by mathematicians for many years

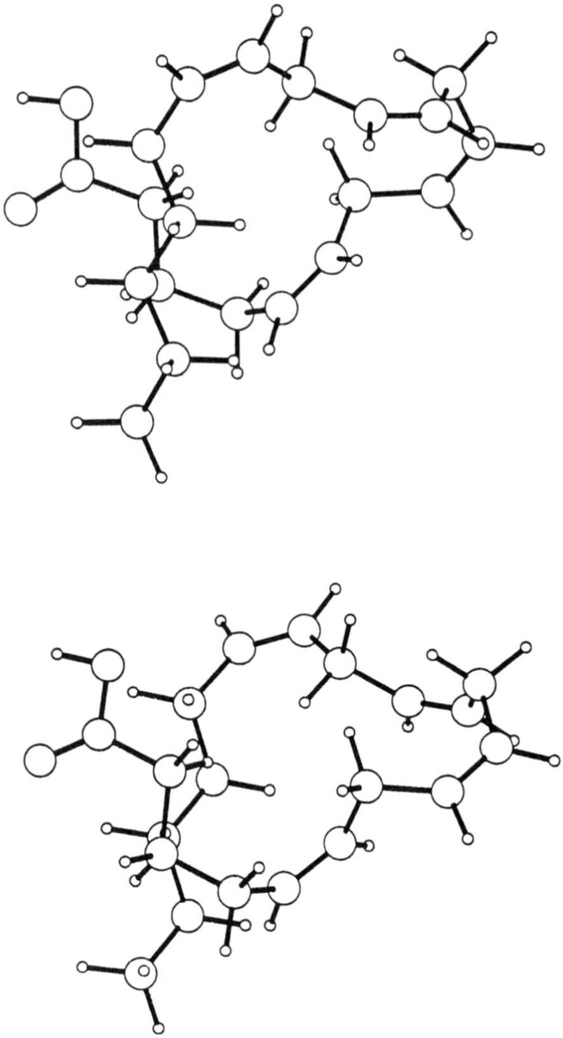

Figure 3.4 The two lowest energy conformations of arachidonic acid found by the COBRA program, after minimisation with MM2 (Allinger, 1977). One of these conformations is the same as found by a simulated annealing method (Wilson *et al.*, 1988). These structures are effectively 'mirror images'. Reproduced from A.R. Leach and K. Prout (1990).

but its application in conformational analysis was first described more recently, by Crippen (1978).

A molecule with N atoms has $(N^2 - N)/2$ interatomic distances. These are most conveniently represented using an $N \times N$ matrix, in which the element

(i, j) of the matrix corresponds to the distance between atoms i and j. Of fundamental importance is that these interatomic distances cannot all adopt arbitrary values and still be consistent with 3D objects. Rather, only very restricted combinations of distances can be used. Moreover, the distances are heavily interdependent.

Distance geometry uses a four-stage process to derive a conformation of a molecule (Crippen, 1981; Crippen and Havel, 1988). First, a matrix of upper and lower interatomic *distance bounds* is calculated. Secondly, trial values are randomly assigned to each interatomic distance between the upper and lower bounds. The resulting distance matrix is then converted into a trial set of Cartesian coordinates, which in the fourth step are refined against the target lower and upper bounds.

Some of the bounds on the interatomic distances come from simple chemical principles. For example, the distance between two bonded atoms is restricted to a very narrow range of values that depends on the atomic number and hybridisation of the atoms involved. The distance between two atoms which are both bonded to a third atom is also very restricted. For two atoms separated by three bonds, the lower bound corresponds to a torsion angle of 0° and the upper bound to one of 180°. For the remaining interatomic distances the remaining lower bound distances are set to the sum of the van der Waals radii and the upper bounds to a large number. A procedure called *triangle smoothing* is then used to refine this initial set of distances. This method uses simple geometrical restrictions on groups of three atoms A, B and C. Using u to represent the upper limit and l the lower limit:

$$u_{AC} \leqslant u_{AB} + u_{BC}$$

$$l_{AC} \geqslant l_{AB} - u_{BC}$$

These inequalities are applied until they are satisfied by all possible triplets of atoms. Having derived the matrix of upper and lower bounds (a process which need only be performed once for each molecule), we proceed to the generation of conformations:

1. Assign a random value to each interatomic distance between the upper and lower bounds to give the distance matrix.
2. Embed the distance matrix. In this procedure we convert from 'distance space' to Cartesian coordinates by performing a series of matrix operations. First, we calculate the *metric matrix*, **G**, each of whose elements (i, j) is equal to the scalar product of the vectors from the origin to atoms i and j:

$$g_{ij} = \mathbf{i} . \mathbf{j} \tag{3.1}$$

The elements g_{ij} can be calculated from the distance matrix using the cosine rule:

$$g_{ij} = (d_{i0}^2 + d_{j0}^2 - d_{ij}^2)/2$$

where d_{i0} is the distance from the origin to atom i, d_{j0} is the distance from the origin to atom j and d_{ij} is the distance between atoms i and j.

It is usual to take the centre of mass of the molecule as the origin. The distances from the centre of mass can be calculated directly from the interatomic distances using the following expression:

$$d_{i0}^2 = \frac{1}{n}\sum_{j=1}^{n} d_{ij}^2 - \frac{1}{n^2}\sum_{j=2}^{n}\sum_{k=1}^{j-1} d_{jk}^2$$

G is a square symmetric matrix and a general property of such matrices is that they can be decomposed as follows:

$$\mathbf{G} = \mathbf{V}\mathbf{L}^2\mathbf{V}^{\mathrm{T}} \tag{3.2}$$

where the diagonal elements of $\mathbf{L}^2$ are the eigenvalues of **G** and the columns of **V** are its eigenvectors. The atomic coordinates can be derived from the metric matrix by rewriting (3.1) as

$$\mathbf{G} = \mathbf{X}\mathbf{X}^{\mathrm{T}} \tag{3.3}$$

where **X** is a matrix containing the atomic coordinates. Equating (3.2) and (3.3) gives

$$\mathbf{X} = \mathbf{V}\mathbf{L}$$

(as **L** has only diagonal entries $\mathbf{L} = \mathbf{L}^{\mathrm{T}}$). Thus to obtain the atomic coordinates the square roots of the eigenvalues are multiplied by the eigenvectors.

3. Refine the resulting coordinates against the starting upper and lower bounds. The initial set of coordinates obtained after step (3) often will fail to satisfy all of the initial distance constraints perfectly, and so the coordinates are refined using a conjugate gradients minimisation algorithm to make them agree more exactly with the distance ranges from triangle smoothing. The final structure may then be subjected to force-field energy minimisation.

Experimental techniques (particularly NMR) can provide information on interatomic distances in a molecule which can be used together with those from theoretical considerations. For example, nuclear Overhauser spectroscopy (NOESY) provides information on the distance between non-bonded atoms. The observation of a nuclear Overhauser effect (nOe) signal between two atoms indicates that they are close together in space, and an approximate value for the distance can be derived (the nOe signal is theoretically proportional to the inverse sixth power of the distance). Correlated spectroscopy (COSY) provides short-range covalent information (for example, backbone torsional angles). These values can then be used to restrict the distances between specific atom pairs in the distance geometry algorithm. As we shall see, it may be desired to generate the conformations in which the distance between specific functional groups is restricted to lie between particular values.

Distance geometry is at heart a random algorithm, and so it may generate the same conformation more than once (indeed, in order to demonstrate that the space has been properly covered with any random algorithm we should generate the same conformation many times). One advantage, however, is that it preferentially samples those regions of distance space more likely to lead to low-energy conformations.

3.9 Genetic algorithms

Genetic algorithms are a class of methods based on biological evolution to find optimal solutions to problems (Goldberg, 1989; Holland, 1992). The basic principle is to create a population of possible solutions to the problem. The 'fitness' of each member of the population is calculated and the genetic information of the fitter members is preferentially carried forward when a new population is created. Each member of the population is coded by a 'chromosome' which is a bit string that can be decoded to give a set of physical parameters. For example, in conformational analysis, the chromosome might be decoded to provide the torsion angles of the rotatable bonds in the molecule. The fitness function can then be calculated for each population; for conformation analysis the function might be the internal energy. Having generated an initial population by randomly assigning bits in the chromosomes, the fitness of each individual is determined. A new population is then generated using operators. A simple scheme involves just three operators: reproduction, crossover and mutation. Application of the reproduction operator simply consists of copying individual chromosomes according to their fitness, such that the highest scoring individuals have a greater probability of being reproduced than their weaker brethren. A simple way to implement this is by creating a biased roulette wheel in which each individual has a slot size that is proportional to its fitness. The crossover operator randomly selects pairs of individuals for 'mating'. A crossover position is then randomly selected at a position i ($1 \leqslant i \leqslant l - 1$, where l is the length of the chromosome). Two new strings are then created by swapping the bits between positions $i + 1$ and l. For example, suppose we have the following two chromosomes:

00100011110001
11000011001100

and the crossover point is chosen to be 6. Then the two new strings are:

00100011001100
11000011110001

The third operator is mutation. Here, bits are randomly selected and changed (0 to 1 and vice versa). The mutation operator is usually assigned a low probability.

Many variants on these basic operations have been devised; for example, it is common practice to use an 'élitist' strategy where the highest ranking individuals are carried forward unchanged in the new population. This ensures that the best individuals are not lost, as would occur if crossover or mutation were applied to every member of the new population.

Judson and colleagues have described how genetic algorithms can be used to search conformational space (Judson *et al.*, 1993; McGarrah and Judson, 1993). In their implementation, six bits are used to code for each torsion angle; this corresponds to a resolution of approximately 5°. Fitness values for each conformation are calculated using an appropriate force field. The method was tested using a set of 72 molecules extracted from the Cambridge Structural Database (Allen *et al.*, 1979; Clark *et al.*, 1989). The molecules contained between 1 and 12 rotatable bonds. The test set was also subjected to a systematic search method in which the torsional increment was 30°. Rings were kept fixed in their crystallographic conformations. The genetic algorithm was found to exhibit very satisfactory performance, particularly for molecules with many rotatable bonds. In a separate investigation, the method was applied to the problem of exploring the conformational space of cyclohexaglycine. One important point to note is that the basic genetic algorithm is a method for performing global optimisation whereas it is often desired to identify a number of low-energy conformations. This can be achieved by premature termination, before any significant convergence of the chromosomes has been achieved.

The genetic algorithm method appears to be a promising approach to exploring conformational space. It has also been used to perform conformationally flexible ligand docking as will be described below. However, research is still required to determine the optimal combination of parameters.

3.10 Molecular dynamics

Molecular dynamics aims to reproduce the time-dependent motional behaviour of a molecule (Allen and Tildesley, 1987; McCammon and Harvey, 1987; Brooks *et al.*, 1988). The force on each atom is determined by differentiating the energy function. Given the force and mass it is then possible to calculate the acceleration. Newton's laws of motion can then be integrated to determine how the positions, velocities and other quantities vary with time. An important feature of molecular dynamics is that it provides a means of overcoming conformational barriers, due to the kinetic energy present in the system. Kinetic energy is directly related to the temperature of the system: the higher the temperature, the higher the kinetic energy. For this reason, it is common practice to use high-temperature molecular dynamics when searching conformational space. Conformations are selected at regular intervals from the trajectory and then minimised to give the associated minimum energy structures. Molecular dynamics is an extremely important technique for determining the structures of large molecules such as proteins when experimental information is available from X-ray crystallography or NMR that allows restraints to be incorporated into the potential function, thereby restricting the range of conformational space that is covered (Brunger *et al.*, 1987). Simulated annealing protocols are often used here. However, the computational resources required mean that molecular dynamics is not currently regarded as a practical method for examining large numbers of molecules. In addition, molecular dynamics requires a force field description of all molecules to be available, giving rise to potential parameterisation problems when applied to a diverse set of molecules.

Having considered the various methods for exploring the conformational space of isolated molecules in order to locate structure for subsequent energy minimisation, we now turn to the problem of incorporating the conformational degrees of freedom when tackling the similarity or complementarity problem.

3.11 Conformational analysis in similarity searching

We will consider two types of similarity search. In the first class, a compound that is active against the receptor has been identified and its active conformation determined. The objective is to determine the optimal similarity between a second molecule and the active compound. In the second type of similarity search, the aim is to determine the optimal similarity between a set of molecules which are all known to be active against the same biological target. The resulting ensemble of structures can then be used to derive a 3D pharmacophore model of the receptor. We are not concerned here with methods to calculate similarity, nor with the factors involved in molecular recognition, but with the way in which the conformational properties of the molecules are taken into account. In all cases, the selected conformation(s) should be 'reasonable', and knowledge of the full conformational space makes it easier to quantify the relative population of the selected conformation. However, for reasons of computational cost it is often desired merely to locate the conformation that maximises the similarity, so long as the conformation does not contain any serious problems (atoms too close, etc.).

In principle, any of the techniques described above for exploring conformational space could be used to determine the conformation in which a given molecule is most similar to the target. However, it is important to recognise that the number of degrees of freedom is increased in a similarity search, for when comparing two molecules it is necessary to specify their relative orientations. Six degrees of freedom are required to specify the relative orientation of two rigid bodies (three translational and three rotational). Having specified the relative orientation of the two molecules and having defined their conformations (if they are not fixed) the similarity between them can be calculated, using the chosen similarity measure. Richards and co-workers have used a similarity measure that is based on electrostatic potentials; the similarity between two molecules A and B is defined (Hodgkin and Richards, 1987) as

$$H_{\mathrm{AB}} = \frac{2\int \rho_{\mathrm{A}}\rho_{\mathrm{B}}\,\mathrm{d}v}{\int \rho_{\mathrm{A}}^2\,\mathrm{d}v + \int \rho_{\mathrm{B}}^2\,\mathrm{d}v}$$

where ρ is the electron density or electrostatic potential as appropriate. The electrostatic potentials can be obtained by fitting atomic charges to the wavefunction (see Chapter 2 and Burt and Rchards, 1990; Burt *et al.*, 1990; Good *et al.*, 1992).

Dean and co-workers have developed a similarity measure in which an appropriate property of the molecule such as electrostatic potential is projected onto the surface of a sphere (Chau and Dean, 1987; Dean and Chau, 1987; Dean and Callow, 1987; Dean *et al.*, 1988). This is known as a *gnomonic projection.* Its particular advantage is that dissimilar molecules can be compared simply by considering the difference between the values of the property on the surface of the two spheres. By rotating one sphere relative to another, the rotational degrees of freedom can be taken into account, and the conformational degrees of freedom can be considered by recalculating the projected values on the surface of the sphere as the conformation changes. In practice, a number of discrete points on the surface of the sphere are considered; in particular, tesselated icosahedra have been used. At each vertex on the icosahedron, a line is drawn to the centroid and the point where it pierces the molecular surface is determined. The value associated with the vertex is then the value of the property at the pierce point. It is possible to match the entire surfaces of the spheres, but a more realistic option is to restrict the comparison to patches on the sphere. The rationale for this is that molecules of any size often interact with their receptors at a specific face. Dean's group has also considered similarity measures based upon interatomic distances: the objective is to find a set of atom pairs (one from each molecule) such that the sum of the elements of the difference distance matrix is a minimum. Given two atoms, i and j, from molecule A which are matched with two atoms k and l from molecule B, the appropriate element of the difference distance matrix is given by:

$$d_{ijkl}^{\mathrm{AB}} = |d_{ij}^{\mathrm{A}} - d_{kl}^{\mathrm{B}}|$$

Conformational flexibility is taken into account using either a systematic search method or a random algorithm based on simulated annealing (Perkins and Dean, 1992). The conformational analysis is performed independently of the similarity procedure and a set of representative conformations selected using cluster analysis. To identify the best matches between a given pair of molecular conformations, both systematic searches and simulated annealing protocols have been employed. Chapter 4 describes these methods in more detail.

Given a set of molecules that are active against a receptor, a variety of methods can be used to determine a three-dimensional model (a 3D pharmacophore) of the receptor site. The models produced may then be used to search '3D databases' for novel lead compounds, as will be discussed below. Several steps can be identified in the elucidation of a pharmacophore. First, the groups required for activity are chosen. At least three non-collinear groups are required to uniquely superimpose two molecules. The problem then is to find low-energy conformations of the molecules in which the matching groups are superimposed. A popular method is the constrained systematic search developed by Dammkoehler and colleagues (Motoc *et al.*, 1986a,b; Dammkoehler *et al.*, 1989). A key feature of their approach is the use of the results for molecules considered early in the procedure to restrict the conformational space considered for later molecules. This is because all of the molecules must fit the receptor model. If it is possible to restrict the conformational space that must be considered then the effects of the combinatorial explosion (which will be particularly pronounced as a large number of molecules may be in the test set) can be mitigated. Thus, if a pair of atoms must lie within some distance range to satisfy the current model, then the set of torsional angles that will enable this constraint to be satisfied can be calculated, and only these values used in the search (rather than the full 360°). Clearly, it is important to choose the most restricted molecules first as these will have a reduced conformational space and so would be expected to have a greater impact on the conformational possibilities of subsequent molecules. The method has been used to predict a model of the receptor for angiotensin converting enzyme inhibitors (Mayer *et al.*, 1987).

Other techniques can be used to determine optimal superpositions of two or more molecules in order to derive a common pharmacophore. In ensemble distance geometry (Sheridan *et al.*, 1986) the dimensions of the bounds and distance matrices are equal to the sum of the number of atoms in the molecules, with elements 1 to N_1 appropriate to the N_1 atoms of molecule 1, elements $N_1 + 1$ to $N_1 + N_2$ appropriate to the N_2 atoms of molecule 2, and so on. Elements (i, j) of these matrices thus represent upper or lower bounds or distances (as appropriate) between atoms i and j, where i and j may or may not be in the same molecule. The upper bound distances of all pairs of atoms are set in the usual way: the distances between bonded atoms, atoms separated by an angle, or atoms separated by a torsion angle can be assigned from geometric considerations but all other elements are set to a large value. The lower bound values for all pairs of atoms (i, j) where i and j are both in the same molecule are set to the sum of their van der Waals radii, but the lower bounds for atoms that are in different molecules are set to zero. This means that different molecules can pass through each other in order to superimpose the appropriate functional groups. The bounds for atoms to be superimposed would be included: here, the lower bound is set to zero and the upper bound to a tolerance parameter. The procedure then follows the usual distance geometry steps, involving smoothing, assignment of random distances, and optimisation against the initial bounds. In the original publication the ensemble distance geometry approach was used to derive a model of the nicotinic pharmacophore from four compounds. The model was then checked against four different compounds; three antagonists and one agonist.

Ensemble molecular dynamics can be used to permit the molecules to explore their conformational degrees of freedom while simultaneously superimposing specific groups by including in the potential a term that forces particular atoms or functional groups to be overlaid (Spellmeyer *et al.*, 1994). Kearsley has described a quaternion-based algorithm that enables different conformations to be superimposed using a penalty function that considers all unique pairs of structures (Kearsley, 1990). The method has also been used to simultaneously maximise electrostatic and steric overlap using atomic partial charges and steric volumes (Kearsley and Smith, 1992). Genetic algorithms can also be used to superimpose molecules and elucidate pharmacophores (Payne and Glen, 1993).

3.12 Conformational aspects of complementarity searching

In a complementarity search the optimal similarity between a given molecule and the receptor is determined using a three-dimensional model of the receptor. The model may be an X-ray structure giving coordinate positions of all the atoms or it may be a pharmacophore which provides just the relative orientations of particular interaction sites. Depending on the type of receptor model available, different techniques can be used to identify molecules complementary to it. We will first consider the problem where a detailed structure of the receptor is available and then examine the situation where only a pharmacophore model is present.

Predicting the structure formed between a putative ligand and a receptor (which we will refer to as the 'ligand docking problem') is a difficult task due to the enormous number of degrees of freedom involved. First, there are the six degrees of freedom (three translational and three rotational) that define the orientation of one molecule relative to the other. There are then the conformational degrees of freedom of the ligand; these may range from zero in the case of a wholly rigid molecule, to perhaps tens of thousands for a very flexible molecule. Finally, there are the conformational degrees of freedom of the receptor. Here, there are no reliable estimates of the number of accessible conformations but molecular dynamics simulations have suggested the presence of many local minima in the region of the X-ray structure of myoglobin, a small protein (Elber and Karplus, 1987).

The first algorithms to tackle the 'ligand docking problem' assumed that both ligand and receptor were rigid, and so only searched the six degrees of translational/rotational freedom. In part this limitation was imposed by the computing resources then available. As computers became more powerful, so it has been possible to use these algorithms to search structural databases containing thousands of compounds. The DOCK algorithm developed by I.D. Kuntz and co-workers (Kuntz *et al.*, 1982) is perhaps the best known example of a rigid-body ligand docking program. The first stage is to calculate the molecular surface of the receptor or of the binding site using Connolly's algorithm (Langridge *et al.*, 1981; Connolly, 1983a,b). The points that define the molecular surface, together with their normals, are then used to characterise the binding site as a collection of overlapping spheres, each of which touches the molecular surface at two points. The spheres, produced by a program called SPHGEN, together form a 'negative image' of the binding site. To suggest a structure of the intermolecular complex, atoms of the ligand (in some predetermined conformation) are matched to the sphere centres. If at least four matching pairs can be found, such that the distance between every pair of atoms equals the corresponding distance between the matching sphere centres within a given tolerance, then the appropriate translation/rotation matrix is used to orient the ligand within the site. The orientation is checked to ensure that there are no unfavourable interatomic contacts (atoms too close) and if acceptable it is scored. A variety of scoring functions have been developed, to measure both the shape complementarity and electrostatic complementarity. Alternative orientations can be generated from different sets of matching atoms and sphere centres. The highest scoring ligands are retained for subsequent analysis. A variety of targets have been tackled by the Kuntz group, including the HIV protease, thymidylate synthase, dihydrofolate reductase and haemagglutinin (Kuntz, 1992). In each case novel inhibitors at the micromolar level were found. The ligands come from a variety of sources: the initial applications used a subset of the Cambridge Structural Database (Allen *et al.*, 1979) but more recently structures from the Fine Chemicals Database[1] (generated by the CONCORD program; Pearlman *et al.*, 1987) were employed.

Although the basic DOCK algorithm does not itself take account of the conformational flexibility of either ligand or receptor, the conformational properties of the ligand can be indirectly considered, via the scoring function. One contribution to the free energy of binding arises from the conformational entropy associated with ligand binding. *In extremis*, a ligand is restricted to

[1] The Fine Chemicals Directory is distributed by MDL Information Systems Inc., 14600 Catalina Street, San Leandro, California 94577.

a single conformation in the binding site and so, other factors being equal, a ligand with a large number of accessible conformations in solution will suffer a greater entropy penalty upon binding than a ligand with only a small number of conformations. If it is assumed that all conformations are equally accessible and that each ligand is restricted to a single conformation in the binding site then the free energy change associated with the loss of conformational entropy is $RT \ln N$ where N is the number of low-energy accessible conformations. The COBRA program (Leach and Prout, 1990) has been used to search the conformational space of the molecules contained in the Fine Chemicals Directory in order to determine the number of accessible conformations for each entry, and thus N. The results of this conformational calculation were used in two ways (Shoichet *et al.*, 1994). First, it provided a useful way to eliminate from the database molecules with a particularly large conformational space; molecules that would suffer a large loss of conformational entropy on binding. An additional advantage of ignoring such molecules altogether is that they require a disproportionate amount of time to be processed by the DOCK algorithm. Secondly, the correction factor can be added to the other contributions in the scoring scheme.

We now consider methods to perform ligand docking in which the conformational degrees of freedom of the ligand are taken into account. Almost all of these approaches assume that the receptor is rigid. A variety of algorithms have been used to effect the conformational search, including systematic and random search methods, genetic algorithms and distance geometry.

The most obvious way to incorporate ligand conformational flexibility into molecular docking is to apply the rigid body docking algorithm to different conformations of the ligand. This approach requires the structure of the ligand to be predetermined (typically conformations at local energy minima are selected), with no possibility for the ligand to adapt to the receptor environment. An alternative approach, based upon the DOCK program, is to dock fragments of the ligand which are then reformed using energy minimization (DesJarlais *et al.*, 1986). This is an attractive approach for ligands that are composed of a small number of relatively large fragments, but it might be difficult to implement for very flexible ligands. Moreover, the intrinsic reliance on a minimization to reform the ligand adds to the computational cost of such an approach. A third approach which also uses the DOCK algorithm was devised by Leach and Kuntz (1992). Here, part of the ligand is oriented in the site using a variant on the DOCK algorithm which takes potential hydrogen bonding interactions between ligand and receptor into account. The orientations so obtained then form the basis for a systematic exploration of the conformational space of the rest of the ligand within the binding site. A novel feature of the conformational search was the development of a scheme whereby the conformation of the ligand could be modified by making small adjustments to torsion angles, in order to account for both unfavourable (e.g.

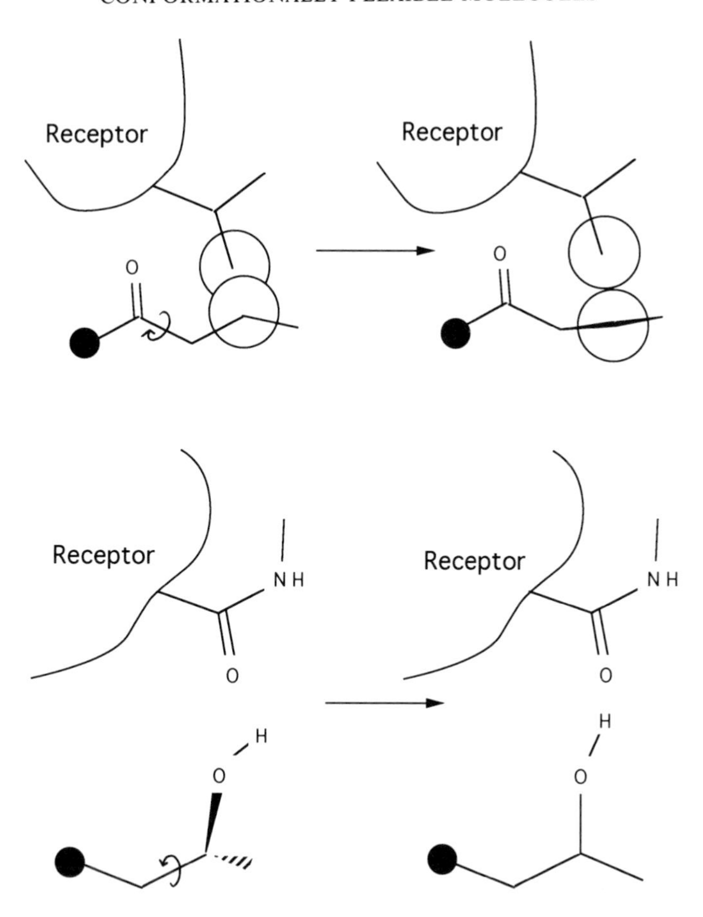

Figure 3.5 Schematic illustration of the way in which the ligand conformation is adjusted to account for steric and hydrogen bonding interactions with the receptor site. Reproduced from A.R. Leach and I.D. Kuntz (1992).

steric) and favourable (e.g. hydrogen bonding) interactions with the binding site, a procedure illustrated in Figure 3.5.

Crippen and co-workers have developed a number of approaches to molecular docking based on distance geometry. Ghose and Crippen used a bounds matrix containing two blocks, one appropriate to the ligand atoms and the other to the site atoms (Ghose and Crippen, 1985). Elements within the ligand block refer to intraligand distances whereas those in the site block refer to intrasite distances (in the case of a rigid receptor the upper and lower bounds for the site atoms are equal). Elements **(i, j)** where **i** and **j** are in different blocks refer to ligand–site interatomic distances. The lower bound of the distances which involve at least one ligand atom (both intra-

and intermolecular) are typically set equal to the sum of the appropriate van der Waals radii and the upper distance constraints are initially set to a large value. After triangle smoothing conformations of the ligand are generated using the standard procedure of randomly assigning distances, embedding, and then optimizing the resulting coordinates against the initial distance bounds. Ligand atoms can be positioned in close proximity to specific site atoms (for example, to form hydrogen bonds) by assigning appropriate bounds. Here, it is first necessary to identify all possible hydrogen bonding donors and acceptors in the binding site and in the ligand. The objective then is to find sets of matching donor/acceptor pairs of ligand atoms and receptor atoms such that the distances or distance ranges between the pairs of ligand atoms are equal to the corresponding distances or distance ranges between the receptor atoms. By using the upper and lower bounds between the pairs of ligand atoms, conformational flexibility can be taken into account. Crippen has used *clique finding algorithms* to automatically determine such sets of hydrogen bonding ligand–site atom pairs. A clique is a maximal set of matching pairs, such that no other pair of matching points (that also satisfy the distance constraints) can be added. Having identified each clique, distance geometry can then be used to generate conformations of the ligand within the site. Identifying all the cliques is a difficult problem (it falls into the class of NP-complete problems) but the method of Bron and Kerbosch (1973) was shown to provide a practical solution by Kuhl *et al.* (1984) and also by Smellie *et al.* (1991), who enhanced the speed of the procedure using a series of filters.

Blaney has developed a distance geometry docking algorithm that does not require the receptor atoms to be directly included in the calculation (Blaney, 1993). Rather, conformations of the ligand are generated within the SPHGEN representation of the docking site. An additional term is included in the penalty function at the optimisation stage. This term has the effect of constraining the ligand to lie within the spheres. The resulting orientations can then be scored using the DOCK scoring functions. Blaney reports that the method is sufficiently fast to make it a practical method for performing conformationally flexible ligand docking on a database, using approximately 100 trials of the distance geometry algorithm.

The simulated annealing approach of Goodsell and Olsen (1990) uses the Metropolis algorithm at a decreasing sequence of temperatures to derive a structure for the interaction complex. The ligand is permitted six degrees of translational/rotational freedom and ligand conformational flexibility is accommodated by rotating about single bonds. Currently, there is no way to deal with conformationally variable rings. At each step a new configuration is generated by making random changes to each variable (the rotatable bonds, $x/y/z$ translations and $x/y/z$ rotations). The energy of the configuration is calculated and it is accepted or rejected according to the Metropolis criterion. This is performed for a series of decreasing temperatures until a final configuration is obtained. The configuration energies are calculated using a grid with parameters from the AMBER force field (Weiner *et al.*, 1984, 1986). It was found necessary to significantly increase the relative contribution of the hydrogen bond term in order to produce the desired results. For flexible molecules, convergence could require a great increase in computational effort (the cases studied, which had a maximum of four rotatable bonds, required up to 1 hour of Convex central processor unit time for each run, each of which only produces a single structure). Nevertheless, simulated annealing should be an efficient method for local sampling and refinement, perhaps using as a starting point a structure from an alternative approach. Hart and Read (1992) have developed a Monte Carlo docking procedure which uses a similar method.

Dixon (1993) has investigated the application of genetic algorithms in ligand docking. The use of genetic algorithms in exploring the conformational space of isolated molecules was described above. In the ligand docking problem, each chromosome contains information not only about the internal conformation of the molecule, but also about its orientation in the receptor site. In Dixon's algorithm, the orientation is defined by matching atoms to the SPHGEN spheres in the same fashion as used by the DOCK program. The DOCK scoring functions are employed to provide a measure of the 'fitness' of each gene. The standard operations of reproduction, crossover and mutation are performed. The method was able to successfully dock methotrexate (which contains approximately 10 rotatable bonds) into the active site of dihydrofolate reductase, among others.

All of the approaches discussed so far use the rigid receptor approximation, in which the binding site is fixed in the conformation determined by X-ray crystallography. This is clearly a limitation, for a change in the conformation of the receptor could significantly alter both the shape and electrostatic character of the binding site. Perhaps the most natural way to take account of the conformation of the binding site would be using molecular dynamics, but at present this approach would take far too long: molecular dynamics simulations are useful for exploring the energy hypersurface in the vicinity of the starting structure, but not for large-scale changes. Indeed, it is commonly regarded as an indication of a stable simulation if the structure stays close to the crystal structure! Leach has described an algorithm which takes the side-chain conformational degrees of freedom into account when performing ligand docking (Leach, 1994b). The backbone is assumed to be fixed and the side-chains are restricted to discrete conformations, obtained from an analysis of high-resolution protein structures; several authors have shown that there are distinct conformational preferences for the side-chains, particularly when X-ray data are considered (Ponder and Richards, 1987). For a given orientation of the ligand within the site, it is possible to determine the lowest energy combination of side-chain rotamers, together with all combinations within a given energy of the lowest energy combination. A combination of two algorithms is used to perform the search: first, dead-end elimination (DEE) (Desmet *et al.*, 1992) and then the A* algorithm (Hart *et al.*, 1968). The DEE algorithm is used to identify side-chain rotamers and pairs of rotamers that cannot possibly be present in the lowest energy conformation or in any conformation within the energy threshold of the global minimum. Application of the DEE algorithm enables the number of possible combinations of rotamers to be significantly reduced. The space remaining is then searched using the A* algorithm. When coupled to a method for exploring the translational and rotational degrees of freedom of the ligand relative to the receptor, the method can be used to perform ligand docking. One interesting result obtained using this algorithm was that the number of combinations of side-chain conformations can be greater in the intermolecular complex than the isolated molecule, and that the conformational entropy of the receptor might thus increase on ligand binding. This counter-intuitive result arises because the ligand modulates the energy surface of the protein, making more conformational states accessible.

3.13 Conformational flexibility in 3D database searches

A common use of a 3D database is to search for compounds that match a three-dimensional pharmacophore. An excellent review of the use of 3D databases in drug design has been published by Martin (1992). There are in essence two different problems to address in a 3D search. The first is to identify whether a given compound contains the appropriate functional groups that are present in the pharmacophore. It is then necessary, *in extremis*, to generate every possible combination of such groups and to determine whether the molecule could position the groups in the appropriate relative orientation. Most 3D pharmacophores are expressed in terms of distance ranges. The first 3D databases used only a single conformation of each molecule, obtained either from X-ray crystallography or from a structure generation program such as CONCORD. The use of a single conformation for each molecule is obviously a potentially serious limitation. There are two possible solutions to this problem. The first is to store multiple conformations for each molecule in the database. This can only be seen as a partial solution as none of the conformations stored may correspond to the 'active' conformation. It is much preferred to be able to consider the entire conformational space.

In recent years a number of methods have been developed that can, in principle at least, take full account of the conformational space of the molecules when searching a 3D database. This is a difficult problem due to the large number of diverse molecules that are contained in a typical database. The aim of these methods is to identify every molecule that could satisfy the 3D pharmacophore in a 'reasonable' conformation (i.e. one without any serious steric interactions). The algorithms that have been described to date fall into two categories. Methods in the first category, exemplified by the CHEMDBS-3D program from Chemical Design[2] (Murrall and Davies, 1990), perform a conformational search on each molecule as it is entered into the database. The algorithm used to perform the conformational search in this program bears some similarities to the WIZARD-II and COBRA programs discussed above, with a particular feature of the approach being the use of conformational 'rules', which enable the search to rapidly eliminate disfavoured conformations. The conformational search is then analysed to derive distance 'keys', one for each pairwise combination of pharmacophoric groups in the molecule. The default set of pharmacophoric groups includes hydrogen bonding donors, hydrogen bonding acceptors, ring centroids and heteroatoms with a formal positive charge. The key consists of a binary number in which each bit corresponds to a distance range between a given type of pharmacophoric group (e.g. donor–donor, donor–acceptor, etc.). For example, the first bit could correspond to the distance being 2.0–2.5 Å, the second bit to the distance range 2.5–3.0 Å and so on. In fact, it is more efficient to use smaller ranges for the more 'common' distances and larger ranges for distances that are less common. This is because the distribution of distances is not uniform (Ash *et al.*, 1985). The key initially contains all zeros. As conformations of the molecule are generated, the distances between the pharmacophoric groups are determined and the appropriate bits in the appropriate keys are set to ones. To search the database, the keys corresponding to the pharmacophore are established. For example, for a pharmacophore containing two donors and one acceptor two keys would be calculated (donor–donor and donor–acceptor). These are then ANDed with the molecular keys, so identifying all molecules in which at least one pair of pharmacophoric groups corresponds to the actual distances. To enhance efficiency, a separate formula screen is also used: this contains information about the number of each type of feature (for example, number of donors, etc.). Clearly, should the pharmacophore contain a larger number of any type of group than is contained in the molecule then it is possible to move immediately to the next entry in the database. Finally, the conformational search is repeated for all 'hits', to regenerate the corresponding conformations which are overlaid for molecular graphics display and subsequent analysis. This final stage makes use of information provided by a substructure search which finds all ways in which the functional groups in the molecule can be matched to the pharmacophore. Chapter 6 describes the use of this software in some practical examples of 3D database searches.

An alternative approach to exploring the conformational space of each molecule when it is entered into the database is to do the conformational search 'on the fly'. Clarke *et al.* have developed a flexible 3D searching system

[2] Chemical Design Ltd, Roundway House, Cromwell Park, Chipping Norton, Oxon OX7 5SR.

based on distance geometry (Clarke *et al.*, 1992, 1993). The first component of their system is a distance screen that is based on distance inequalities obtained from triangle smoothing. The molecules that pass through the filter are then subjected to a distance geometry conformational search to try and generate a conformation that satisfies all the distance constraints. It was found that, using distance geometry at least, the incorporation of conformational flexibility had a severe effect on the time taken when compared to a search using rigid conformations. One way to partially ameliorate these problems under circumstances where only a given number of hits are required is to rank the molecules using a flexibility index. The conformational search is thus applied to the most rigid molecules first (those expected to have the fewest conformations).

The 3D searching systems developed by MDL[3] (Moock *et al.*, 1994) and by Tripos Associates[4] employ a somewhat different strategy. A single conformation of each molecule is stored in the database. The search algorithm then attempts to 'adjust' the conformation, by rotating about single bonds, to try and force it to satisfy the requirements of the pharmacophore. If such an adjustment is possible, then the compound is added to the list of 'hits'. A variety of methods could, in principle, be used to perform the adjustment. For example, the distances between the pharmacophoric groups could be incorporated as restraints into a molecular dynamics simulation, in much the same way that NMR distance constraints are incorporated in the simulated annealing approaches to structure determination. However, for reasons discussed above, molecular dynamics is not currently regarded as a practical method for database searching, due to the time taken and the parameterisation difficulties. The Tripos system uses the 'tweak' algorithm of Fine and colleagues (Fine *et al.*, 1986; Shenkin *et al.*, 1987). This algorithm gives the changes in dihedral angle that are required to enable a given pair of atoms to adopt a specified distance. It was originally devised to model antibody hypervariable loops. An advantage of this particular method over other chain-closure algorithms is that it scales with the number of constraints, rather than the length of the chain and is applicable to any length of chain. The MDL system uses a similar approach that minimises a function that gives the variation in distance with torsion angle.

3.14 Conclusions

As this volume indicates, similarity and complementarity searching are areas of very active research and will continue to be so in the foreseeable future.

[3] MDL Information Systems Inc., 14600 Catalina Street, San Leandro, California 94577.
[4] Tripos Associates, St Louis, Missouri.

A significant number of novel inhibitors of therapeutically important receptors have now been discovered by the methods described in this and other chapters. However, a proper treatment of the problem must take conformational properties into account. There are now a variety of promising ways in which the conformational space of the ligand can be considered; the next stage will be to evaluate the validity of the 'rigid receptor' approximation and to develop algorithms that can take these degrees of freedom into account. This will be of increasing importance given that the number of X-ray and NMR structures is increasing at an exponential rate, a trend that seems likely to continue.

Acknowledgements

The author acknowledges funding by the Science and Engineering Research Council (UK).

References

Allen, F.H., Bellard, S.A., Brice, M.D., Cartwright, B.A., Doubleday, A., Higgs, H., Hummelink, T., Hummelink-Rogers, B.G., Kennard, O., Motherwell, W.D.S., Rodgers, J.R. and Watson, D.G. (1979) The Cambridge Crystallographic Data Centre: computer-based search, retrieval, analysis and display of information. *Acta Crystallographica*, **B35**, 2231–2339.

Allen, M.P. and Tildesley, D.J. (1987) *Computer Simulation of Liquids*. Oxford University Press, Oxford.

Allinger, N.L. (1977) Conformational analysis 130. MM2, a hydrocarbon force field utilising V1 and V2 torsional terms. *Journal of the American Chemical Society*, **99**, 8127–8134.

Ash, J.E., Chubb, P.A.,Ward, S.E., Welford, S.M. and Willett, P. (1985). In *Communication Storage and Retrieval of Chemical Information*, pp.160–167. Ellis Horwood, Chichester.

Beveridge, D.L. and DiCapua, F.M. (1989) Free energy via molecular simulation: a primer. In *Computer Simulation of Biomolecular Systems* eds W.F. van Gunsteren and P.K. Weiner. Escom, Leiden.

Blaney, J.M. (1993) A distance geometry-based approach for docking conformationally flexible molecules from 2D or 3D-chemical datases. Personal communication, in preparation.

Bron, C. and Kerbosch, J. (1973). Finding all cliques of an undirected graph. *Communications of the American Chemical Society Meeting*, **16**, 575–577.

Brooks, C.L. III, Karplus, M. and Pettitt, B.M. (1988) A theoretical perspective of dynamics, structure and thermodynamics. *Advances in Chemical Physics* LXXI. John Wiley, New York.

Brunger, A.T., Kuriyan, J. and Karplus, M. (1987). Crystallographic *R*-factor refinement by molecular-dynamics. *Science* **235**, 458–460.

Burkert, U. and Allinger N.L. (1982) Molecular mechanics. *ACS Monograph* 177. American Chemical Society.

Burt, C. and Richards, W.G. (1990) Molecular similarity: the introduction of flexible fitting. *Journal of Computer-Aided Molecular Design* **4**, 231–238.

Burt, C., Richards, W.G. and Huxley, P. (1990) The application of molecular similarity calculations. *Journal of Computational Chemistry*, **10**, 1139–1146.

Chang, G., Guida, W.C. and Still, W.C. (1990) An internal coordinate Monte Carlo method for searching conformational space. *Journal of the American Chemical Society*, **111**, 4379–4386.

Chau, P.L. and Dean, P.M. (1987) Molecular recognition: 3D surface structure comparison by gnomonic projection. *Journal of Molecular Graphics*, **5**, 97–100.

Clark, D.E., Willett, P. and Kenny, P.W. (1992) Pharmacophoric pattern matching in files of three-dimensional chemical structures: use of bounded distance matrices for the representation and searching of conformationally flexible molecules. *Journal of Molecular Graphics*, **10**, 194–204.

Clark, D.E., Willett, P. and Kenny, P.W. (1993) Pharmacophoric pattern matching in files of three-dimensional chemical structures: implementation of flexible searching. *Journal of Molecular Graphics*, **11**, 146–156.

Clark, M., Cramer, R.D. and Opdenbosch, N.V. (1989) Validation of the general-purpose Tripos 5.2 force-field. *Journal of Computational Chemistry*, **10**, 982–1012.

Clark, T. (1985) *A Handbook of Computational Chemistry: A Practical Guide to Chemical Structure and Energy Calculations*. John Wiley, New York.

Connolly, M.L. (1983a) Analytical molecular surface calculation. *Journal of Applied Crystallography*, **16**, 548–558.

Connolly, M.L. (1983b) Solvent-accessible surfaces of proteins and nucleic-acids. *Science* **221**, 709–713.

Crippen, G.M. (1981) *Distance Geometry and Conformational Calculations*. Chemometrics Research Studies Series 1. John Wiley, New York.

Crippen, G.M. (1978). Rapid calculation of coordinates from distance matrices. *Journal of Computational Physics*, **26**, 449–452.

Crippen, G.M. and Havel, T.F. (1988) *Distance Geometry and Molecular Conformation*. Chemometrics Research Studies Series 15. John Wiley, New York.

Dammkoehler, R.A., Karasek, S.F., Shands, E.F.B. and Marshall, G.R. (1989) Constrained search of conformational hypersurface. *Journal of Computer-Aided Molecular Design*, **3**, 3–21.

Dean, P.M. and Chau, P.-L. (1987) Molecular recognition: optimized searching through rotational 3-space for pattern matches on molecular surfaces. *Journal of Molecular Graphics*, **5** , 152–158.

Dean, P.M. and Callow, P. (1987) Molecular recognition: identification of local minima for matching in rotational 3-space by cluster analysis. *Journal of Molecular Graphics* **5**, 159–164.

Dean, P.M., Callow, P. and Chau, P.-L. (1988) Molecular recognition: blind searching for regions of strong structural match on the surfaces of two dissimilar molecules. *Journal of Molecular Graphics*, **6**, 28–34.

DesJarlais, R.L., Sheridan, R.P., Dixon, J.S., Kuntz, I.D. and Venkatarghavan, R. (1986) Docking flexible ligands to macromolecular receptors by molecular shape. *Journal of Medicinal Chemistry*, **29**, 2149–2153.

Desmet, J., DeMaeyer, M., Hazes, B. and Lasters, I. (1992) The dead-end elimination theorem and its use in protein side-chain positioning. *Nature*, **356**, 539–542.

Dixon, J.S. (1993) Flexible docking of ligands to receptor sites using genetic algorithms. Abstract of a paper presented at the *9th European Symposium on Structure–Activity Relationships*. In press.

Dolata, D.P., Leach, A.R. and Prout, K. (1987) WIZARD: AI in conformational analysis. *Journal of Computer-Aided Molecular Design*, **1**, 73–85.

Elber, R. and Karplus, M. (1987) Multiple conformational states of proteins — a molecular-dynamics analysis of myoglobin. *Science*, **235**, 318–321.

Ferguson, D.M. and Raber, D.J. (1989) A new approach to probing conformational space with molecular mechanics: random incremental pulse search. *Journal of the American Chemical Society*, **111**, 4371–4378.

Fine, R.M., Wang, H., Shenkin, P.S., Yarmush, D.L. and Levinthal, C. (1986) Predicting antibody hypervariable loop conformations. II: Minimization and molecular dynamics studies of MCPC603 from many randomly generated loop conformations. *Proteins: Structure, Function and Genetics*, **1**, 342–362.

Ghose, A.K. and Crippen, G.M. (1985) Geometrically feasible binding modes of a flexible ligand molecule at the receptor-site. *Journal of Computational Chemistry*, **6**, 350–359.

Goldberg, D.E. (1989) *Genetic Algorithms in Search, Optimization and Learning*. Addison Wesley, Reading, Massachusetts.

Good, A.C., Hodgkin, E.E. and Richards, W.G. (1992) Utilization of Gaussian functions for the rapid evaluation of molecular similarity. *Journal of Chemical Information and Computer Sciences*, **32**, 188–191.

Goodsell, D.S. and Olson, A.J. (1990) Automated docking of substrates to proteins by simulated annealing. *Proteins*, **8**, 195–202.

Hart, P.E., Nilsson, N.J. and Raphael, B. (1968) A formal basis for the heuristic determination of minimum cost paths. *IEEE Transactions on SSC*, **4**, 100–114.

Hart, T.N. and Read, R.J. (1992) A multiple-start Monte-Carlo docking method. *Proteins: Structure–Function and Genetics*, **13**, 206–222.

Hehre, W.J., Radom, L., Schleyer, P. v. R. and Pople, J.A. (1986) *Ab Initio Molecular Orbital Theory*. John Wiley, New York.

Hodgkin, E.E. and Richards, W.G. (1987) Molecular similarity based on electrostatic potential and electric-field. *International Journal of Quantum Chemistry*, **14**, 105–110.

Holland, J.H. (1992 Genetic algorithms. *Scientific American*, July, 66–72.

Howard, A.E. and Kollman, P.A. (1988) An analysis of current methodologies for conformational search of complex molecules. *Journal of Medicinal Chemistry*, **31**, 1669–1675.

Kearsley, S.K. (1990) An algorithm for the simultaneous superposition of a structural series. *Journal of Computational Chemistry*, **10**, 1187–1192.

Kearsley, S.K. and Smith, G.M. (1992) An alternative method for the alignment of molecular structures: maximizing electrostatic and steric overlap. *Tetrahedron Computer Methodology*, **3**, 615–633.

Kirkpatrick, S., Gelatt, C.D. Jr and Vecchi, M.P. (1983) Optimization by simulated annealing. *Science*, **220**, 671–680.

Kuhl, F.S., Crippen, G.M. and Friesen, D.K. (1984) A combinatorial algorithm for calculating ligand binding. *Journal of Computational Chemistry*, **5**, 24–34.

Kuntz, I.D. (1992) Structure-based strategies for drug design and discovery. *Science*, **257**, 1078–1082.

Kuntz, I.D., Blaney, J.M., Oatley, S.J., Langridge, R. and Ferrin, T.E. (1982). A geometric approach to macromolecule-ligand interactions. *Journal of Molecular Biology*, **161**, 269–288.

Judson, R.S., Jaeger, E.P., Treasurywala, A.M. and Peterson, M.L. (1993) Conformational searching methods for small molecules. II. Genetic algorithm approach. *Journal of Computational Chemistry*, **14**, 1407–1414.

Langdrige, R., Ferrin, T.E., Kuntz, I.D. and Connolly, M.L. (1981) Real-time color graphics in studies of molecular-interactions. *Science*, **211**, 661–666.

Leach, A.R. (1991) A survey of methods for searching the conformational space of small and medium-sized molecules. In *Reviews in Computational Chemistry*, Vol. II (eds K.B. Lipkowitz and D.B. Boyd, pp. 1–55. VCH Publishers, New York.

Leach, A.R. (1994a) An algorithm to directly identify a molecule's 'Most different' conformations. *Journal of Chemical Information and Computer Sciences*. **34**, 661–670.

Leach, A.R. (1994b) Ligand docking to proteins with discrete side-chain flexibility. *Journal of Molecular Biology*, **235**, 345–356.

Leach, A.R. and Prout, K. (1990) Automated conformational analysis: directed conformational search using the A* algorithm. *Journal of Computational Chemistry*, **11**, 1193–1205.

Leach, A.R. and Kuntz, I.D. (1992) The conformational analysis of flexible ligands in macromolecular receptor sites. *Journal of Computational Chemistry*, **13**, 730–748.

Leach, A.R., Prout, K. and Dolata, D.P. (1990) The application of artifical intelligence to the conformational analysis of strained molecules. *Journal of Computational Chemistry*, **11**, 680–693.

Lipton, M. and Still, W.C. (1988) The multiple minimum problem in molecular modelling. Tree searching internal coordinate conformational space. *Journal of Computational Chemistry*, **9**, 343–355.

McCammon, J.A. and Harvey, S.C. (1987) *Dynamics of Protein and Nucleic Acids*. Cambridge University Press, Cambridge.

McGarrah, D.B. and Judson, R.S. (1993) Analysis of the genetic algorithm method of molecular conformation determination. *Journal of Computational Chemistry*, **14**, 1385–1395.

Martin, Y.C. (1992) 3D database searching in drug design. *Journal of Medicinal Chemistry*, **35**, 2145–2154.

Mayer, D., Naylor, C.B., Motoc, I. and Marshall, G.R. (1987) A unique geometry of angiotensin-converting enzyme consistent with the structure-activity studies. *Journal of Computer-Aided Molecular Design*, **29**, 453–462.

Moock, T.E., Henry, D.R., Ozkabak, A. and Alamgir, M. (1994) Conformational searching in ISIS/3D databases. *Journal of Chemical Information and Computer Sciences*. **34**, 184–189.

Motoc, I., Dammkoehler, R.A., Mayer, D. and Labanowski, J. (1986a) Three-dimensional quantitative structure–activity relationships. I. General approach to the pharmacophore model validation. *Quantitative Structure–Activity Relationships*, **5**, 99–105.

Motoc, I., Dammkoehler, R.A. and Marshall, G.R. (1986b) Three-dimensional structure–activity relationships and biological receptor mapping. In *Mathematical and Computational Concepts in Chemistry* (ed. N. Trinajstic), pp. 222–257. Ellis Horwood, Chichester.

Murrall, N.W. and Davies, E.K. (1990) Conformational freedom in 3-D databases. 1. Techniques. *Journal of Chemical Information and Computer Sciences*, **30**, 312–316.

Payne, A.W.R. and Glen, R.C. (1993) Molecular recognition using a binary genetic search algorithm. *Journal of Molecular Graphics*, **11**, 74–91.

Pearlman, R.S., Ruskino, A., Skell, J.M. and Balducci, R. (1987) CONCORD is available from Tripos Associates Inc., St Louis, Missouri.

Perkins, T.D.J. and Dean, P.M. (1992) An exploration of a novel strategy for superposing several flexible molecules. *Journal of Computer-Aided Molecular Design*, **7**, 155–172.

Ponder, J.W. and Richards, F.M. (1987). Tertiary templates for proteins — use of packing criteria in the enumeration of allowed sequences for different structural classes. *Journal of Molecular Biology*, **193**, 775–791.

Pople, J.A. and Beveridge, D.L. (1970) *Approximate Molecular Orbital Theory*. McGraw-Hill, New York.

Saunders, M. (1987) Stochastic exploration of molecular mechanics energy surfaces. Hunting for the global minimum. *Journal of the American Chemical Society*, **109**, 3150–3152.

Saunders, M. (1989) Stochastic search for the conformations of bicyclic hydrocarbons. *Journal of Computational Chemistry*, **10**, 203–208.

Shenkin, P.S., Yarmush, D.L., Fine, R.M., Wang, H. and Levinthal, C. (1987) Predicting antibody hypervariable loop conformation. I. Ensembles of random conformations for ringlike structures. *Biopolymers*, **26**, 2053–2085.

Sheridan, R.P., Nilakantan, R., Dixon, J.S. and Venkataraghavan, R. (1986) The ensemble approach to distance geometry: application to the nicotinic pharmacophore. *Journal of Medicinal Chemistry*, **29**, 899–906.

Shoichet, B.K., Leach, A.R., Corwin, C. and Kuntz, I.D. (1994) Ligand solvation and conformational effects in molecular docking. In preparation.

Smellie, A.S., Crippen, G.M. and Richards, W.G. (1991) Fast drug–receptor mapping by site-directed distances. A novel method of predicting new pharmacological leads. *Journal of Chemical Information and Computer Sciences*, **31**, 386–392.

Spellmeyer, D., Blaney, J.M. and Ripke, W. (1994) Ensemble molecular dynamics. In preparation.

Weiner, S.J., Kollman, P.A., Case, D.A., Singh, U.C., Ghio, C., Alagona, G., Profeta, S. and Weiner, P. (1984) A new force-field for molecular mechanical simulation of nucleic-acids and proteins. *Journal of the American Chemical Society*, **106**, 765–784.

Weiner, S.J., Kollman, P.A., Nguyen, D.T. and Case, D.A. (1986) An all atom force-field for simulations of proteins and nucleic acids. *Journal of Computational Chemistry*, **7**, 230–252.

Wilson, S.R. and Cui, W. (1990) Applications of simulated annealing to peptides. *Biopolymers*, **29**, 225–235.

Wilson, S.R., Cui, W., Moskowitz, J.W. and Schmidt, K.E. (1988) Conformational analysis of flexible molecules: location of the global minimum energy conformation by the simulated annealing method. *Tetrahedron Letters*, **29**, 4373–4376.

4 Exploiting similarity between highly flexible and dissimilar molecular structures

T.D.J. PERKINS

4.1 Introduction

There are many examples of chemically unrelated and apparently dissimilar molecules that bind to a common site at a receptor and elicit the same pharmacological response. Ideally, these drug–receptor interactions would be examined using X-ray crystallography or nuclear magnetic resonance (NMR) spectroscopy to provide a detailed atomic description of the interaction. This information could then be used as the basis for the direct design of ligands to fit this site. However, despite the dramatic increase in the number of crystal structures of enzyme–ligand complexes in recent years, it is still the case that there are no crystal data for the vast majority of potential drug molecules in their sites. Membrane-bound receptors are a notable example of the problem since they are often pharmacological and therapeutic targets, and have provided no crystal structures to date. In this type of situation, a less direct approach must be used; an obvious tactic is to compare a series of ligands known to bind to the site and attempt to determine their common features and potential pharmacophores. If the molecules can be superposed in three-dimensional (3D) space, their properties can be studied using 3D quantitative structure–activity relationships (QSAR) using procedures such as CoMFA (Cramer *et al.*, 1988; see also Chapter 12). This, in turn, may allow the design of novel molecules that incorporate some or all of the common features and that should therefore be capable of binding to the same receptor site.

Most drug molecules are flexible to some extent (i.e. they have a number of single bonds that are able to rotate), making their overall conformation and 3D structure uncertain. Small peptides, which are of growing interest as tools in the search for novel lead compounds in drug design (e.g. Jacobs and Fodor, 1994) have a very large number of potential conformations, many of which exist simultaneously in solution, since they have a flexible linear backbone with flexible side-chains. The number of possible conformations of any molecule increases exponentially with the number of rotatable bonds, and thus the computer time required to search through these conformers also increases exponentially.

If two molecules A and B are to be matched, each conformation of molecule

A must be matched with each conformation of molecule B to ensure a thorough search. If both molecules are even moderately flexible, the number of matches that have to be performed becomes unmanageably large. An additional problem is that many of the conformers generated would be energetically unstable when bound in the receptor site; only a subset of the many possible conformations should therefore be included in the matching process and no simple criteria are available for this selection. Apparently obvious choices such as the global-minimum-energy conformation or the crystal structure of the isolated molecule may not be appropriate, since the receptor environment may favour quite different conformations (see Jorgensen, 1991). Any conformation that the molecule could feasibly adopt in the receptor site should therefore be included, increasing the number of conformations to be matched.

In some cases, one molecule of interest is rigid or relatively so, and it can be used as a template onto which the remaining flexible molecules are forced (Hopfinger and Burke, 1990). This is a useful procedure if a rigid analogue of an active compound is available, but is clearly not applicable as a general method.

The other factor to consider in the search for molecular similarity is the choice of superposition method. For clearly similar molecules like bio-isosteres the superposition is obvious and can be performed by hand, although it should be borne in mind that there may be more than one good superposition. For matches between dissimilar molecules, and for those involving larger molecules, manual matching is not an option as the number of possibilities becomes infeasibly large. The superpositions can be made in terms of any one of a number of molecular features such as atomic superposition, potential pharmacophoric atoms, or surface properties. The choice will be problem dependent, but in all cases it should ideally search through all feasible superpositions without operator bias.

This chapter examines the difficulties involved in matching flexible and apparently dissimilar molecules, in terms of both the conformational analysis of each molecule, and the optimum overlay or superposition of pairs (and larger numbers) of molecules. A possible approach to the problem is outlined, and is illustrated with a set of angiotensin II antagonists. This is followed by a discussion on the usefulness of the matches obtained, and the ways in which they can be used in the design of novel compounds.

4.2 Molecular conformation

The determination of molecular conformation is described in more detail in Chapter 3; this section deals with those aspects particularly relevant to molecular matching. There are two major problems in determining conformation: firstly, the combinatorial complexity of searching through conformational space and finding the global-minimum-energy conformation in a particular environment (presumably the gas phase); secondly, the relative stability of conformers in the receptor site, compared to the gas phase.

Once these problems have been solved, a fundamental choice has to be made in the way conformations are generated for molecular matching. If there are efficient rules for searching conformational space, it may be possible to apply them to molecules during the matching process. Alternatively, the conformational search and molecular matching aspects of the process can be separated; this may be particularly useful if the same molecules are used in more than one match, as the conformational analysis can be re-used.

4.2.1 Combinatorial complexity and the multiple minima problem

If a molecule has k rotatable bonds each with j_k rotameric states, the total number of conformations C can be written as

$$C = \prod_{i=1}^{k} j_i \tag{4.1}$$

For example, a typical drug molecule with eight rotatable bonds, each considered in 1° steps, has a total of $360^8 \approx 10^{20}$ conformations available to it. It is clearly infeasible to enumerate each of these conformations, without even assessing their relative energies.

The primary aim in conformational analysis is to determine the global minimum-energy conformation of the molecule under consideration. With such a large conformational space to search, it is clear that approximations will have to be made and efficient strategies devised to give a reasonable chance of discovering the global minimum. In addition, the conformational space of a molecule typically possesses a highly complex landscape, with many local minima of approximately equal depth. This complicates the search and renders some optimization methods of little value. Three classes of methods, each with their own variants, can be envisaged that allow a search through the conformational hypersurface: 'classical' minimization, a systematic search, or a random-based method.

Classical minimization methods. These methods, which include Newton–Raphson and simplex procedures, are not suitable for a complete conformational analysis of a molecule because of the combination of landscape topography and optimization procedures. The methods generally use a steepest descent strategy, making them very efficient at finding the nearest local minimum in a surface and thus ideal for the optimization of bond lengths and angles in modelled structures. However, once this minimum has been found, there is no mechanism to escape and search for a deeper minimum. These methods are thus of little use for conformational analysis unless the starting structure is very close to the global-minimum-energy conformation or many random starting positions are used.

Systematic search methods. In systematic or grid-search methods, a step size (typically 60°) is chosen and a complete search of the conformational space is made; the number of steps required is given by equation (4.1). Note that this implies that, however coarse the step size, the time required to complete the search still increases exponentially with the number of rotatable bonds. There are a number of variants of the method that essentially aim to slow this increase. For example, in peptides, torsion angles can be considered by residue rather than individually (Moult and James, 1986), or the search can attempt to ignore non-productive regions of the conformational space (Goodman and Still, 1991). In general, however, this class of methods is suitable only for relatively small molecules.

Random methods. Random methods are necessary if larger structures are to be examined, although they do not guarantee a complete coverage of conformational space. The algorithm visits conformations at random, subject to rules specific to each procedure. Random searches may be completely stochastic (Saunders, 1987), based on genetic algorithms (Davis, 1991), Monte Carlo methods (Metropolis *et al.*, 1953), or optimized using simulated annealing (Kirkpatrick and Smith, 1983).

One of the most successful of these methods is simulated annealing, based on methods originally used in statistical mechanics (Metropolis *et al.*, 1953), and applied to molecular conformation by Wilson *et al.* (1988). Small changes are made to the configuration of the system (typically a change in the conformation by rotation around a single bond) and the difference in potential energy ΔE between the current and previous conformations is assessed. If the change is negative, it is accepted immediately; otherwise it is accepted with a probability of $\exp(-\Delta E/T)$ where T is an annealing control temperature. The value of T is initially large (most changes are accepted and the entire conformational surface is explored) but is gradually reduced so that the algorithm settles in the lowest position in the deepest valley. In principle, simulated annealing should find the same global minimum each time it is run, although this is often not the case in practice.

4.2.2 *Determination of relative values of structures*

At each stage in the search through the conformational states of a molecule, its energy must be calculated in some way. For most drug molecules, the only practical method available is an empirical calculation of the potential energy in the gas phase. This energy has components that represent bond length, bond angles, torsion angles, van der Waals and coulombic interactions. More accurate (and generally slower) methods of calculation are also available (including semi-empirical methods) but if a large number of calculations are involved, speed is clearly important. The absolute energies of conformers are not important in a conformational analysis since the aim is merely to rank conformers of interest. In addition, comparisons do not need to be made between molecules (merely between conformers), so that a molecular-mechanics force field is probably sufficiently accurate.

The major flaw with *in vacuo* (or even solution) molecular-mechanics methods is that the conformation of interest is that of the ligand bound to its receptor. The minimum-energy conformation of the bound ligand is thus dependent on the ligand–receptor complex as a whole; the ligand itself need not adopt a low-energy conformation of the isolated molecule. Unfavourable conformations can be compensated for by favourable ligand–receptor interactions. This is a crucial difference that also implies that experimental evidence from X-ray crystal data of the isolated molecule and NMR spectroscopic data of the ligand in solution cannot be relied upon to give a good estimate of the binding conformation of a ligand. If the 3D structure of the receptor is unknown (and this is very likely if molecular similarity methods are being used), then it will obviously be impossible to calculate any of the crucial interaction energies.

Gas-phase energies will be useful in spotting conformations that are infeasible (e.g. poor steric contacts) and will provide some kind of pointer to stability at the receptor. They should probably be considered as a rather approximate measure of receptor bound energy and treated with caution. Probably the best use of force-field-derived energies is to consider any conformer within some energy cutoff (perhaps 25 kJ mol^{-1}) to be a candidate for the receptor-bound conformation.

4.2.3 Methods for generating conformers

There are two fundamentally opposing methods that can be used for generating conformers to be used in molecular matching: conformational search during matching or using predetermined conformers from a prior conformational analysis step.

Conformer generation during matching. In this type of search, conformational analysis is combined with molecular matching. This can be considered as the addition of further dimensions to those of the matching process. For example, if matching is performed by rigid body transformations (three rotation and three translation axes) and there are n_A and n_B torsion angles in molecules A and B, respectively, the total number of dimensions for the search is $6 + n_A + n_B$. The total number of states to examine (i.e. the number of matches to undertake) is simply the product of the number of states in each dimension; this can easily become a very large number. However, the method may be feasible if only one or two torsion angles have a significant effect on the conformation, so that the remainder can be neglected.

If new conformers are generated during the search for molecular similarity, this allows improved conformations to be produced for the curentlyconsidered superposition. This has the advantage that final matches should have a better correspondence, apparent as a lower root mean square (rms) deviation between matched points. Calculation of conformation together with matching may, however, be inefficient if the same molecule is to be matched a number of times; the conformational analysis will have to be repeated for each match.

Conformer generation prior to matching. In this style of search, a thorough analysis is made of the conformational hypersurface of the molecule before any matching is undertaken. The aim is to select a small number of conformers at this stage to reduce the combinatorial problems at the matching stage. The selection may favour low-energy conformers or, preferably, attempt to cover the entire range of feasible conformers. There is considerable choice in the way that conformers can be selected and duplicates omitted, and a number of methods are described in the next section. The remaining problem is then reduced to matching a number of combinations of rigid molecules. This simplifies the matching since there are already a number of algorithms suitable for matching rigid structures. The major disadvantage of making this approximation is that the quality of the fits obtained is limited by the range of conformations used in the matching.

4.3 Similarity and conformation

4.3.1 Molecular matching methods

The simplest form of molecular matching involves the manual selection of equivalent atoms in two molecules and their superposition by minimizing the rms difference in equivalent atom positions. This approach is suitable if the molecules to be superposed are very similar (as is the case for bio-isosteres or for molecules within a homologous series). For less obviously similar molecules, or where there may be more than one possible solution to the choice of atoms to match, this method is highly dependent on operator bias. This is a particular problem with larger molecules because of the increase in the possible number of matches. For an atom-centred match, the number of possible superpositions S for two n-atom molecules can be written as

$$S = n! \tag{4.2}$$

If only a subset of all possible atoms are included, then there is additional complexity in the choice of atoms to omit.

An alternative approach is to use a molecular graphics system to superpose the molecules manually on screen, by rotation and translation of one molecule onto the other. This has the advantage that overlaps can be compared visually and followed by a quantitative assessment of the goodness of fit of the match, perhaps by measuring molecular overlap (Itai *et al.*, 1993) or electrostatic potential similarity (Blaney *et al.*, 1993). This procedure is often simplified by initially superposing the centres of gravity of the two molecules and considering only rotation around *x*-, *y*- and *z*-axes. The method is subject to bias in the choice of superposition and the problem of searching through a 6D superposition space. Potentially good matches may be completely missed, and there is no means of optimizing those close to a local minimum. The only truly unbiased approach is to use a matching procedure that is fully automated and as thorough as possible. The remainder of the methods discussed fall into this category.

Rigid matching. There are a number of methods available that examine molecular similarity and perform molecular matching, although the majority can be applied only to rigid molecules and do not take account of flexibility. Molecules have been matched on the basis of a number of different features, and each method has its own advantages.

Hydrogen bonding is one of the most important determinants of receptor–ligand binding since it uniquely combines strength and directionality. If only selected hydrogen-bonding atoms together with hydrophobic centres are considered, then the combinatorial complexity can be reduced sufficiently to allow enumeration of all possible matches by systematic methods (Danziger and Dean, 1985). This type of method also has the advantage that the selected atoms/hydrophobic centres are potential pharmacophores.

Molecular matches have also been performed on the basis of automatically-selected atomic superpositions (Barakat and Dean, 1990a,b; 1991). The method uses simulated annealing to minimize the difference-distance matrix between two molecules by reordering the matched atoms. This has the advantage that no rigid body translations or rotations are performed and that rms differences do not need to be calculated. This is a shape-only comparison; an atom can be matched onto any other, regardless of atomic type or chemical character.

Matching on the basis of electron densities (Meyer and Richards, 1991) has become a popular method, combining steric and electronic effects. However, these methods tend to be better at describing differences between a series of related molecules, rather than matching dissimilar structures.

Molecular volume matches (Masek *et al.*, 1993a) and molecular surface matches (Dean *et al.*, 1988; Masek *et al.*, 1993b) are both useful approaches. In the former, the entire molecular interiors are superposed, whilst in the latter only a thin shell or skin, such as that between the van der Waals and accessible surfaces, around each molecule is superposed. These methods are discussed in more detail elsewhere in this volume (see Chapter 7). This surface shell is expected to be particularly useful for drug design, since it defines the position of the receptor for those parts that interact with the ligand, and will be common for all ligands that bind to the same receptor.

The inability of molecular matching programs to cope with flexible structures does indeed cause problems for those wishing to examine molecular similarity. For example, Diana *et al.* (1993) used the Barakat and Dean (1990a,b; 1991) algorithms re-implemented in SYBYL; these authors were able to match molecules successfully if crystal structures were known, but found the algorithms far less helpful when the molecular conformations were unknown.

Flexible matching. There have been a number of examples of the incorporation of flexibility into molecular matching. Burt and Richards (1990) introduced flexibility into molecular fitting based on electron density; they used a simplex algorithm to optimize molecular overlap whilst constraining conformational energy to within a small range above the global minimum using a Boltzmann function. The potentially enormous configurational space explored by this algorithm means that only relatively small structures are suitable candidates. In addition, the use of simplex minimization makes the method useful only if the optimal molecular superposition is near the start configuration.

Fantucci *et al.* (1992) used a simple and fast energy evaluation technique to randomly search through torsion-angle space to select low-energy conformers; these were then subjected to a full energy minimization, and the lower-energy conformers were used in electrostatic potential similarity calculations.

Borea *et al.* (1992) attempted to find low-energy conformers by partitioning torsion angles into isolated, non-interacting pairs, each of which was searched for energy minima. The conformers involved in the matching were limited to those within 15° of the *cis* and *trans* minima for each pair in 1° steps; this comprised those within 30 kJ mol^{-1} of the global minimum. All combinations of conformers were then matched on the basis of potential hydrogen-bonding atoms and hydrophobic centres. They were assessed on the basis of the quality of the superposition, and of the similarity between their surface electrostatic potentials.

Payne and Glen (1993) used a genetic search algorithm to optimize both conformation and relative orientation. The algorithm uses bitstrings to describe the two molecules and the techniques of mutation, crossover, and selection are employed to optimize the match. However, the match they obtained between two benzodiazepine receptor antagonists was very different from that predicted by structure–activity studies (Codding and Muir, 1985) and other superposition studies (Borea *et al.*, 1992).

Kato *et al.* (1992) have described an algorithm AUTOFIT that matches molecules on the basis of their hydrogen bonds. Conformational flexibility was included by considering only conformationally significant bonds. Matches were generated for all combinations of conformers (in 30° steps) together with all combinations of hydrogen-bond pairs. The total number of rotatable bonds was limited to two, or in later studies (Itai *et al.*, 1993), three.

A number of the examples of matching flexible molecules described above have used *ad hoc* short-cuts possible because of characteristics specific to the molecules being compared. These include decisions to ignore or partition torsion angles, required to reduce the combinatorial complexity. Whilst these measures allow a particular problem to be studied, it is not always clear how they could be applied to the general case of matching two flexible, dissimilar molecules.

4.3.2 *The FMATCH algorithm*

A possible approach to matching flexible molecules has been described (Perkins and Dean, 1993; Dean and Perkins, 1993) using an algorithm termed FMATCH (Flexible Molecules Annealed To a Consensus Hierarchy). The approach circumvents many of the combinatorial problems present in matching flexible molecules by its combined use of simulated annealing and cluster analysis. It attempts to be completely general and can also be used to investigate similarity between more than two molecules. The following section describes the stages in the FMATCH algorithm and presents an example of its use in the superposition of a set of angiotensin II receptor antagonists.

The method separates the process of matching into a number of discrete steps (a–d) (see Figure 4.1): determination of low-energy conformers; selection of representative conformers; combinatorial matching of rigid representative conformers; and analysis of matches (particularly important when more than two molecules are involved).

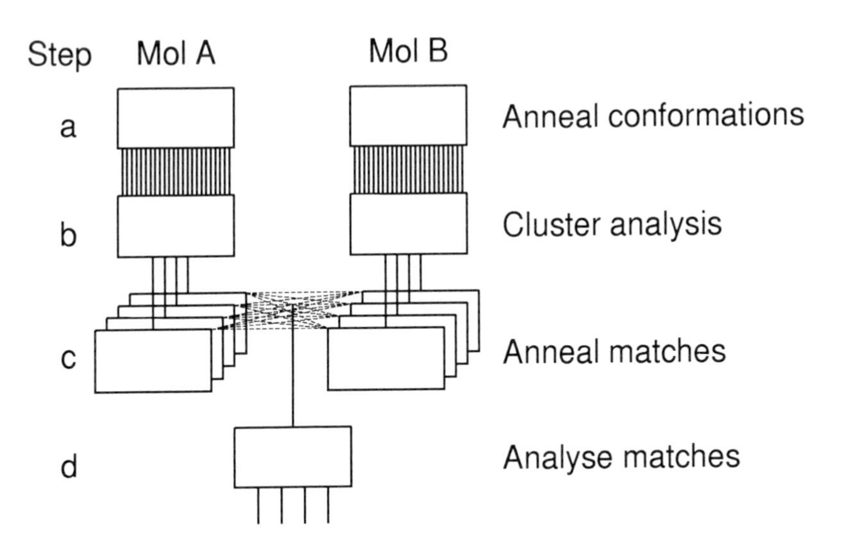

Figure 4.1 Schematic overview of the matching process in FMATCH between two molecules, Mol A and Mol B. Each step (a–d) corresponds with a section in the text.

Determination of low-energy conformers. Simulated annealing is used to explore the conformational space and determine the global-minimum-energy conformation of a molecule. The system is simplified by modifying only torsion angles for rigid rotations around acyclic bonds (ring torsions, bond angles and lengths remain fixed as found in the initial structure). An empirical molecular mechanics force field from the COSMIC routines (Vinter *et al.*, 1987) is used to calculate the potential energy of conformers in terms of van der Waals, coulombic, and torsion parameters during the conformational optimization. This approach to conformational analysis is very similar to that of Wilson *et al.* (1988) and Wilson and Cui (1990).

A record of all conformations visited is stored, in terms of torsion angles, in a history file. This can be used to study the behaviour of the annealing at a graphics terminal, but far more importantly it also acts as a record of the pseudo-random search through conformational space. The global-minimum-energy conformation determined by the annealing is not considered important in itself in the FMATCH algorithm. However, its energy provides a useful absolute zero to which all other conformers can be compared. There are two reasons for the lack of importance of this conformation: firstly, any gas-phase molecular mechanics energy is of dubious relevance to the receptor environment; secondly, optimization using only rigid rotors adds another level of approximation, since a full energy minimization may be able to relax bond lengths and torsion angles, thereby reducing the potential energy without markedly altering the overall conformation.

As a consequence of these problems, the energy band that designates allowed conformers should be rather generous. In the FMATCH algorithm, allowed low-energy conformers are those accessible at room temperature and thus feasibly stable when bound at the receptor. In practice, this has been defined as a probability of at least 10^{-6}: for a Boltzmann distribution at 290K this is equivalent to conformations within $\approx 33\,\mathrm{kJ\,mol^{-1}}$ of the global-minimum-energy conformation. This filter step can still leave a large number of allowed conformers, and thus to simplify the housekeeping, a random subset of these (typically 500–2000) is selected for the next stage of the analysis.

Cluster low-energy structures. Cluster analysis is a useful tool for examining similarity between objects. For conformers, there are at least two different ways in which a 'distance' (i.e. a measure of dissimilarity) between conformers can be assessed: torsion angles and real-space measures.

Torsion angles are an intuitive measure of conformation and particularly relevant if the conformational search is based on these internal coordinates. Essentially, the torsion distance T between two conformers of a molecule with n rotatable bonds is written

$$T^{jk} = \sum_{i=1}^{n} (\tau_i^j - \tau_i^k)^2 \qquad (4.3)$$

where τ is the torsion angle value, subscript i is the sequential number of the torsion angle and superscripts j and k refer to the conformer. Equation (4.3) must be modified to take account of the rotational degeneracy of torsion angles (i.e. $-180° = +180°$), and the symmetry present, for example, in terminal methyl groups. T^{jk} thus provides a measure of the difference in the conformation between conformers j and k.

There are some potential problems with this measure of dissimilarity between conformers, since all torsion angles are given equal weighting in the calculation whereas they do not all have equal effects on the 3D space occupied by the molecule. A shift in a central torsion angle of say 30° will alter the positions of a larger number of atoms and move them further than the same size change at a peripheral torsion.

An alternative method in Cartesian space uses differences between equivalent elements in the distance matrices of the two conformers. The distance D^{jk} between conformers j and k is

$$D^{jk} = \sum_{l=1}^{n-1} \sum_{m=l+1}^{n} |d_{lm}^{k} - d_{lm}^{j}| \tag{4.4}$$

where d_{lm} is the distance between atoms l and m. This method gives a better impression of the overall shape, rather than conformation in terms of individual torsion angles. In the selection of conformers to match, this may be a better measure to detect significantly different shapes. However, Perkins and Dean (1993) found little difference between the conformers produced by the two methods for a number of angiotensin II receptor antagonists. This, however, may be due to the topology of these molecules, which essentially consisted of a rigid core with side-chains. The results may be very different for a peptide chain since the essentially linear structure means that a 60° change in a side-chain torsion will contribute the same 'distance' as a 60° change to a central main-chain torsion angle.

Variations are possible to both methods; these could include a weighting system to alter the distribution of importance between torsion angles in an attempt to overcome the greater influence of central bonds on overall shape, or the inclusion only of torsions/atoms that contribute to displacements of atoms present in a predetermined pharmacophore.

The distances between conformers (T^{jk} or D^{jk}) produced by these methods allow the construction of a distance matrix (T or D) that includes all interconformer distances. This is a square, symmetric matrix with zeros along one diagonal which represent self-comparisons. This information can then be subjected to cluster analysis; a number of techniques are available that process the distance in different ways but one of the most useful is the minimum variance method (Ward, 1963), which tends to generate spherical clusters (Everitt, 1980).

The aim of cluster analysis, within the FMATCH algorithm, is to reduce the dataset of all selected low-energy conformers to a small number of clusters that span the entire conformational space, and to select a representative from each. Cluster analysis by itself provides no way of determining the optimum number of clusters, merely the distribution of conformers among clusters for any given number of clusters. However, the stopping rules proposed by Mojena (1977) can be used to automatically select a particular number of clusters, such that the clusters are significantly different from each other. Ward's method is an agglomerative method of cluster analysis: for N conformers, $(N - 1)$ fusions are performed between pairs of conformers (these may already be part of a cluster), and each fusion i has a criterion value α_i, related to the 'distance' between the conformers involved in the fusion. The mean ($\bar{\alpha}$) and standard deviation (σ_α) of the fusion values are determined, and number of deviates (k_i) calculated as

$$k_i = (\alpha_{i+1} - \bar{\alpha})/\sigma_\alpha \tag{4.5}$$

This is simply a measure of how far the next fusion $(i + 1)$ lies from the mean value of α. The agglomeration is continued until the subsequent fusion (α_{i+1}) is significantly different from the mean ($\bar{\alpha}$) (generally using a t-test with the probability $p < 0.05$), leaving i fusions and $(N - i)$ clusters. The conformer closest to the centroid of each cluster (measured in D- or T-space according to equations (4.3) or (4.4) as appropriate) is selected as the single representative of that cluster. This method provides an objective way of selecting a number of significantly different clusters that provide an even spread through the available conformational space.

An example of a dendrogram produced from cluster analysis of the angiotensin II antagonist DuP 753 is shown in Figure 4.2, with the solid horizontal line representing the optimum division of 500 low-energy conformers into eight clusters. The eight representative conformers are shown superposed with one five-membered ring in common in Figure 4.3, illustrating the diversity of conformational space occupied.

Match representative conformers. Once representatives have been determined, all combinations of conformers of pairs of molecules can be matched using any method available for rigid molecules. The choice will be governed by the time available, size, and chemical character of the ligands being matched. The orginal FMATCH algorithm used the Barakat and Dean (1990a,b, 1991) atom-based method which ignores any distinction between atom types.

Analysis of matches. Methods of molecular matching based on simulated annealing have significant random components and although the algorithm should, in theory, always find the same global minimum, this is not generally true in practice. The standard way to overcome this problem is to perform replicate matches (typically 5–20) for each pair of conformations. For

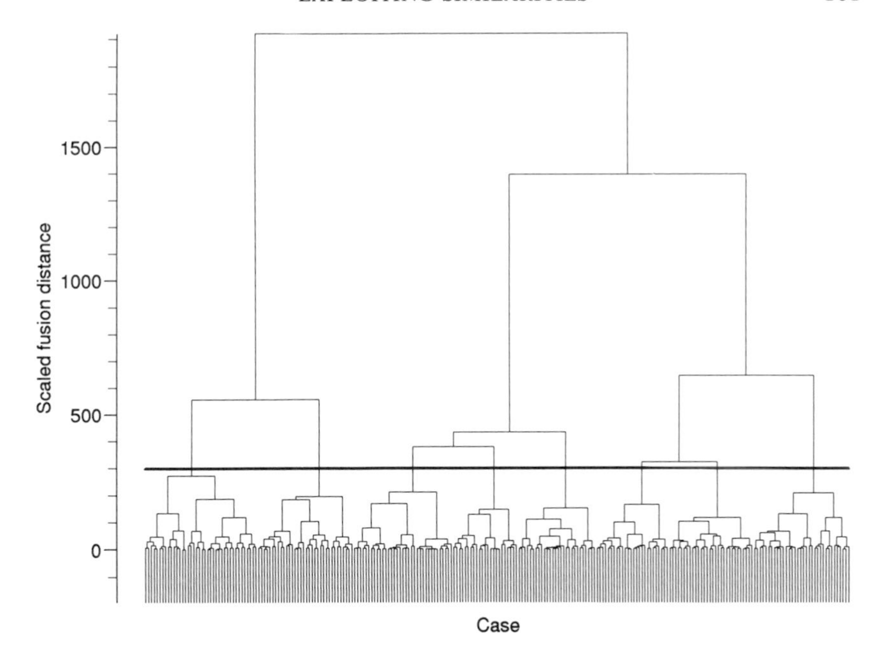

Figure 4.2 Dendrogram illustrating the cluster analysis of 250 low-energy conformers of DuP 753, after clustering with Ward's method according to torsion angle distance. The thick horizontal line represents the fusion at which clustering is stopped according to Mojena's stopping rules with $p < 0.05$.

atom-based matches, the matches can be distinguished and analysed by the generation of a 'distance' between matches in match space on the basis of equivalent matched atoms (see Table 4.1). This describes the difference between both replicate matches between the same conformers, and matches

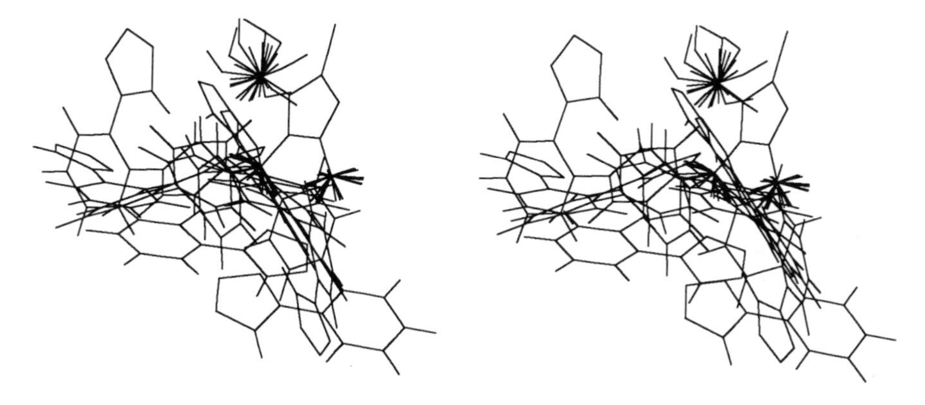

Figure 4.3 Stereoview of the superposition of eight representative DuP 753 conformers.

Table 4.1 Binary scoring scheme for measuring the 'distance' between replicate matches between two molecules. For each match, pairs of numbers represent index numbers of equivalent atoms in molecules A and B. Identical matches score zero and non-identical matches one; the 'distance' between the matches is the sum of these scores

Match 1		Match 2		Score
A	B	A	B	
1	4	1	4	0
2	3	2	6	1
3	1	3	1	0
4	7	4	7	0
6	2	5	2	1
			Distance:	2

between different conformers. This distance information can then be used to create a symmetric distance matrix in match space between all matches, analogous to the conformation-space distance matrix decribed below. Cluster analysis can then be applied in the same way as for conformational analysis. This allows the generation of a small number of representative, significantly different matches. Papadopoulos and Dean (1991) used a match 'distance' based on the rotation and translation matrix required to superpose one molecule onto the other, and performed cluster analysis on these data. This latter scheme has the advantage that it can be applied to non-atom-based matches, although it cannot be used if multiple matches also cover multiple conformations.

So far, matches have only been considered between two molecules. In most cases, however, more than two molecules bind to the same site, and the drug designer would like to be able to match all of these molecules at the same time. If an attempt is made to match a number of flexible molecules simultaneously, the combinatorial problems become enormous; the only feasible solution is to carry out all pairwise matches and then analyse the results. There are significant organizational problems even with pairwise matching: for example, for six molecules, each in ten representative conformations with ten replicate matches, the total number of comparisons is 15 000. Considerable thought needs to be given to the organization and storage of this number of results, as well as to the computer time required for the matching.

In FMATCH, the analysis between more than two molecules is carried out in two different ways: a base-reference method and a consensus method. Both methods hunt through all combinations of representative conformers for all molecules and use the superpositions generated for pairwise matches. The quality of the multi-molecule match is determined using a modified difference-distance matrix measure. In the atom matching procedures, the sum of the elements in the difference-distance matrix is minimized, whereas in this analysis it is the average element of the matrix that is used. This normalizes matches between molecules with differing numbers of atoms.

In the base-reference method, the conformers are chosen such that the match statistic of all the molecules onto a selected base molecule is a minimum. This is analogous to a series of elastic bands joining one atom on each molecule to an equivalent atom on the base molecule, with the proviso that as the conformations of the molecules change, the equivalent matched atoms change (the elastic bands are swapped onto different atoms). The best base-reference match for six angiotensin II antagonists (onto DuP 753) is shown in Figure 4.4. There is a fairly good overall shape similarity, although few atoms are directly superposed. Equivalent atoms for a small fragment of this superposition are shown in Figure 4.5, illustrating the principles involved. This method is clearly dependent on the choice of base molecule; quite different superpositions are generated for different base molecules. In addition, much of the information generated from matching all pairs of molecules is lost if only a base-reference method is used.

In the consensus method, the approach is similar to the base-reference method, but the number of atom pairs involved in the calculations is increased; all pairs of equivalent atoms are included. Thus, there are monitored distances from each atom in each molecule to all other molecules in the match set. The consensus superposition for the same six angiotensin II molecules is shown in Figure 4.6. On visual inspection, the overall match appears tighter than the base-reference match with better atomic superposition. This is probably preferable to the first method, since it is not biased towards any base molecule. In either case, a ranked list of best matches is generated to allow matches to be viewed and further analysed.

The visual appearance and quality of both styles of matches suffer from the use of only rigid representative conformers. It should be possible to improve the matches by allowing the torsion angles to relax with say $\pm 30°$

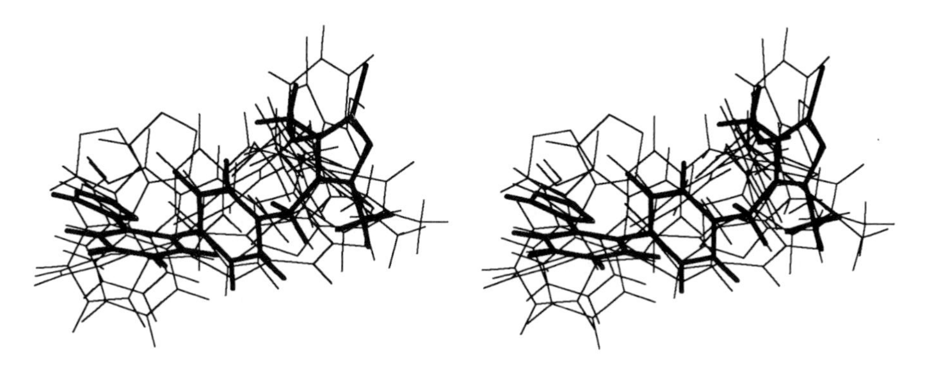

Figure 4.4 Stereoview of base-reference superposition of five angiotensin II antagonists onto DuP 753 (shown in bold).

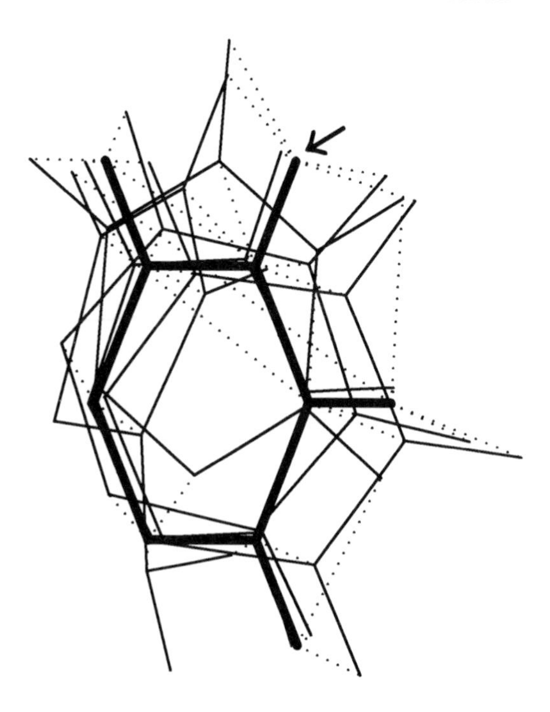

Figure 4.5 Detail of the base-reference superposition shown in Figure 4.4. Dotted lines link equivalent atoms in DuP 753 and each of the other structures. The five dotted lines can be seen most clearly at the atom marked with an arrow.

whilst retaining the equivalent atom pairs determined by the matching algorithms. If successful, this would allow the representative conformers to be transient tools which are expanded to the full set of low-energy conformations after matching and improve the quality of the superpositions.

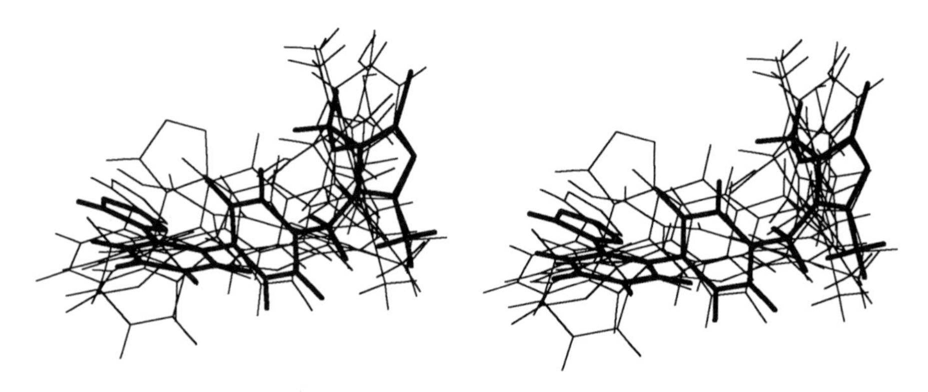

Figure 4.6 Stereoview of consensus superposition of six angiotensin II antagonists.

4.4 Exploitation of matches

The algorithms described in the previous section enable a drug designer to take a set of molecules that bind to a common receptor site and to create a superposition in 3D space of the molecules. The next question is how good and how useful are these matches, and in what ways can they be used in drug design?

4.4.1 Assessment of matches

The quality and usefulness of the matches generated are very closely related, and will depend on the method chosen to determine the superposition and its final use. Matches performed on the basis of steric features alone (such as the atom position superpositions in FMATCH) necessarily fulfil steric criteria for matching, but need to be assessed for chemical similarity (i.e. hydrogen bonding, hydrophobic features and electrostatic potential). This separation may turn out to be useful, since partitioning matching into a primary match and subsequent assessment reduces combinatorial complexity in the same way that the separation of conformation and matching reduces the number of matches that need to be performed.

Molecular shape matching, in some form, is an ideal candidate for this primary matching; steric complementarity between receptor and ligand is a prerequisite for binding and/or activation of the ligand. Molecular surface matching methods (e.g. Masek, 1993b) may be particularly relevant as they concentrate on the surface that the receptor presents to the ligand. These methods may also be suitable in cases where a series of ligands bind at overlapping parts of the receptor surface.

The surfeit of matches between flexible molecules produced by stochastic methods can then each be assessed for hydrogen bonding, hydrophobic features and electrostatic potential. This can be used to rank matches, either independently, or in combination with steric match statistics.

Primary matching need not be based on steric similarity, although other chemical features may not be important for all matches (for example, if all the molecules are hydrophobic, then a hydrogen-bond potential is probably not helpful). In any case, the results need to be assessed according to features not present in the original matching.

An alternative approach is to match on the basis of a combination of factors simultaneously. This will require a weighting scheme and for a general matching algorithm it is difficult to predict the importance of each feature; they are likely to vary between chemically different ligands. If steric features are not given a high enough weighting, then it may be possible to generate matches with good hydrogen-bonding similarity but without a common steric surface. These may not be physically meaningful if they require ligands to occupy quite different parts of receptor space.

Another use of the multiple matches generated is the prediction of potential multiple-binding modes. These can occur in a number of ways including different binding orientations of similar ligands in one site, identical ligands in similar sites, and identical ligands in the same site (see Mattos and Ringe, 1993). The use of cluster analysis within FMATCH to illustrate significantly different matches may also allow the prediction of multiple-binding modes, so long as there is chemical as well as steric similarity.

4.4.2 *Use of generated pharmacophores*

The superpositions generated by matching programs can be used directly to predict pharmacophores, particularly if they contain hydrogen bond, hydrophobic or electrostatic potential features. This can be formalized by the use of 3D QSAR, typically using the procedure from CoMFA (see Chapter 12). These common features can then form a search query as a distance matrix between pharmacophoric points; this can be used to search a 3D structural database for novel compounds that fit the query (see Chapter 6). It should be possible to use the information from flexible searching to map out a pharmacophore with upper and lower bounds on the distance matrix elements to provide a more general search query. This approach can discover existing compounds or fragments that satisfy the constraints input (see Mason, 1993; Martin, 1992).

4.4.3 *Envelope-directed drug design*

The use of pharmacophoric points alone to search a 3D database ignores potentially valuable information that describes the 3D surface of the molecule. It also has the disadvantage that generated structures are limited by the size and diversity of the database searched. An alternative is to use an approach analogous to that for *de novo* generation of ligands to fit known receptor sites. A superposition of a set of molecules can be used to generate a common molecular surface and hence an image of the map of the proposed receptor, complete with pointers to hydrogen bond, electrostatic and hydrophobic features. The same procedures used to design ligands to fit a site can then be used for *de novo* design of a set of matched ligands (see Rotstein and Murcko, 1993).

4.4.4 *Unresolved problems and future directions*

Molecular similarity is a useful tool for drug design but suffers from a number of problems that it cannot hope to attack in its current form. The principal of these is the underlying assumption that the receptor molecule is rigid and invariant; all the methods described depend on the assumption that there is a common cavity into which all the ligands under review bind. It is a commonly-observed feature of crystal structures that proteins deform (and sometimes by a large amount), for example, to accommodate a ligand. The solution to this problem is not yet clear; proponents of the use of molecular similarity must hope that all ligands will deform the receptor site in the same way.

Another problem, particularly important for apparently dissimilar molecules, is the assumption that each ligand binds to the same site, and to the same part of the site on the receptor. It may be possible, for example with molecular surface methods, to cope with ligands that have overlapping binding sites. These potential superpositions may then be testable by the design and synthesis of ligands that incorporate parts of both molecules to increase the binding surface area.

One interesting example of independent binding sites is provided by Moore (1994) who has described separate activation and desensitization sites on the angiotensin II receptor; these require different conformations of the angiotensin II molecule for binding. This raises the possibility of a more selective approach to therapy, particularly the ability to desensitize a receptor without first activating it, an alternative to the classical antagonist. However, if this is a common occurrence it provides a difficult challenge for drug design and molecular similarity.

4.5 Conclusions

This chapter has described a number of the problems associated with matching a set of apparently dissimilar and flexible molecules and illustrated ways in which these may be overcome. Particular attention has been focused on the FMATCH algorithm.

Combinatorial complexity of matching flexible molecules can be reduced by pruning the search space at each stage in the analysis. This is achieved for the FMATCH algorithm with combination of simulated annealing and cluster analysis. The selection of a small number of representative conformers provides the single most important improvement. A range of potential matches are generated which can be used as ideas for a drug designer or can be input directly to algorithms that search for or design ligands to bind the same site.

Acknowledgements

The author wishes to thank Rhône-Poulenc Rorer for personal financial support. Part of this work was carried out in the SERC Cambridge Centre for Molecular Recognition.

References

Barakat, M.T. and Dean, P.M. (1990a) Molecular structure matching by simulated annealing. I. A comparison between different cooling schedules. *Journal of Computer-Aided Molecular Design*, **4**, 295–316.

Barakat, M.T. and Dean, P.M. (1990b) Molecular structure matching by simulated annealing. II. An exploration of the evolution of configuration landscape problems. *Journal of Computer-Aided Molecular Design*, **4**, 317–330.

Barakat, M.T. and Dean, P.M. (1991) Molecular structure matching by simulated annealing. III. The incorporation of null correspondences into the matching problem. *Journal of Computer-Aided Molecular Design*, **5**, 107–117.

Blaney, F., Finn, P., Phippen, R. and Wyatt, M. (1993) Molecular surface comparison: application to drug design. *Journal of Molecular Graphcis*, **11**, 98–105.

Borea, P.A., Dean, P.M., Martin, I.L. and Perkins, T.D.J. (1992) The structural comparison of two benzodiazepine receptor antagonists: a computational approach. *Molecular Neuropharmacology*, **2**, 261–268.

Burt, C. and Richards, W.G. (1990) Molecular similarity: the introduction of flexible fitting. *Journal of Computer-Aided Molecular Design*, **4**, 231–238.

Codding, P.W. and Muir, A.K.S. (1985) Molecular structure of Ro 15-1788 and a model for the binding of benzodiazepine receptor ligands. *Molecular Pharmacology*, **28**, 178–184.

Cramer, R.D., III, Patterson, D.E. and Bunce, J.D. (1988) Comparative molecular field analysis (CoMFA). 1. Effect of shape on binding steroids to carrier proteins. *Journal of the American Chemical Society*, **110**, 5959–5967.

Danziger, D.J. and Dean, P.M. (1985) The search for functional correspondences in molecular structure between two dissimilar molecules. *Journal of Theoretical Biology*, **116**, 215–224.

Davis, L. (1991) *Handbook of Genetic Algorithms*. Van Nostrand Reinhold, New York.

Dean, P.M. and Perkins, T.D.J. (1993) Searching for molecular similarity between flexible molecules. In *Trends in QSAR and Molecular Modelling 92* (ed. C.G. Wermuth), pp. 207–215. Escom, Leiden.

Dean, P.M., Callow, P. and Chau, P.-L. (1988) Molecular recognition: blind-searching for regions of strong structural match on the surface of two dissimilar moleules. *Journal of Molecular Graphics*, **6**, 28–34.

Diana, G., Jaeger, E.P., Patterson, M.L. and Treasurywala, A.M. (1993) The use of an algorithmic method for small molecule superimpositions in the design of antiviral agents. *Journal of Computer-Aided Molecular Design*, **7**, 325–335.

Everitt, B. (1980) *Cluster Analysis*, 2nd edn. Heinemann Educational, London.

Fantucci, P., Mattioli, E., Villa, A.M. and Villa, L. (1992) Conformational behaviour and molecular similarity of some β_1-adrenergic ligands. *Journal of Computer-Aided Molecular Design*, **6**, 315–330.

Goodman, J.M. and Still, W.C. (1991) An unbounded systematic search of conformational space. *Journal of Computational Chemistry*, **12**, 1110–1117.

Hopfinger, A.J. and Burke, B.J. (1990) Molecular shape analysis: a formalism to quantitatively establish spatial molecular similarity. In *Concepts and Applications of Molecular Similarity* (eds M.A. Johnson and G.M. Maggiora), pp. 173–209. John Wiley, New York.

Itai, A., Tomioka, N., Yamada, M., Inoue, A. and Kato, Y. (1993) Molecular superposition for rational drug design. In *3D QSAR in Drug Design: Theory, Methods and Applications* (ed. H. Kubinyi), pp. 200–225. Escom, Leiden.

Jacobs, J.W. and Fodor, S.P.A. (1994) Combinatorial chemistry: applications of light-directed chemical synthesis. *Trends in Biotechnology*, **12**, 19–26.

Jorgensen, W.L. (1991) Rusting of the lock and key model for protein–ligand binding. *Science*, **254**, 954–955.

Kato, Y., Inoue, A., Yamada, M., Tomioka, N. and Itai, A. (1992) Automatic superposition of drug molecules based on their common receptor site. *Journal of Computer-Aided Molecular Design*, **6**, 475–486.

Kirkpatrick, S.K. and Smith, G.M. (1983) Optimization by simulated annealing. *Science*, **220**, 671–680.

Masek, B.B., Merchant, A. and Matthew, J.B. (1993a) Molecular shape comparison of angiotensin-II receptor antagonists. *Journal of Medicinal Chemistry*, **36**, 1230–1238.

Masek, B.B., Merchant, A. and Matthew, J.B. (1993b) Molecular skins: a new concept for quantitative shape-matching of a protein with its small molecule mimics. *Proteins: Structure, Function and Genetics*, **17**, 193–202.

Martin, Y.C. (1992) 3D database searching in drug design. *Journal of Medicinal Chemistry*, **35**, 2145–2154.

Mason, J.S. (1993) Experiences with conformationally flexible 3D databases. In *Trends in QSAR and Molecular Modelling 92* (ed. C.G. Wermuth), pp. 252–255. Escom, Leiden.

Mattos, C. and Ringe, D. (1993) Multiple binding modes. In *3D QSAR in Drug Design: Theory, Methods and Applications* (ed. H. Kubinyi), pp. 226–254. Escom, Leiden.

Metropolis, N., Rosenbluth, A.W., Rosenbluth, M.N., Teller, A.H. and Teller, E. (1953) Equation of state calculations by fast computing machines. *Journal of Chemical Physics*, **21**, 1087–1092.

Meyer, A.Y. and Richards, W.G. (1991) Similarity of molecular shape. *Journal of Computer-Aided Molecular Design*, **5**, 427–439.

Mojena, R. (1977) Hierarchical grouping methods and stopping rules: an evaluation. *Computer Journal*, **20**, 359–363.

Moore, G.J. (1994) Designing peptide mimetics. *Trends in Pharmacological Sciences*, **15**, 124–129.

Moult, J. and James, M.N.G. (1986) An algorithm for determining the conformation of polypeptide segments in proteins by systematic search. *Proteins: Structure, Function and Genetics*, **1**, 146–163.

Papadopoulos, M.C. and Dean, P.M. (1991) Molecular matching by simulated annealing. IV. Classification of atom correspondences in sets of dissimilar molecules. *Journal of Computer-Aided Molecular Design*, **5**, 119–133.

Payne, A.W.R. and Glen, R.C. (1993) Molecular recognition using a binary genetic search algorithm. *Journal of Molecular Graphics*, **11**, 74–91.

Perkins, T.D.J. and Dean, P.M. (1993) An exploration of a novel strategy for superposing several flexible molecules. *Journal of Computer-Aided Molecular Design*, **7**, 155–172.

Rotstein, S.H. and Murcko, M.A. (1993) GroupBuild: a fragment-based method for *de novo* drug design. *Journal of Medicinal Chemistry*, **36**, 1700–1710.

Saunders, M. (1987) Stochastic exploration of molecular mechanics. Energy surfaces. Hunting for the global minimum. *Journal of the American Chemical Society*, **109**, 3150–3152.

Vinter, J.G., Davis, A. and Saunders, M.R. (1987) Strategic approaches to drug design. I. An integrated software framework for molecular modelling. *Journal of Computer-Aided Molecular Design*, **1**, 31–51.

Ward, J.H. (1963) Hierarchical grouping to optimize an objective function. *Journal of the American Statistical Association*, **58**, 236–244.

Wilson, S.R. and Cui, W. (1990) Applications of simulated annealing to peptides. *Biopolymers*, **29**, 225–235.

Wilson, S.R., Cui, W., Moskowitz, J.W. and Schmidt, K.E. (1988) Conformational analysis of flexible molecules: location of the global minimum energy conformation by the simulated annealing method. *Tetrahedron Letters*, **29**, 4373–4376.

5 Similarity-searching and clustering algorithms for processing databases of two-dimensional and three-dimensional chemical structures

P. WILLETT

5.1 Introduction

Databases of chemical structures play an increasingly important role in the fine-chemicals industry, e.g. for the development of novel pharamaceuticals and agrochemicals (Ash *et al.*, 1991). These databases contain tens or hundreds of thousands of chemical substances, either in two-dimensional (2D) or in three-dimensional (3D) form, and several different searching mechanisms have been developed to provide access to the data that is stored in them. The most common mechanisms are *structure searching*, which involves the retrieval of a single specific molecule, and *substructure searching*, which involves the retrieval of all of those molecules that contain a user-defined partial structure, e.g. a putative pharmacophoric pattern. An extended programme of research in the University of Sheffield has sought to develop a complementary means of access, called *similarity searching*, and this chapter provides an overview of some of the algorithms that have been developed for this purpose since the programme commenced in the early 1980s. Specifically, we are interested in techniques that will allow a user of a chemical database to input a target structure of interest, and then to retrieve those molecules in the database that are structurally most similar to the target molecule. Our programme of research has also considered how cluster-analysis methods can be used for the processing of chemical databases.

The molecules in a chemical database are represented in a machine-readable form that will facilitate a range of searching and database-processing applications. This chapter discusses measures for the structural similarity between a pair of molecules that are defined in terms of the molecules' database representations (or characteristics that can be derived automatically from these representations). The design of a similarity-searching system is thus strongly dependent on the structural representation that is used, and hence the next three paragraphs provide a very brief introduction to the representational and searching methods used in chemical database systems; more detailed accounts are provided by Ash *et al.* (1985, 1991) and by Willett (1987a, 1992).

The primary means of representation for a 2D chemical structure diagram is by means of a *connection table*. A connection table contains a list of all of the (typically non-hydrogen) atoms within a structure, together with bond information that describes the exact manner in which the individual atoms are linked together. An important characteristic of a connection table is that it can be regarded as a *graph*, a mathematical construct that describes a set of objects, called *nodes* or *vertices*, and the relationships, called *edges* or *arcs*, that exist between pairs of these objects. Chemical graphs are examples of *labelled graphs*, since the atoms and bonds in a connection table are characterised by their elemental and bond types, respectively. A range of *isomorphism* algorithms are available that identify structural relationships between pairs of graphs, e.g. a subgraph-isomorphism algorithm determines whether a graph is completely contained within another, larger graph (which corresponds to the chemical problem of substructure searching). Graph-theoretic algorithms and data structures provide the basis for most modern 2D chemical-information systems.

Gund (1977) noted that the nodes and edges in a graph could also be used to represent the atoms and the interatomic distances, respectively, in a 3D chemical molecule, so that the IJth edge in such a graph represents the distance separating the Ith and the Jth atom in the molecule. The edge labels are thus real numbers, i.e. distances in Å, rather than the integer bond-type labels used to denote the edges of a 2D chemical graph. Such a 3D chemical graph is an example of a *fully-connected graph*, i.e. one in which there is an edge between every pair of nodes (since there is an interatomic distance between every pair of atoms in a molecule), so that the graph representing a structure that contains N atoms will contain $O(N^2)$ edges. This is in marked contrast to a 2D chemical graph, where an individual node will typically be linked by an edge to only a small number of other nodes and there will hence be $O(N)$ edges. The chemical graphs considered in 3D database systems are thus far more complex than those considered in 2D systems; even so, graphs based on atoms and interatomic distances provide a simple and effective way of describing the geometries of small 3D molecules, and a range of chemical database systems are now available for searching files of 3D chemical substances (Willett, 1992; Bures *et al.*, 1994). Thus far, we have considered only rigid 3D molecules, but Clark *et al.* (1992) have recently demonstrated that it is also possible to represent a flexible 3D molecule in graph-theoretic terms. Specifically, the IJth edge in the graph describing a 3D structure is characterised by a distance range, which represents the minimum-possible and the maximum-possible separations of the Ith and Jth atoms in all of the energetically-feasible conformations that can be adopted by that molecule. There is currently substantial interest in the development of efficient algorithms for flexible 3D substructure searching that can operate on databases of such representations. To date, however, there has been no work on flexible 3D similarity searching in large databases.

Graph-based matching algorithms are effective in operation but are also highly demanding of computational resources. *Fragment screens* have thus been developed that provide a rapid way of eliminating the great majority of the structures in a database from the time-consuming graph-matching search. The screens that are used for 2D substructure searching are small, atom-centred or bond-centred fragment substructures (see for example Dittmar *et al.*, 1983), while those that have been developed for 3D substructure searching (both rigid and flexible) consist of pairs of atoms and interatomic distance ranges (see for example Cringean *et al.*, 1990).

The use of structural representations that were originally developed for substructure searching means that we can draw upon the large body of algorithmic techniques that have been developed to handle such representations when designing systems for similarity searching, and this has formed the basis for much of our work in Sheffield. Specifically, we have sought to develop methods for similarity searching that are *effective*, in the sense that they result in the retrieval of structures that are related to a user's target structure (as discussed in detail in section 5.2 below), and that are also *efficient*, in the sense that they are sufficiently fast in operation to permit the searching of databases of non-trivial size. It is important that both characteristics are considered when designing a system, since there is often a trade-off between them, with the most effective procedures often demanding substantial computing resources. An example of this important point is provided by the *maximal common substructure*, or MCS. The MCS represents the maximal superimposition of one molecule upon the other, thus providing a very precise measure of the degree of similarity between them. MCS algorithms have, however, been little used for similarity searching in 2D databases, owing to their computational requirements. The identification of the MCS for a pair of molecules belongs to the class of computational problems known as *NP-complete* (Harel, 1987), for which no polynomial-time algorithms are believed to exist. Accordingly, rapid, MCS-based similarity searching in large databases is feasible only if the user is prepared to accept similarity rankings that are based on large, but not necessarily on the largest, common substructures, since submaximal common substructures can be identified much more rapidly than can the true MCS (Hagadone, 1992). MCS-based searching is feasible, however, if the database that is to be searched does not contain very large numbers of molecules and if a realtime response is not necessary; in such cases, the MCS provides an extremely powerful way of measuring the similarities between pairs of structures, as we demonstrate in section 5.5 below.

Much of our early work on similarity searching was concerned with the development of techniques for processing databases of 2D structures. This took as its basis a study by Adamson and Bush (1973), who published one of the very first papers to suggest that it was possible to derive quantitative measures of structural resemblance from machine-readable representations of chemical molecules. Specifically, these authors suggested that the similarity between a pair of molecules could be calculated by identifying the fragment substructures that they had in common and then using this common-fragment information to calculate a similarity coefficient of some sort. Our studies have used the atom-centred and bond-centred fragments that are used for the screening stage of substructure searches and we have been able to demonstrate that measures based on such screens provide a reasonably effective, and computationally efficient, way of effecting a similarity search of a database of 2D structures: these studies are described in detail by Willett (1987b). The techniques that were developed in Sheffield have now been widely adopted and they figure in many operational chemical database systems.

The use of fragment-based similarity searching is discussed in Chapter 6. In this chapter, we shall discuss more recent work that has focused on the use of such 2D similarities for clustering chemical databases (rather than for carrying out similarity searches upon them) and review our progress thus far on the development of similarity searching methods for databases of 3D structures. Specifically, section 5.2 describes the approach we have developed to evaluate the effectiveness of similarity searching and clustering procedures when applied to databases of chemical structures. We illustrate this approach by means of a recent comparison of standardisation methods for 2D similarity searching. Section 5.3 then summarises our studies of clustering chemical databases, focusing on those methods that we have found to be most appropriate to the clustering of chemical-structure databases for the selection of molecules for inclusion in biological screening programmes. In section 5.4, we describe an approach to 3D similarity searching that has been developed in Sheffield and that is sufficiently fast in operation to allow in-house searching of corporate databases. This section also discusses the use of upperbound algorithms as a general technique for increasing the efficiency of similarity searching in large databases. Section 5.5 then discusses a graph-theoretic approach to similarity searching in databases of 3D macromolecules that can reveal previously-unknown resemblances between protein structures. We illustrate this characteristic of the method by means of a recent study that has demonstrated such a resemblance between the β-sandwich domains of prealbumin, protocatechuate 3,4-dioxygenase and thaumatin. The chapter concludes with some thoughts about the future development of similarity searching and clustering in large chemical databases.

5.2 Evaluation of effectiveness

It is relatively simple to evaluate the efficiency of a similarity-searching procedure, either by carrying out a complexity analysis or by timing an operational implementation. It is less obvious how one can evaluate the effectiveness of such a procedure, especially if there is a need to carry out a

(a)

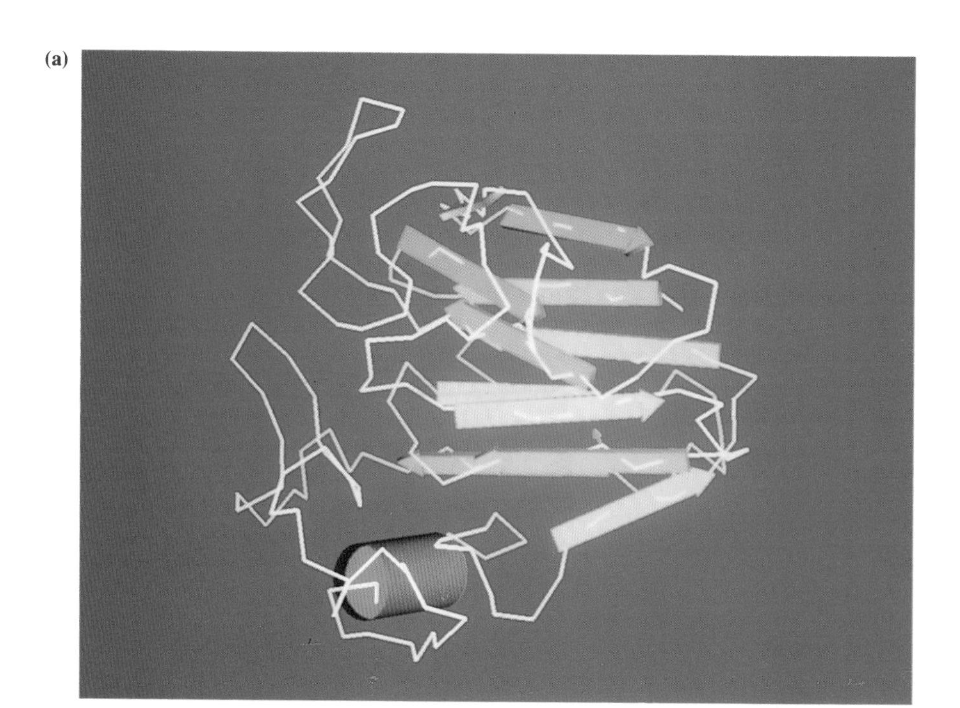

(b)

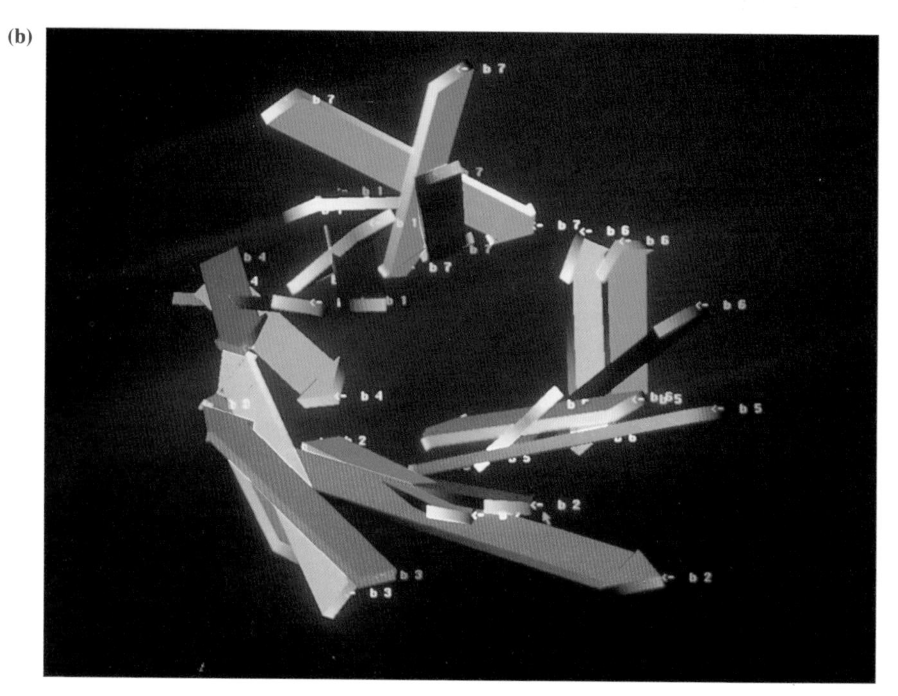

Plate 5.2(a) Three-dimensional structure of thaumatin (1THI) (Kim *et al.*, 1988).

Plate 5.2(b) Common ß-sandwich motif in 2PAB, 1PCD and 1THI identified using the PROTEP program (Artymiuk *et al.*, 1994b). The matched ß-strands from 2PAB, 1PCD and 1THI are shown in red, blue and green, respectively.

as a way of identifying potential lead compounds (Johnson and Maggiora, 1990; Ash *et al.*, 1991).

Each molecule in these datasets was characterised by its constituent *augmented atoms* (Willett, 1987b). An augmented atom is a substructural fragment that consists of a non-hydrogen atom, together with the immediately-adjacent atoms and bonds. The small size of an augmented atom means that a particular type can occur several, or many, times in a molecule, thus making standardisation methods of potentially greater importance than when larger, more detailed types of fragment are used to characterise a molecule for similarity searching. Assume that a given dataset contains a total of M different types of augmented atom. Then a molecule, A, may be characterised by a vector

$$\{A(1), A(2), A(3), \ldots, A(I), \ldots, A(M)\}$$

where $A(I)$ contains the frequency of occurrence of the Ith fragment in that molecule. The similarity between A and another molecule, B, is then calculated by the Tanimoto coefficient, after standardisation of each of the raw frequencies by one of the standardisation methods mentioned above. The Tanimoto coefficient was chosen to measure the intermolecular similarities following previous studies that have demonstrated its appropriateness for database-searching applications (Willett *et al.*, 1986a; Moock *et al.*, 1988). The Tanimoto coefficient for the similarity between two molecules A and B is given by

$$\frac{\sum_{I=1}^{M} A(I)B(I)}{\sum_{I=1}^{M} A(I)^2 + \sum_{I=1}^{M} B(I)^2 - \sum_{I=1}^{M} A(I)B(I)}$$

Our evaluation procedure is based on a 'leave-one-out' approach. An active molecule in a dataset is selected and its similarity calculated with each of the other molecules in that dataset, using one of the standardisation methods in the calculation of the Tanimoto coefficient. The molecules are then ranked in decreasing order of similarity, and a cut-off applied to retrieve some fixed number of the top-ranking compounds. These are then checked to determine whether they have been classified as being active or inactive in the particular biological test system associated with that dataset. This procedure is repeated using each of the actives in turn as the target structure, and the resulting rankings then form the basis for the calculation of a percentage measure of search efficiency. The measure is calculated by taking the mean numbers of active molecules at the cut-off position, C, when averaged over all of the active compounds that have been used to generate a ranking, normalising this mean number by dividing by C, and then converting to a percentage. Thus, if there are ACT actives in a dataset and if the use of the Ith active as the target identifies AI actives above the chosen cut-off (5, 10 or 20 compounds in our experiments), then the overall effectiveness of the current standardisation method is given by

$$\frac{100 \sum_{I=1}^{ACT} AI}{ACT \times C} \tag{5.1}$$

The relative merits of the various standardisation methods can then be assessed on the basis of these percentage search efficiencies.

A typical set of results, those for the top-ranked five compounds, are shown in Table 5.1; other results are presented by Bath *et al.* (1993). In this table, S1–S7 refer to the seven different standardisation methods that were studied and S0 refers to the unstandardised data, while A–H refer to the eight datasets that were used. An inspection of these results suggests that there is no obvious difference in the effectivenesses of the various standardisation methods, a conclusion that is confirmed by a statistical analysis using the Kendall coefficient of concordance, W (Siegel and Castellan, 1988). Kendall's W measures the extent to which k rankings of the same set of L objects are in agreement with each other, and the relative ordering of the L objects if a statistically-significant measure of agreement does, in fact, exist. In the present context, both k (the number of datasets that are being used to rank the standardisation methods) and L (the number of different methods) are 8. W is used to test the null hypothesis, H_0, that the k sets of rankings of the L standardisation methods are independent of each other, and thus that there is no significant agreement in these rankings. The calculated value of W for the experiments using the top five molecules (as detailed in Table 5.1) is 0.08. The significance of this value can be determined from a modified χ^2 test, which shows that the calculated value of W is not significant at the 0.05 level of statistical significance for $L - 1$ (i.e. 7 in the present context) degrees of freedom. This result implies that there is no significant difference in the

Table 5.1 Comparison of standardisation methods. The figures in the main body of the table are the percentage search efficiencies that were obtained when the top-ranked five structures were retrieved from the eight datasets (A–H) using the eight standardisation methods S0–S7 (Bath *et al.*, 1993)

Method	A	B	C	D	E	F	G	H
S0	60.0	67.2	92.4	89.4	84.2	79.6	74.6	78.2
S1	55.6	66.0	88.6	89.0	78.6	84.4	83.4	74.2
S2	52.0	66.0	88.6	90.2	78.6	84.4	83.4	74.2
S3	52.0	65.0	88.6	88.6	78.6	84.4	83.4	77.0
S4	52.0	65.0	87.0	89.6	77.4	83.8	82.6	82.4
S5	52.4	60.0	83.2	87.6	80.2	83.0	83.4	81.8
S6	52.4	62.2	81.0	88.4	78.8	82.8	86.0	78.2
S7	56.2	62.6	87.0	88.8	78.6	77.2	75.4	79.4

performance of the various standardisation methods. Similar conclusions can be drawn from a consideration of the top 10 and top 20 searches, and Bath *et al.* (1993) hence concluded that there was little obvious benefit to be gained from the standardisation of fragment-occurrence data when it is used for similarity searching in databases of 2D chemical structures.

The study reported here is typical of many that we have carried out over the years, in its use of activity-based similarities, to provide a quantitative method for comparing structurally-based similarities and, though to a lesser extent, in its use of the Kendall test for the analysis of the results. This particular study has involved discrete activity data and a single component (data standardisation) of a single similarity-searching procedure. Very similar approaches can be used if continuous data is available, such as inhibitor concentration levels, LD_{50} values or partition coefficients (Willett and Winterman, 1986), or if one wishes to compare different clustering (rather than similarity-searching) procedures (Downs and Willett, 1991). For example, the comparison of clustering methods that is described in the following section involves datasets for which quantitative property data was available. The property value of a molecule, A, within a dataset is assumed to be unknown, and the classification resulting from the use of some particular clustering method is scanned to identify the cluster that contains the molecule A. The predicted property value for A, P(A), is then set equal to the arithmetic mean of the observed property values of the other compounds in that cluster. This procedure results in the calculation of a P(A) value for each of the N structures in a dataset, and an overall figure of merit for the classification is then obtained by calculating the product moment correlation coefficient between the sets of N observed and N predicted values (Willett, 1987b). The most generally useful clustering methods will be those that give high correlation coefficients across as wide a range of datasets as possible, and it is these that have been selected for discussion in the next section.

5.3 Clustering databases of 2D structures

5.3.1 Introduction

Clustering is the process of subdividing a group of entities into more homogeneous subgroups on the basis of some measure of similarity between the entities (Sneath and Sokal, 1973; Everitt, 1993). Two main applications of clustering to chemical-database processing have been suggested (Willett *et al.*, 1986b; Barnard and Downs, 1992), both of which involve the selection of compounds from each of the groupings that have been revealed by the clustering procedure. The first application is the clustering of the hits resulting from a substructure search, with compound selection being carried out to give an overview of the range of structural classes present in the search output; the second is the clustering of an entire database, e.g. a corporate structure file, with compound selection being carried out to identify candidates for inclusion in a biological screening programme. To date, much greater interest has been shown in the use of cluster-analysis methods to select compounds for screening (Downs and Willett, 1994).

Very many different clustering methods have been described in the literature, and there are no *a priori* guidelines as to which will be most appropriate for a particular application domain. We have thus carried out an extended series of experiments to evaluate the effectivenesses of these methods when they are used for clustering files of 2D chemical structures (Willett, 1982a; Rubin and Willett, 1983; Willett, 1984; Willett *et al.*, 1986b). The intermolecular similarities in these experiments were all calculated in the same way as we have discussed in section 5.2, i.e. using fragment-occurrence data and a similarity coefficient such as the Tanimoto coefficient. Recently, we have extended our comparative studies to encompass structures characterised by physical properties (Downs *et al.*, 1994).

Clustering methods can produce *overlapping clusters*, in which each object may be in more than one cluster, or *non-overlapping clusters*, in which each object occurs in only one cluster. All of the methods that we have investigated have been of the latter type; an example of an overlapping method that has been used for compound selection is discussed by Hodes (1989). There are two main classes of non-overlapping cluster methods — *hierarchical methods* and *non-hierarchical methods* — and our studies have encompassed a large number of methods from both of these classes.

5.3.2 *Hierarchical clustering methods*

A hierarchical clustering method produces a classification in which small clusters of very similar molecules are nested within larger and larger clusters of less closely-related molecules. Hierarchical *agglomerative methods* generate a classification in a bottom-up manner, by a series of agglomerations in which small clusters, initially containing individual molecules, are fused together to form progressively larger clusters. Conversely, hierarchical *divisive methods* generate a classification in a top-down manner, by progressively subdividing the single cluster which represents the entire dataset.

Hierarchical agglomerative methods are the most widely used type of clustering procedure, and have been the subject of intense study since their initial development and use for taxonomic applications in the early 1960s. All of these methods can be implemented by means of the algorithm shown in Figure 5.1, which is known as the *stored-matrix algorithm* since it involves random access to the intermolecular similarity matrix throughout the entire cluster-generation process. It is completely general in nature since it permits the implementation of all of the different hierarchical agglomerative methods, which differ merely in the ways in which the most similar pair of points is

1. Calculate the similarity matrix, which contains the similarities between all pairs of molecules in the dataset that is to be clustered.
2. Find the most similar pair of points (where a point denotes either a single molecule or a cluster of molecules) and merge them into a cluster to form a new single point.
3. Calculate the similarity between the new point and all remaining points.
4. Repeat Steps 2 and 3 until only a single point remains, i.e. until all of the entities have been merged into one cluster.

Figure 5.1 Stored-matrix algorithm for the implementation of hierarchic agglomerative clustering methods. Different methods are implemented by choosing the appropriate similarity definitions in Steps 2 and 3 (Lance and Williams, 1967).

defined and in the representation of the merged pair as a single point (Lance and Williams, 1967).

Murtagh (1983) classifies hierarchic agglomerative methods into *graph-theoretic* or *linkage methods* and *geometric* or *cluster-centre methods.* Graph-theoretic methods include the *single-linkage*, *complete-linkage*, *weighted-average* and *group-average methods*, whilst geometric methods include the *centroid*, *median* and *Ward's* or *minimum-variance methods.* An important concept which Murtagh discusses is that of the *reducibility property.* If a method satisfies the reducibility property then agglomerations can be done in restricted areas of the similarity space and the results amalgamated to form the overall hierarchy of relationships. Satisfaction of the property also means that *reversals* or *inversions* of the hierarchy cannot occur. Reversals are a problem associated with geometric methods in which a cluster may end up being more similar to its parent cluster than to any of its constituent entities. Ward's is the only one of the geometric methods mentioned above that satisfies the reducibility property and thus does not suffer from this problem, whereas reversals can be a serious problem with the median and centroid methods. The reducibility property is not generally applicable to graph-theoretic methods, with the sole exception of the group average method under certain conditions (Voorhees, 1986; El-Hamdouchi and Willett, 1989).

The stored matrix algorithm has expected time and storage complexities of $O(N^3)$ and $O(N^2)$, respectively, for a dataset of N molecules, which makes it infeasible for the processing of anything but the smallest datasets. The benefit of isolating methods which satisfy the reducibility property is that the stored-matrix algorithm can be replaced by the *reciprocal nearest-neighbour (RNN) algorithm*, which has expected time and storage complexities of $O(N^2)$ and $O(N)$, respectively. In this, a path is traced through the similarity space until two points are reached that are more similar to each other than they are to any other points, i.e. they are reciprocal nearest neighbours. These RNN points are fused to form a single new point, and the search continues until the last unfused point is reached. The basic RNN algorithm is thus as shown in Figure 5.2. The RNN method is applicable to clustering methods in which the most similar pair at each stage is defined by a distance measure, i.e. geometrically. In Ward's method, the intercluster variance is maximised

1. Mark all molecules, I, as being 'unused'.
2. Starting at an unused molecule, A, trace a path of unused nearest neighbours until a pair of reciprocal nearest-neighbours (RNNs) is encountered; i.e. trace a path of the form $B := NN(A)$, $C := NN(B)$, $D := NN(C)$, etc. until a pair of points is reached for which $Q = NN(P)$ and $P = NN(Q)$.
3. Add the RNNs P and Q to the list of RNNs along with the distance between them, mark Q as 'used', and replace the centroid of P with the combined centroid of P and Q.
4. Continue the chain of nearest neighbours from the point in the path prior to P, or choose another unused starting point if P was a starting point.
5. Repeat Steps 2–4 until only one unused point remains.

Figure 5.2 Reciprocal nearest-neighbours algorithm for the implementation of hierarchic agglomerative methods that satisfy the reducibility property (Murtagh, 1983). A point here refers to either an individual molecule or to a cluster of molecules.

as the intracluster variance is minimised, with the Euclidean distance being used to determine distances between points and hence to define a cluster centroid. Thus, if the Euclidean distance is used in the RNN algorithm, the clusters obtained are those that would be obtained from the stored-matrix algorithm using the update formula appropriate to Ward's method (Lance and Williams, 1967).

Hierarchical divisive methods are generally much faster in operation than the corresponding agglomerative methods. However, Rubin and Willett (1983) found that they gave worse levels of performance than the hierarchical agglomerative methods for clustering chemical structures. This they ascribed to the fact that the divisive methods that they tested were all *monothetic* in character, i.e. the divisions are based on just a single attribute. This is in marked contrast to all of the other clustering methods described in this section, which are *polythetic* in character, i.e. all of the attributes are considered simultaneously during the cluster-generation process.

Polythetic divisive clustering methods are extremely rare, and those that have been described tend to have very large computational requirements. One of the few successful polythetic divisive methods is the *minimum-diameter method* (Guenoche *et al.*, 1991), a general outline of which is given in Figure 5.3. In this figure, the diameter of a cluster is defined as the largest dissimilarity between any two of its members, and does not necessarily reflect the size (number of members) of the cluster; the diameter of a singleton cluster is defined as zero. One of the potential benefits of polythetic hierarchical divisive clustering is that users typically want only a few clusters. For hierarchical

1. For the set of N molecules to be clustered, produce an input list of all $N(N-1)/2$ dissimilarities, sorted into decreasing order of magnitude.
2. The top two molecules in the sorted dissimilarity list become the focus of the first bipartition of the dataset. Assign all other molecules to the least dissimilar of these initial cluster centres.
3. Recursively select the cluster with the largest diameter and partition it into two clusters such that the larger cluster has the smallest possible diameter.
4. Repeat Step 3 for a maximum of $N-1$ bipartitions.

Figure 5.3 The hierarchic divisive minimum-diameters method of Guenoche *et al.* (1991).

agglomerative clustering this requires the production of most of the hierarchy, with the accompanying problem that erroneous assignments made early in the hierarchy are not corrected and become compounded. For hierarchical divisive clustering only the first part of the hierarchy need be produced, with consequently less risk of compounding erroneous assignments. Some recent experiments with the minimum-diameter method have shown that it gave an excellent level of performance in clustering molecules characterised by physical properties, rather than substructural fragments as in the other methods described here (Downs *et al.*, 1994). It is, however, computationally expensive in that it requires random access to the full dissimilarity matrix (to calcuate the diameter in Step 3 of the algorithm shown in Figure 5.3) if the time requirements are not to become unacceptably large; specifically, the method has expected time and space complexities of $O(N^2 \log N^2)$ and $O(N^2)$, respectively.

5.3.3 *Non-hierarchical clustering methods*

A non-hierarchical method generates a classification by partitioning a dataset, giving a set of (generally) non-overlapping groups having no hierarchical relationships between them. A systematic evaluation of all possible partitions is quite infeasible, and many different heuristics have thus been described to allow the identification of good, but possibly suboptimal, partitions; such methods are generally much less demanding of computational resources than the hierarchical methods. The three main categories of non-hierarchical method are the *single-pass*, *relocation* and *nearest-neighbour methods*. Examples of all of these methods have been studied for the clustering of chemical structures (Willett, 1984; Willett *et al.*, 1986b), with the best level of performance being given by the nearest-neighbour method due to Jarvis and Patrick (1973).

Nearest-neighbour methods assign structures to the same cluster as some number of their nearest neighbours. User-defined parameters determine how many nearest neighbours need to be considered, and the necessary level of similarity between nearest neighbour lists. The Jarvis–Patrick method involves the use of a list of the top K nearest neighbours for each of the N molecules in a dataset, where the nearest neighbours are identified using a fast inverted-file algorithm (Willett, 1982b) for the calculation of the Tanimoto similarities. Once the lists of top K nearest neighbours for each molecule have been produced, a second stage is used to create the clusters. Here, structures are grouped together on the basis of their nearest-neighbour lists (rather than on the basis of common attribute values as is normally the case); this state is detailed in Figure 5.4. The partition that is produced by the method is governed by the choice of K and of K_{min}, so that it is necessary to experiment with a range of K and K_{min} values until roughly the required number of clusters is obtained.

1. Create an N-element label array that contains a cluster label for each of the N molecules in the dataset. Initialise this array by setting each element to its array position; this assigns each molecule to its own initial cluster.
2. Compare the nearest-neighbour lists for all pairs of molecules, I and J ($I < J$) to determine whether all three of the following clustering conditions are obeyed:
   I is in the top K nearest-neighbour list of J;
   J is in the top K nearest-neighbour list of I; and
   I and J have at least K_{min} of their top K nearest neighbours in common, where K_{min} is a user-defined parameter in the range $1 \leqslant K_{min} \leqslant K$.
   If all of these conditions are satisfied then replace the label-array entry for J, and all other occurrences of the label-array entry for J in the label array, with the label-array entry for I.
3. The label array contains the lowest array entries for the molecules in each cluster, so that the members of a given cluster all have the same array entry. Scan the array to extract the members of each cluster.

Figure 5.4 Implementation of the nearest-neighbour clustering method of Jarvis and Patrick (1973).

The three methods that we have detailed above—Ward's, minimum-diameter and Jarvis–Patrick—are those that we have found to be most effective for clustering files of chemical structures. Of these, the Jarvis–Patrick is by far the fastest (although it has the same time complexity as the RNN implementation of Ward's method (Downs and Willett, 1994) and we thus believe that it is currently the most appropriate for the clustering of large files of chemical molecules. This method forms the basis for the clustering package of the chemical-database software produced by Daylight Chemical Information Systems, Inc. and for the CLASS routine in the CAVEAT package for molecular design that is distributed by the University of California, as well as for the selection of compounds in several in-house clustering systems, e.g. the system at Pfizer Central Research (Willett et al., 1986b). That said, it must be emphasised that other clustering procedures can, of course, be used for compound selection if the processing requirements are computationally feasible; examples of such procedures have been described by Hodes (1989) and by the group at Upjohn (Lajiness *et al.*, 1989; Lajiness, 1991).

5.4 Similarity searching in databases of 3D structures

5.4.1 *Introduction*

The two previous sections have described the use of similarity-searching and clustering procedures with databases of 2D structures. While there is much developmental work still to be done, such procedures are now well established and are increasingly being used in both commercial and in-house chemical information systems. With the development of substructure searching techniques for databases of 3D structures over the last few years, there is now interest in the development of complementary systems for 3D similarity searching (and, one assumes in the future, in systems for 3D clustering). A summary of current research is provided by Bures *et al.* (1994); in this section, we discuss some of our own work in this area.

The molecules in a 3D database can be represented by their interatomic distance matrices and the similarity measures that we have investigated have thus all been based on information that can be generated from such matrices. Specifically, we have compared four different ways of processing interatomic distance information, using the approach detailed in section 5.2 to compare the methods' effectivenesses. These experiments are described by Pepperrell and Willett (1991), who showed that the *atom-mapping method* was the most cost-effective of those tested, in that the atom-mapping method and a maximal common-substructure (MCS) method gave the best levels of effectiveness but the former was much less demanding of computational resources than the MCS method (which, as has been noted previously, belongs to the class of NP-complete problems). A subsequent study demonstrated the general robustness of the atom-mapping method and its use with a wide range of types of structural descriptor (Pepperrell *et al.*, 1991). In this section, we describe the atom-mapping method and then use it to illustrate how upper-bounding strategies can be employed to increase the efficiency of nearest-neighbour algorithms when they are used to search large databases.

5.4.2 The atom-mapping method

The atom-mapping method provides a quantitative measure of the similarity between a pair of 3D molecules, A and B, that are represented by their interatomic distance matrices. The measure is calculated in two stages. Firstly, the geometric environment of each atom in A is compared with the corresponding environment of each atom in B to determine the similarity between each possible pair of atoms. The resulting interatomic similarities are then used to identify pairs of geometrically-related atoms and these equivalences allow the calculation of the overall, intermolecular similarity.

Assume that the molecule A contains $N(\mathrm{A})$ non-hydrogen atoms. The distance matrix for A, DA, is then an $N(\mathrm{A}) \times N(\mathrm{A})$ matrix such that the IJth element, $DA(I, J)$, contains the distance between the Ith and Jth atoms in A (and similarly for the matrix DB representing the $N(\mathrm{B})$-atom molecule B, where we assume that $N(\mathrm{A}) \leqslant N(\mathrm{B})$). The first stage of the procedure compares each atom from A with each atom from B to identify the extent to which each pair of atoms lies at the centre of comparable volumes of 3D space, as defined by the geometric arrangement of the atoms that are contained within these volumes.

Consider the Ith atom in A, $\mathrm{A}(I)$: then the Ith row of the distance matrix DA contains the distances from I to all of the other atoms in A. This set of distances is compared with each of the rows, J, from the distance matrix DB to identify the matching distances, where a matching distance is one that is the same in the two molecules to within some user-defined tolerance, e.g.

± 0.5 Å, and that separates atom pairs that are of the same elemental type(s) in A and B. Let there be $C(I, J)$ such distance matches when A(I) is compared with B(J). Then the similarity S(I, J), between A(I) and B(J) is given by the Tanimoto coefficient, which in this case has the form

$$S(I, J) = \frac{C(I, J)}{N(A) + N(B) - C(I, J)}$$

S(I, J) is the IJth element of an $N(A) \times N(B)$ matrix, S, that contains the similarities between all pairs of atoms, A(I) and B(J), from A and B. This matrix is referred to subsequently as the *atom-match matrix*.

Once the atom-match matrix has been created, the interatomic similarities contained within it are used to establish a set of equivalences of the form $A(I) \Leftrightarrow B(J)$, where B($J$) is that atom in B that lies at the centre of the most similar area of 3D space, as defined by the atoms surrounding it, as does the atom A(I). A *greedy algorithm*, i.e. one that selects the largest-available value for some function or variable without consideration of any subsequent selections that may take place, is used to establish the set of equivalences, and is as shown in Figure 5.5. The overall degree of similarity between A and B is then calculated as the mean of the similarities over all of the atoms in A, using the information in the sets of values that are stored in Step 3 of the algorithm shown in Figure 5.5. Thus, the intermolecular similarity is given by

$$\frac{1}{N(A)} \sum_{I=1}^{N(A)} S(I, J)$$

Ideally, we would like to map the atoms from A onto those from B so as to give the largest possible value for the intermolecular similarity coefficient; however, as discussed by Pepperrell and Willett (1991), the identification of the maximal value of the coefficient involves a combinatorial procedure that is far too demanding of computational resources for a practicable similarity-searching system. The greedy algorithm used here provides an efficient and effective heuristic that usually, but not invariably, identifies the most similar pairs of atoms.

The description thus far assumes that some atom, A(I), in the target structure is considered for mapping to some atom, B(J), in a database structure if, and only if, both A(I) and B(J) are of the same elemental type.

1. Sort the elements of the atom-match matrix into order of decreasing similarity.
2. Scan the atom-match matrix to find the remaining pair of atoms, one from A and one from B, that have the largest calculated value for S(I, J).
3. Store the resulting equivalences as a set of values of the form $\{A(I) \Leftrightarrow B(J); S(I, J)\}$.
4. Remove A(I) and B(J) from further consideration.
5. Return to Step 2 if it is possible to map further atoms in A to atoms in B.

Figure 5.5 A greedy algorithm for identifying the sets of equivalent atoms in the atom-mapping algorithm of Pepperrell and Willett (1991).

However, there is no *a priori* reason why this should be so, and it is thus possible, if the user so wishes, for the program to match molecules on the basis of the arrangement of some specific atomic property (or properties) in 3D space (Pepperrell *et al.*, 1990). Specifically, methods have been developed that allow an atom to be characterised by its hydrogen-bonding characteristics, its partial charge, its atomic radius, or a combination of these three types of property (rather than the elemental type as in the basic form of the method). Alternative approaches to property-based 3D similarity searching have been reported by Fisanick *et al.* (1992) and by Perry and van Geerestein (1992).

5.4.3 *Upperbound strategies*

In similarity searching, the user may know little about the structures that are to be retrieved, other than that they are likely to be related, in some imprecisely defined way, to the target structure. Such an imprecise approach to query formulation necessitates a browsing-like retrieval mechanism in which the user is able to inspect output (i.e. the top-ranked structures) and then to refine the search in an iterative manner until an appropriate set of structures has been obtained. A rapid response is required if the browsing is to be effective, and it is thus important that efficient algorithms are available if one wishes to implement similarity searching on databases of non-trivial size. This is true of any type of nearest-neighbour search procedure, whatever the application area, and there is an extensive literature that discusses a range of approaches to this very general problem: the reader is referred to the recent review by Murtagh (1993) for an introduction to this subject. One approach that we have found to be of particular applicability in our work has been the use of *upperbound strategies*.

The nearest-neighbour searching problem may be defined as follows. Assume that the target structure, Q, is to be matched against each of the N structures in a database, and that the search is to retrieve the M nearest neighbours ($M \ll N$) for Q. At a given point in the search of the set of N structures, let the similarity of the current Mth nearest neighbour (i.e. the database record that is the Mth most similar to Q of those that have been tested thus far) be Sim(M). If this is the case, then none of the remaining, untested records need be considered further for inclusion amongst the set of M nearest neighbours unless there is a non-zero probability that their similarity could be greater than Sim(M). An upperbound strategy involves calculating the maximum possible similarity, MaxPossSim, for each record that has not yet been matched against Q. Each such match then needs to be carried out if, and only if, MaxPossSim > Sim(M), since the full match cannot affect the current list of M nearest neighbours if this inequality is not satisfied. Such a strategy can bring about substantial reductions in the computational cost of a nearest-neighbour search if the upperbound calculation can be carried out far more cheaply than can the full similarity match; if this is not the case, then one might just as well carry out the full match for each of the N structures in the database. We shall exemplify this approach to similarity searching in chemical databases by describing an upperbound strategy that we have developed for an operational implementation of the atom-mapping method described in the previous section. Other examples of our use of upperbound strategies for chemical similarity searching are described by Brint and Willett (1988), by Downs *et al.* (1994) and by Willett (1983).

A complexity analysis of the atom-mapping algorithm shows that it has an expected time complexity of no less than $O(N(\mathrm{A})^3)$ if $N(\mathrm{A}) \approx N(\mathrm{B})$, with the computation being dominated by the calculation of the elements of the atom-match matrix (Pepperrell *et al.*, 1990). It is thus clear that extremely efficient algorithms are required if atom mapping is to be used for similarity searching in large files of 3D structures; this we have sought to achieve by means of an upperbound procedure that is based on the molecular formulae of the target structure and of the current database structure with which it is being compared. If these molecular formulae are, for example, $C_aN_bO_c$ and $C_xN_yO_z$, respectively, then an upperbound to the number of atoms that can be mapped onto each other during the similarity calculation, UpbMappedAtom, is given by

$$\mathrm{UpbMappedAtom} = \mathrm{Min}\{a, x\} + \mathrm{Min}\{b, y\} + \mathrm{Min}\{c, z\}$$

where $\mathrm{Min}\{p, q\}$ denotes the lesser of p and q. More generally, if we assume that there are X different atomic types in the two molecules that are being compared, then UpbMappedAtom is given by

$$\mathrm{UpbMappedAtom} = \sum_{K=1}^{X} \mathrm{Min}\{N\mathrm{A}(K), N\mathrm{B}(K)\}$$

where $N\mathrm{A}(K)$ and $N\mathrm{B}(K)$ are the numbers of atoms of element K in molecules A and B, respectively. The upperbound to the number of distances that can possibly be in common between any two atoms, including the zero-valued distances to the atoms themselves, is also UpbMappedAtom. Substituting into the Tanimoto coefficient formula, an upperbound to the interatomic similarity, UpbAtomSim, is hence given by

$$\mathrm{UpbAtomSim} = \frac{\mathrm{UpbMappedAtom}}{N(\mathrm{A}) + N(\mathrm{B}) - \mathrm{UpbMappedAtom}}$$

The maximum possible value for the intermolecular similarity will occur when there are UpbMappedAtom atom mappings, each with an interatomic similarity value of UpbAtomSim, i.e. when every possible atom has the largest possible interatomic similarity. Thus an upperbound to the overall intermolecular similarity, UpbMolSim, is given by

$$\text{UpbMolSim} = \frac{1}{N(\text{A})} \times \text{UpbMappedAtom} \times \text{UpbAtomSim}$$

$$= \frac{\text{UpbMappedAtom}^2}{N(\text{A})(N(\text{A}) + N(\text{B}) - \text{UpbMappedAtom})}$$

When the target structure is to be compared with a structure in the file, their two molecular formulae are used to evaluate UpbMappedAtom, and thence UpbAtomSim, as described above. If the resulting calculated value of UpbMolSim is not greater than the Mth best similarity found so far, then the atom-mapping algorithm need not be invoked for the current molecule, since the actual similarity then cannot possibly be sufficiently large for that molecule to be included in the list of M nearest neighbours that is presented to the user at the end of the search.

The description thus far assumes that the upperbound calculation is carried out for each molecule in the database in turn, to determine whether the full atom-mapping calculation needs to be carried out. In fact, it is possible to improve the efficiency of our upperbound algorithm still further by carrying out the upperbound calculation for all of the molecules in the database before the search commences and before any atom-mapping calculations have been carried out. Specifically, the UpbMolSim values are used to provide a rapid, but approximate, ranking of the database in decreasing order of the maximum possible similarity that will be calculated if, or when, the full atom-mapping procedure is invoked. This ranking of the database will increase the likelihood that molecules which are highly similar to the target structure will be identified early on in a database search, thus maximising the numbers of structures that can be eliminated from the full atom-mapping procedure. Thus, UpbMolSim is calculated for all of the molecules in the database, and the molecules are sorted into decreasing order of their upperbound values. The atom-mapping calculations then need to be carried out only until the calculated value for the Mth nearest neighbour becomes equal to the UpbMolSim value for the next molecule in the sorted file. The search terminates at this point since there is no chance that any of the subsequent molecules can possibly belong to the final list of nearest neighbours. The overall search algorithm is shown in Figure 5.6.

Pepperrell *et al.* (1990) suggest that this procedure can eliminate up to ca. 80% of the database in a typical atom-mapping search. These authors also describe a second, less precise upperbound algorithm that can be used to reduce the computation for those molecules where the atom-mapping calculation does, indeed, need to be carried out. An implementation of these ideas is now used for similarity searches of the Zeneca Agrochemicals corporate database of ca. 180 000 structures. The algorithm is coded in C and runs on a Unix workstation, with a response time for a typical search

1. The user submits a target molecule and specifies M, the number of nearest-neighbour structures that are to be retrieved by the search.
2. Initialise an M-element data structure that will contain the identifiers and calculated similarities for the M nearest neighbours.
3. Calculate UpbMolSim for each molecule in the database with respect to the target structure.
4. Rank the database in decreasing order of the UpbMolSim values.
5. If the value of UpbMolSim for the next molecule in the sorted database is not greater than for the Mth nearest neighbour (or if there are no more structures to be processed), then go to Step 7.
6. Carry out the full atom-mapping calculation; if the resulting similarity is greater than the similarity for the Mth nearest neighbour then update the list of nearest-neighbour structures accordingly. Go to Step 5.
7. Display the top M structures to the user in order of decreasing similarity.

Figure 5.6 Use of an upperbound calculation to minimise the number of full calculations that need to be carried out in an atom-mapping search (Pepperrell *et al.*, 1990).

being in the range 20–60 minutes (the precise time depending primarily upon the size of the target structure). It is also possible to use hardware, rather than software, means to increase the speed of the algorithm, and we have recently discussed several ways in which atom mapping can be implemented on massively-parallel processors (Wild and Willett, 1994).

5.5 Similarity searching in databases of 3D protein structures

Thus far, we have considered databases of small molecules, typically containing an average of 25–30 non-hydrogen atoms per structure. It is also necessary to develop appropriate search algorithms for the 3D information in the Brookhaven Protein Data Bank (Bernstein *et al.*, 1977), which contains atomic coordinate data for all of the macromolecules (overwhelmingly proteins) for which a 3D structure has been obtained (normally by X-ray crystallography but there are also several NMR structures in the database). We have noted previously that both 2D and 3D small molecules may be represented for searching purposes by labelled graphs, and we are now engaged in a project to evaluate the utility of graph-based methods for the representation and searching of the 3D structures in the Protein Data Bank.

The main focus of interest to date in our work has been the representation and searching of *tertiary-structure motifs*, i.e. patterns of secondary-structure elements in 3D space. The graph representation of a protein that we have adopted makes use of the fact that the two most common types of secondary-structure element, the α-helix and the β-strand, are both approximately linear, repeating structures, which can hence be represented by vectors drawn along their major axes. The set of vectors corresponding to the secondary-structure elements in a protein can then be used to describe that protein's 3D structure, with the secondary-structure elements and the interelement angles and distances corresponding to the nodes and to the edges, respectively, of a graph (Mitchell *et al.*, 1990). More precisely, each edge in such a labelled graph is a three-part data element that contains the angle between a pair of vectors (i.e. a pair of secondary-structure elements), the distance of closest approach of the two vectors and the distance between their midpoints. This labelled-graph representation provides a compact, yet highly precise specification of the geometric arrangement of the secondary-structure elements in a protein that can be used for similarity searching by means of a maximal common subgraph isomorphism algorithm (Grindley *et al.*, 1993).

The maximal common subgraph procedure that we have used in this work is based on the *clique-detection algorithm* of Bron and Kerbosch (1973), where a clique is a subgraph of a graph in which every node is connected to every other node and which is not contained in any larger subgraph with this property. The input to the clique-detection procedure is a *correspondence graph* or *compatibility table.* Given a pair of graphs A and B, i.e. a pair of protein structures in the present context, the correspondence graph, C, is formed in two stages. In the first stage, the set of all pairs of nodes is created, one from A and one from B, such that the nodes of each pair are of the same type. C is then the graph whose nodes are the pairs from the first step. Two nodes $(A(I), B(X))$, $(A(J), B(Y))$ are marked as being connected in C if the values of the edges from $A(I)$ to $A(J)$ and $B(X)$ to $B(Y)$ are the same. Maximal common subgraphs then correspond to the cliques in the correspondence graph, so that the identification of the maximal common substructure for a pair of 3D chemical molecules is equivalent to the identification of the largest clique in the correspondence graph linking the two molecules. This largest clique then represents the maximal overlap of one structure on the other, subject to any geometric tolerances that have been defined by the user.

Clique detection was first studied in Sheffield in the context of finding the maximal substructure common to pairs of small 3D molecules (Brint and Willett, 1987), where it was shown that the clique-detection algorithm due to Bron and Kerbosch was the most efficient for this application. A subsequent report discussed the use of the algorithm for the implementation of similarity searching in files of 3D small molecules (Brint and Willett, 1988) and this led to its implementation in the program described here, called PROTEP (for PROtein Topographic Exploration Program), where it is used to identify patterns of α-helices and β-strands in 3D space that are common to a pair of proteins. Specifically, a target protein is matched against each of the proteins in the Protein Data Bank in turn to identify the structural similarities that are present, subject to the user's search parameters. The clique-detection algorithm has been implemented so that it identifies not only the largest common pattern but also all of the common patterns that are larger than some user-defined threshold size. The precise nature of the output from a search is determined by the tolerances that are used. The angular tolerance is specified in terms of numbers of degrees, while the distance tolerances are specified either in Ångströms or as a percentage of the distance in the target structure (where the distance can refer to the distance of closest approach of the vectors representing two secondary-structure elements, to the distance between the vector midpoints, or to a combination of the two). It is also possible to specify that the matching secondary-structure elements in a database protein are in the same sequence order as in the target; alternatively, the sequence order does not need to be the same. The user thus has a very large degree of control over the number and the quality of the matches that are identified by the program.

The main use of PROTEP is to identify previously unrecognised structural resemblances. We now report such a resemblance that has been identified recently (Artymiuk *et al.*, 1994b), this being the structural relationship between the β-sandwich domains of three otherwise-unrelated proteins: prealbumin (2PAB) (Blake and Oatley, 1977; Blake *et al.*, 1978), protocatechuate 3,4-dioxygenase (1PCD) (Ohlendorf *et al.*, 1988) and thaumatin (1THI) (Kim *et al.*, 1988).

The query pattern consisted of eight helices and 32 strands from 1PCD. This pattern was searched against the Protein Data Bank using an angular tolerance of 30°, and a distance tolerance that was the lesser of 4 Å and of 40% of the distances in the target protein. Only sequence-ordered hits were retrieved, and matching cliques had to contain at least seven secondary-structure elements for that clique to be retrieved. Searches using this pattern resulted in the identification of a size-seven clique in common with 2PAB and with 1THI. The common fold in the three proteins is a β-sandwich consisting of one three-stranded and one four-stranded β-sheet, these motifs comprising a major part of the mixed β-sandwich found in each protein. The individual proteins are shown in Plates 5.1(a,b) and 5.2(a) and the common sandwich motif in Plate 5.2(b). In these plates, the cylinders and the arrows represent the positions of the α-helices and the β-strands, respectively, in a structure.

The three proteins are very different in character. Human prealbumin is a hormone-binding protein which has two thryoxine-binding sites and also independent binding sites for retinol-binding protein (Blake and Oatley, 1977), the specific plasma carrier of vitamin A. The crystal structure has been solved to 1.8 Å resolution and detailed binding studies have revealed the nature of the substrate-specificity sites. The molecule is a tetramer of identical subunits, each containing 127 amino acid residues and consisting of an eight-stranded β-barrel structure organised into two four-stranded β-sheets, together with a short stretch of α-helix. Protocatechuate 3,4-dioxygenase belongs to a family of bacterial dioxygenases that catalyse the cleavage of molecular oxygen with the subsequent incorporation of both oxygen atoms into organic substrates. The molecule is an oligomer of dissimilar subunits, the α-subunit containing 200 residues and the β-subunit containing 238 residues. Both subunits share a common structural core, which consists of a flattened nine-stranded β-barrel and three short α-helices, but they differ in their termini and in several surface loops. Finally, thaumatin is a monomeric (207 residues), intensely sweet protein of unknown function, which is extracted from the fruit of an African plant *Thaumatococcus daniellii* (van der Wel and Loeve, 1972). The crystal structure has been solved at 3.0 Å resolution and reveals a flattened 11-stranded β-barrel, surrounded by many loops closed by disulphide bonds.

The 3D resemblance in the common region is good, with the rms differences over the equivalenced residues ranging from 0.91 Å (for the three-stranded motif in 2PAB and 1PCD) to 2.35 Å (for the four-stranded motif in 1THI and 1PCD). The superpositions show an identical organisation and order of the strands, which also appear to be conserved in position, although the four-stranded motif in 1THI/1PCD superposed less well than the others. A detailed multiple-alignment study failed to identify any statistically-significant sequence homologies between the residues comprising the common structural feature in the three proteins (Ujah, 1992) and the question then arises as to why there is such a strong structural resemblance despite the lack of any sequence resemblance. Three possible explanations suggest themselves.

Firstly, it is possible that the proteins may be evolutionarily related, but so remotely that all significant sequence holology has been lost, and only a common substructure remains. An alternative possibility is that some members of the three families may bind to structurally similar receptors. This explanation could arise by evolutionary processes, and would thus be compatible with the previous suggestion. However, the three proteins are extremely diverse and are not known to bind to structurally similar receptors. Thus, in this case, there seems no reason to believe that there could be any evolutionary similarity; accordingly, the previous explanations are unlikely. The most likely explanation is that the observed 3D resemblance may simply reflect a particularly-stable folding motif, especially as the common fold between 2PAB, 1PCD and 1THI resembles the β-sandwich found in another unrelated protein, azurin (1AZA). The only difference is that the outer pair of parallel β-strands in the three-stranded sheet of azurin is *N*-terminal to the rest of the fold, whereas these parallel strands are *C*-terminal to the rest of the fold in the three proteins studied here. In addition, five of the β-strands in the common fold between 2PAB, 1PCD and 1THI are structurally very similar to five of the β-strands in a type 1 jelly roll (Stirk *et al.*, 1992).The jelly-roll motif is a stable structural feature that is found in a wide variety of unrelated molecules, and it is hence quite possible that the common fold identified in this work is simply a common modification of a particularly stable folding motif, rather than an example of an evolutionary relationship.

The identification of this common β-sandwich domain is the third 'discovery' made by PROTEP since we started to use it on a routine basis for the analysis of protein structures, and we thus believe that it provides an effective means of exploring the structural information that is contained within the Protein Data Bank. Thus, we have discovered a strong resemblance between leucine aminopeptidase (1LAP) and carboxypeptidase A (5CPA), where there is a common substructure containing no less than eight β-strands in a sheet and five of the ten α-helices in the *C*-terminal domain of 1LAP (Artymiuk *et al.*, 1992). More recently, we have identified a strong similarity between the ribonuclease H domain of HIV-1 reverse transcriptase (RT) and the ATPase folds of hexokinase, heat-shock cognate protein and actin, and an intermolecular resemblance between the same ribonuclease H domain and the two other domains in RT (Artymiuk *et al.*, 1993). In addition to providing an effective means of measuring inter-protein similarity, PROTEP is also highly efficient in operation. A typical search of the entire Protein Data Bank takes a few tens of minutes on a Unix workstation, thus allowing the user a sufficiently-rapid response to encourage the use of a range of search parameters to maximise the retrieval of relevant material. At present, no screening or upperbound procedures are needed to ensure a rapid response; however, the current rapid growth in the size of the Protein Data Bank suggests that it may be necessary to develop such procedures in the future.

5.6 Conclusions

This chapter has provided an overview of some of the work that has been carried out at the University of Sheffield over the last few years in an extended programme of research to develop techniques for using similarity searching and clustering in databases of chemical structures. When the project started in the early 1980s, only 2D structure and substructure searching were available in chemical database systems. Current systems additionally provide facilities for 2D similarity searching and cluster-based compound selection, and for both rigid and flexible 3D substructure searching, with algorithms for rigid 3D similarity searching starting to appear in the literature. In this context, we have already described the atom-mapping approach that has been developed in collaboration with Zeneca Agrochemicals (Pepperrell *et al.*, 1990); other such distance-based algorithms have been reported recently by Bemis and Kuntz (1992), Fisanick *et al.* (1992), Nilakantan *et al.* (1993) and Perry and van Geerestein (1992), while Bath *et al.* (1994) have compared distance-based and angle-based measures of 3D similarity.

A characteristic of many of the 3D similarity measures that have been described is that they are much more demanding of computational resources than the fragment-based measures that are used for 2D similarity searching, even though they consider only rigid 3D molecules and take no account of molecular flexibility. Accordingly, some of these measures may not be used on a large scale unless, or until, new software techniques and substantially faster processors become available, and their extension to encompass databases of flexible 3D molecules (as has recently been achieved in the case of substructure searching; Clark *et al.*, 1992) will undoubtedly require still further computational resources. This being so, it may be some time before it is possible to apply clustering methods to databases of 3D structures, and thence to determine whether the resulting classifications are superior to those resulting from the current generation of 2D clustering approaches.

The procedures that have been described in this chapter are all based on the measures of structural similarity that can be computed using information that can readily be generated from a connection table. However, such structural information (e.g. atom-centred or bond-centred fragments generated from a 2D connection table or interatomic distance functions generated from a 3D connection table) does not provide an explicit representation of the steric, electrostatic and hydrophobic fields around a molecule that are the prime determinants of its activity at a biological receptor site. Accordingly, there is a substantial need for the development of techniques that will permit field-based similarity searching, in which a molecule is represented by field values in the elements of a 3D grid that surrounds it and in which the similarity is calculated by a comparison of the sets of corresponding grid points. Several workers have reported the use of similarity measures based on the molecular electrostatic potential (Burt *et al.*, 1990; Manaut *et al.*, 1991; Petke, 1993). However, the techniques that have been described thus far are not appropriate for database searching owing to the need for some efficient mechanism for the maximal alignment of the fields representing a target structure and each of the structures in a large database. We believe that the development of such a mechanism is a necessary prerequisite if field-based searching is to be feasible on files of anything but trivial size, and are currently evaluating a range of strategies by which this could be achieved.

Another area that is currently under investigation is the extension of our work on protein-similarity searching to encompass the geometries of the protein side-chains, rather than the secondary-structure elements as in the work described in section 5.5. We have already developed methods for substructure searching by means of a graph-theoretic representation of the side-chain structure (Artymiuk *et al.*, 1994a) and hope to extend this to the similarity-searching context in the near future. In the longer term, we hope to be able to integrate the various approaches that have been, and are being, described for similarity searching in databases of 3D small molecules and in databases of 3D macromolecules.

Acknowledgements

My grateful thanks are due to Frank Allen, Peter Artymiuk, Peter Bath, David Bawden, Andrew Brint, Geoff Downs, Bill Fisanick, Helen Grindley, Trevor Heritage, Carol Morris, Catherine Pepperrell, Andrew Poirrette, David Rice, Robin Taylor, David Thorner, David Turner, Elizabeth Ujah, Peter Walshe, David Wild, Vivienne Winterman and Matt Wright for their contributions to the work that has been carried out in Sheffield on similarity searching and clustering, to Geoff Downs and Andrew Poirrette for a careful reading of this manuscript, and to Peter Artymiuk for the provision of the colour plates. Funding has been provided by the British Library Research and Development Department, Cambridge Crystallographic Data Centre, Chemical Abstracts Service, the European Communities Joint Research Centre, the James Black Foundation, Pfizer Central Research (UK), the Science and Engineering Research Council, Shell Research Centre, Tripos Associates and Zeneca Agrochemicals. This chapter is a contribution from the Krebs Institute for Biomolecular Research, which has been designated as a centre for biomolecular sciences by the Science and Engineering Research Council.

References

Adamson, G.W. and Bush, J.A. (1973) A method for the automatic classification of chemical structures. *Information Storage and Retrieval*, **9**, 561–568.

Adamson, G.W. and Bush, J.A. (1975) A comparison of the performance of some similarity and dissimilarity measures in the automatic classification of chemical structures. *Journal of Chemical Information and Computer Sciences*, **15**, 55–58.

Artymiuk, P.J., Grindley, H. M., Park, J.E., Rice, D.W. and Willett, P. (1992) Three-dimensional structural resemblance between leucine aminopeptidase and carboxypeptidase A revealed by graph-theoretical techniques. *FEBS Letters*, **303**, 48–52.

Artymiuk, P.J., Grindley, H.M., Kumar, K., Rice, D.W. and Willett P. (1993) Three-dimensional structural resemblance between the ribonuclease H and connection domains of HIV reverse transcriptase revealed using graph theoretical techniques. *FEBS Letters*, **324**, 15–21.

Artymiuk, P.J., Grindley, H.M., Poirrette, A.R., Rice, D.W., Ujah, E.C. and Willett, P. (1994a) Identification of β-sheet motifs, of φ-loops and of patterns of amino-acid residues in three-dimensional protein structures using a subgraph-isomorphism algorithm. *Journal of Chemical Information and Computer Sciences*, **34**, 54–62.

Artymiuk, P.J., Grindley, H.M., MacKenzie, AB., Rice, D.W., Ujah, E.C. and Willett, P. (1994b) PROTEP: a program for graph-theoretic similarity searching of the 3D structures in the Protein Data Bank. In *Molecular Similarity and Reactivity: From Quantum Chemical to Phenomenological Appoaches* (ed. R. Carbo). In press.

Ash, J.E., Chubb, P.A., Ward, S.E., Welford, S.M. and Willett, P. (1985) *Communication, Storage and Retrieval of Chemical Information*. Ellis Horwood, Chichester.

Ash, J.E., Warr, W.A. and Willett, P. (eds) (1991) *Chemical Structure Systems*. Ellis Horwood, Chichester.

Barnard, J.M. and Downs, G.M. (1992) Clustering of chemical structures on the basis of two-dimensional similarity measures. *Journal of Chemical Information and Computer Sciences*, **32**, 644–649.

Bath, P.A., Morris, C.A. and Willett, P. (1993) Effect of standardisation on fragment-based measures of structural similarity. *Journal of Chemometrics*, **7**, 543–550.

Bath, P.A., Poirrette, A.R., Willet, P. and Allen, F.H. (1994) Similarity searching in files of three-dimensional chemical structures: comparison of fragment-based measures of shape similarity. *Journal of Chemical Information and Computer Sciences*, **34**, 141–147.

Bemis, G.W. and Kuntz, I.D. (1992) A fast and efficient method for 2D and 3D molecular shape description. *Journal of Computer-Aided Molecular Design*, **6**, 607–628.

Bernstein, F.C., Koetzle, T.F., Williams, G.J.B., Meyer Jnr, E.F., Brice, M.D., Rodgers, J.R., Kennard, O., Shimanouchi, M. and Tasumi, M. (1977) The Protein Data Bank: a computer-based archival file for macromolecular structures. *Journal of Molecular Biology*, **112**, 535–542.

Blake, C.C.F. and Oatley, S.J. (1977) Protein–DNA and protein–protein hormone interactions in prealbumin: a model of the thyroid hormone nuclear receptor. *Nature*, **268**, 115–120.

Blake, C.C.F., Geisow, M.J. and Oatley, S.J. (1978) Structure of prealbumin: secondary, tertiary and quarternary interactions characterised by Fourier refinement at 1.8 Å. *Journal of Molecular Biology*, **121**, 339–356.

Brint, A.T. and Willett, P. (1987) Algorithms for the identification of three-dimensional maximal common substructures. *Journal of Chemical Information and Computer Sciences*, **27**, 152–158.

Brint, A.T. and Willett, P. (1988) Upperbound procedures for the identification of similar three-dimensional chemical structures. *Journal of Computer-Aided Molecular Design*, **2**, 311–320.

Bron, C. and Kerbosch, J. (1973) Algorithm 457. Finding all cliques of an undirected graph. *Communications of the ACM*, **16**, 575–577.

Bures, M.G., Martin, Y.C. and Willett, P. (1994) Searching techniques for databases of three-dimensional chemical structures. *Topics in Stereochemistry*. **21**, 467–511.

Burt, C., Richards, W.H. and Huxley, P. (1990) The application of molecular similarity calculations. *Journal of Computational Chemistry*, **11**, 1139–1146.

Clark, D.E., Willett, P. and Kenny, P.W. (1992) Pharmacophoric pattern matching in files of three-dimensional chemical structures: use of bounded-distance matrices for the representation and searching of conformationally-flexible moelcules. *Journal of Molecular Graphics*, **10**, 194–204.

Cringean, J.K., Pepperrell, C.A., Poirrette, A.R. and Willett, P. (1990) Selection of screens for three-dimensional substructure searching. *Tetrahedron Computer Methodology*, **3**, 37–46.

Dittmar, P.G., Farmer, N.A., Fisanick, W., Haines, R.C. and Mockus, J. (1983) The CAS ONLINE search system. 1. General system design and selection, generation and use of search screens. *Journal of Chemical Information and Computer Sciences*, **23**, 93–102.

Downs, G.M. and Willett, P. (1991) The use of similarity and clustering techniques for the prediction of molecular properties. In *Applied Multivariate Analysis in SAR and Environmental Studies* (eds J. Devillers and W. Karcher), pp. 247–279. European Communities, Brussels.

Downs, G.M. and Willett, P. (1994) Clustering of chemical-structure databases for compound selection. In *Chemometric Methods in Molecular Design* (ed. H. van de Waterbeemd). VCH, New York. In press.

Downs, G.M., Willett, P. and Fisanick, W. (1994) Similarity searching and clustering of chemical-structure databases using molecular property data. *Journal of Chemical Information and Computer Sciences* (in press).

Edelbrock, C. (1979) Comparing the accuracy of hierarchical clustering algorithms: the problem of classifying everybody. *Multivariate Behavioural Research*, **14**, 367–384.

El-Hamdouchi, A. and Willett, P. (1989) Comparison of hierarchic agglomerative clustering methods for document retrieval. *Computer Journal*, **32**, 220–227.

Everitt, B.S. (1993) *Cluster Analysis*, 3rd edn. Edward Arnold, London.

Fisanick, W., Cross, K.P. and Rusinko, A. (1992) Similarity searching on CAS Registry substances. 1. Global molecular property and generic atom triangle geometric searching. *Journal of Chemical Information and Computer Sciences*, **32**, 664–674.

Grindley, H.M., Artymiuk, P.J., Rice, D.W. and Willett, P. (1993) Identification of tertiary structure resemblance in proteins using a maximal common subgraph isomorphism algorithm. *Journal of Molecular Biology*, **229**, 707–721.

Guenoche, A., Hansen, P. and Jaumard, B. (1991) Efficient algorithms for divisive hierarchical clustering with the diameter criterion. *Journal of Classification*, **8**, 5–30.

Gund, P. (1977) Three-dimensional pharmacophoric pattern searching. *Progress in Molecular and Subcellular Biology*, **5**, 117–143.

Hagadone, T. (1992) Molecular substructure similarity searching: efficient retrieval in two-dimensional structure databases. *Journal of Chemical Information and Computer Sciences*, **32**, 515–521.

Harel, D. (1987) *Algorithmics: the Spirit of Computing*. Addison-Wesley, Reading, Massachusetts.

Hodes, L. (1989) Clustering a large number of compounds. I. Establishing the method on an initial sample. *Journal of Chemical Information and Computer Sciences*, **29**, 66–71.

Jarvis, R.A. and Patrick, E.A. (1973) Clustering using a similarity measure based on shared nearest neighbours. *IEEE Transactions on Computers*, **C-22**, 1025–1034.

Johnson, M.A. and Maggiora, G.M. (eds) (1990) *Concepts and Applications of Molecular Similarity*. John Wiley, New York.

Kim, S.H., de Vos, A. and Ogata, C. (1988) Crystal structures of two intensely sweet proteins. *Trends in Biochemical Science*, **13**, 13–15.

Lajiness, M.S. (1991) An evaluation of the performance of dissimilarity selection. In *QSAR:*

Rational Approaches to the Design of Bioactive Compounds (eds C. Silipo and A. Vittoria), pp. 201–204. Elsevier Science Publishers, Amsterdam.

Lajiness, M.S., Johnson, M.A. and Maggiora, G. (1989) Implementing drug screening programs using molecular similarity methods. In *QSAR: Quantitative Structure–Activity Relationships in Drug Design* (ed. J.L. Fauchere), pp. 173–176. Alan R. Liss Inc., New York.

Lance, G.N. and Williams, W.T. (1967) A general theory of classificatory sorting strategies. I. Hierarchical systems. *Computer Journal*, **9**, 373–380.

Manaut, F., Sanz, F., Jose, J. and Milesi, M. (1991) Automatic search for maximum similarity between molecular electrostatic potential distributions. *Journal of Computer-Aided Molecular Design*, **5**, 371–380.

Milligan, G.W. (1980) An examination of the effect of six types of error perturbation on fifteen clustering algorithms. *Psychometrika*, **45**, 325–342.

Milligan, G.W. and Cooper, M.C. (1988) A study of standardisation of variables in cluster analysis. *Journal of Classification*, **5**, 181–204.

Mitchell, E.M., Artymiuk, P.J., Rice, D.W. and Willett, P. (1990) Use of techniques derived from graph theory to compare secondary structure motifs in proteins. *Journal of Molecular Biology*, **212**, 151–166.

Moock, T.E., Grier, D.L., Hounshell, W.D., Grethe, G., Cronin, K., Nourse, J.G. and Theodosiou, J. (1988) Similarity searching in the organic reaction domain. *Tetrahedron Computer Methodology*, **1**, 117–128.

Murtagh, F. (1983) A survey of recent advances in hierarchical clustering algorithms. *Computer Journal*, **26**, 354–359.

Murtagh, F. (1993) Search aglorithms for numeric and quantitative data. In *Intelligent Information Retrieval: The Case of Astronomy and Related Space Sciences* (eds A. Heck and F. Murtagh), pp. 29–48. Kluwer Academic Publishers, Dordrecht.

Nilakantan, R., Bauman, N. and Venkataraghavan, R. (1993) A new method for rapid characterisation of molecular shape: applications in drug design. *Journal of Chemical Information and Computer Sciences*, **33**, 79–85.

Ohlendorf, D.H., Lipscomb, J.D. and Weber, P.C. (1988) Structure and assembly of protocatechuate 3,4-dioxygenase. *Nature*, **336**, 403–405.

Pepperrell, C.A. and Willett, P. (1991) Techniques for the calculation of three-dimensional structural similarity using inter-atomic distances. *Journal of Computer-Aided Molecular Design*, **5**, 455–474.

Pepperrell, C.A., Willett, P. and Taylor, R. (1990) Implementation and use of an atom-mapping procedure for similarity searching in databases of 3D chemical structures. *Tetrahedron Computer Methodology*, **3**, 575–593.

Pepperrell, C.A., Poirrette, A.R., Willett, P. and Taylor, R. (1991) Development of an atom-mapping procedure for similarity searching in databases of three-dimensional chemical structures. *Pesticide Sciences*, **33**, 97–111.

Perry, N.C. and van Geerestein, V. (1992) Database searching on the basis of three-dimensional molecular similarity using the SPERM program. *Journal of Chemical Information and Computer Sciences*, **32**, 607–616.

Petke, J.D. (1993) Cumulative and discrete similarity analysis of electrostatic potentials and fields. *Journal of Computational Chemistry*, **14**, 928–933.

Rubin, V. and Willett, P. (1983) A comparison of some hierarchical monothetic divisive clustering algorithms for structure property correlation. *Analytica Chimica Acta*, **151**, 161–166.

Siegel, S. and Castellan, N.J. (1988) *Non-Parametric Statistics for the Social Sciences*. McGraw-Hill, London.

Sneath, P.H.A. and Sokal, R.R. (1973) *Numerical Taxonomy*. W.H. Freeman, San Francisco.

Stirk, H.J., Woolfson, D.N., Hutchinson, E.G. and Thornton, J.M. (1992) Depicting topology and handedness in jellyroll structures. *FEBS Letters*, **308**, 1–3.

Stouch, T.R. and Jurs, P.C. (1986) Computer-aided studies of the structure–activity relationships between the structure of some steroids and their anti-inflammatory activity. *Journal of Medicinal Chemistry*, **29**, 2125–2136.

Todeschini, R. (1989) k-nearest neighbour method: the influence of data transformations and metrics. *Chemometrics and Intelligent Laboratory Systems*, **6**, 213–220.

Ujah, E.C. (1992) A Study of β-Sheet Motifs at Different Levels of Structural Abstraction Using Graph-Theoretic and Dynamic Programming Techniques. PhD thesis, University of Sheffield.

van der Wel, H. and Loeve, K. (1972) Isolation and characterization of thaumatin I and II, the sweet-tasting proteins from *Thaumatococcus danielli* Benth. *European Journal of Biochemistry*, **31**, 221–225.

Voorhees, E.M. (1986) Implementing agglomerative hierarchical clustering algorithms for use in document retrieval. *Information Processing and Management*, **22**, 465–476.

Wild, D.J. and Willett, P. (1994) Similarity searching in files of three-dimensional chemical structures: implementation of atom mapping on the Distributed Army Processor DAP-610, the MasPar MP-1104 and the Connection Machine CM-200. *Journal of Chemical Information and Computer Sciences*, **34**, 224–231.

Willett, P. (1982a) A comparison of some hierarchical agglomerative clustering algorithms for structure property correlation. *Analytica Chimica Acta*, **136**, 29–37.

Willett, P. (1982b) The calculation of inter-molecular similarity coefficients using an inverted file algorithm. *Analytica Chimica Acta*, **138**, 339–342.

Willett, P. (1983) Some heuristics for nearest neighbour searching in chemical structure files. *Journal of Chemical Information and Computer Sciences*, **23**, 22–25.

Willett, P. (1984) An evaluation of relocation clustering algorithms for the automatic classification of chemical structures. *Journal of Chemical Information and Computer Sciences*, **24**, 29–33.

Willett, P. (1987a) A review of chemical structure retrieval systems. *Journal of Chemometrics*, **1**, 139–155.

Willett, P. (1987b) *Similarity and Clustering in Chemical Information Systems*. Research Studies Press, Letchworth.

Willett, P. (1992) A review of three-dimensional chemical structure retrieval systems. *Journal of Chemometrics*, **6**, 289–305.

Willett, P. and Winterman, V. (1986) A comparison of some measures for the determination of inter-molecular structural similarity. *Quantitative Structure–Activity Relationships*, **5**, 18–25.

Willett, P., Winterman, V. and Bawden, D. (1986a) Implementation of nearest-neighbour searching in an online chemical structure search system. *Journal of Chemical Information and Computer Sciences*, **26**, 36–41.

Willett, P., Winterman, V. and Bawden, D. (1986b) Implementation of non-hierarchic cluster analysis methods in chemical information systems: selection of compounds for biological testing and clustering of substructure search output. *Journal of Chemical Information and Computer Sciences*, **26**, 109–118.

6 Experiences with searching for molecular similarity in conformationally flexible 3D databases

J.S. MASON

6.1 Introduction

A key step within a modern drug design strategy is the generation of a lead compound. This process can be greatly enhanced by the use of three-dimensional (3D) database searching, where compounds with molecular similarity or complementarity can be extracted from large databases of 3D structures. The use of 3D databases in drug design is the subject of much current interest, and several new commercial software packages for 3D database searching are now available. Recent reviews are available for this topic (Martin, 1992), and more specifically for the generation and use of 3D molecular structures in 3D searching (Pearlman, 1994); software for general molecular modelling, often needed for defining the search criteria or analysing the results, is catalogued in a recent compendium by Boyd (1994).

Large databases of compounds from any source (company registries, external databases, active ligands or hypothetical structures) can now be searched in 3D, taking into account the full conformational flexibility of the structures. 3D database searches can play an important role in the lead generation and optimisation steps in the drug design process: small screening sets tailored to a specific biological target can be generated from large company databases; the idea-based design process can be optimised; *de novo* ligand design is possible through the identification of complementary fragments and new molecular scaffolds; automated pharmacophore identification is possible from a set of compounds of known activity. This chapter presents an experience-based discussion of the design, construction and application of 3D databases of conformationally flexible molecules for lead generation in medicinal chemistry applications.

Two main types of 3D database searching in drug design can be distinguished, based on either pharmacophoric distances or shape and electrostatic complementarity. The former is an example of the extraction of molecular similarity using as search criteria only those features of molecules hypothesised to be essential for biological activity. The latter involves the identification of entire molecules with complementary shape and electrostatic properties to a defined site. This method is ideally used when the 3D structure of an enzyme or receptor is known, although a site can be defined by projecting complementary properties from a ligand, resulting in a search for molecular similarity. Published drug design applications of this type of 3D search include the structure-based design of inhibitors of thymidylate synthase (Shoichet *et al.*, 1993). Conformational flexibility is, however, much less readily handled in this approach. The most common approach is 3D searching of large databases of structures to identify those which can present a particular pharmacophore. Here the idea of the search is to identify structures in a database with similar 3D geometric relationships between essential features. The criteria for similarity can be quite wide, and as the connectivity between the centres defining the 3D query is not considered, diverse structures which are not at all related in a two-dimensional chemical structural sense to the original query can be found. The pharmacophore is often described only by a set of distance constraints between atom centres or centroids of a defined environment (e.g. acidic, basic, hydrogen bond acceptor or donor, aromatic ring centroid or other hydrophobe centre). The query can also contain other geometric constraints such as angles and other features such as exclusion volumes, substructures and geometric objects (e.g. planes, lines, points and normal vectors). An important recent advance has been the ability to allow for the conformational flexibility of structures in the database, as discussed below; some experiences and applications have already been published (Haraki *et al.*, 1990; Mason, 1993; Mason *et al.*, 1994).

3D searching of a single low-energy conformation does not generally give a realistic appraisal of the ability of a flexible molecule to fit a particular pharmacophore geometry. An alternative low-energy conformation may match the query, as has been demonstrated recently (Haraki *et al.*, 1990; Mason 1992, 1993; Mason *et al.*, 1994). Indeed the bioactive (bound) conformation may not be a low energy conformation, and receptor interactions may involve higher energy activated structures (Jorgenson, 1991). Conformational flexibility can now be readily taken into account for distance-based searches, requiring the storage of only a single conformation for each structure in the database. The flexibility is taken into account at search time by either generating new conformations, for example a systematic rule-based conformational analysis or random conformer generation, or by torsional fitting ('tweaking') where a minimisation in torsional space uses the query as a constraint; this is discussed further in section 6.2.2. For shape-based searches, the problem of flexibility is generally approached by storing multiple conformations, and methods to achieve this more efficiently are discussed in section 6.4.3.

6.2 Creation of a viable 3D database

The creation of a 3D database involves the generation of a reasonable 3D structure and the registration of this single conformation into the database.

During or after this registration various 'keys' (bit patterns) are generated to speed up subsequent searches by the very rapid elimination of all 'unsuitable' compounds, defined by 2D and/or 3D distance criteria. Some methods determine and store atom type information, as normally used in molecular modelling studies (e.g. ChemDBS-3D (Chemical Design)),[1] whereas in others only element type is stored (e.g. ISIS/3D, MACCS-3D (MDL)[2] and UNITY (Tripos Associates)[3]) and any differentiation of atom environment must be defined at search time. Associated with the various atom types (e.g. for oxygen: 16A, O–R; 16B, 0=; 16C, O–aromatic; 16D, O–carboxylic) in ChemDBS-3D is the concept of generic centre types. 3D keys can then be generated based on conformationally accessible distances between these centre types. There are four centre types currently used: hydrogen bond donor, hydrogen bond acceptor, aromatic ring centroids and positively charged centres; these are automatically assigned, with total user customisation possible.

6.2.1 Obtaining the 3D structures

The conversion of a 2D database of chemical structures to 3D structures involves the creation of a suitable '2D' connectivity table (CT) file with atom and bonding descriptions for each structure, which is used as input for an automated 3D structure generation program. Several different methods for 3D structure generation are available, the most popular and rapid being bond or fragment-based. Distance geometry-based methods are becoming more widely available, and although slower, are particularly powerful with fused ring systems and macrocycles. Methods for 3D structure generation were recently reviewed by Pearlman (1994).

2D structure representation. The most compact and easily handled 2D structure representation is the SMILES line notation (Weininger, 1989); this is a single line description of the structure in which aromatic atoms are differentiated, for example phenol is 'c1ccccc1O' and cyclohexanol 'C1CCCCC1O'. The SMILES code can be made unique for a given structure, and offers a very convenient means of comparing different databases of structures. Extensions to the code have been developed to allow stereochemistry to be defined using both relative and absolute indicators. The former is the 'official'

[1] Chemical Design Ltd, Roundway House, Cromwell Park, Chipping Norton, Oxfordshire, OX7 5SR, UK.
[2] MDL Information Systems Inc., 14600 Catalina Street, San Leandro, California 94577, USA.
[3] Tripos Associates, 1699 Hanley Road, Suite 303, St Louis, Missouri 63144, USA.

method, developed by Daylight Chemical Information Systems,[4] whereas the latter, termed 'conversational' SMILES, and developed for CONCORD (Pearlman, 1987; Pearlman *et al.*, 1988) allows the convenient use of the atom chirality indicators *R* and *S* and bond stereochemistry indicators *E*, *Z*, C(*cis*) and T(*trans*). With these absolute stereochemistry identifiers it is possible to readily verify stereochemistry relative to a chemical name and modify if necessary. SMILES codes can be used as input for many programs, and can be generated automatically from the other main 2D format, MDL MOL or SDfiles (Dalby *et al.*, 1992). Programs such as GEMINI and CCT (Daylight Chemical Information Systems) can be used for this, and many software programs can also write SMILES codes directly. The SDfile CT file format was developed by MDL for the structural database program MACCS, and is now a universal file transfer format for both 2D and 3D structure representations (Dalby *et al.* 1992). In this format *x*-, *y*- and *z*-coordinates are stored for each atom, and all bonds are explicitly defined; this gives rise to large databases or very large unwieldy files which cannot easily be edited, unlike their SMILES equivalents, where only one line per structure is needed. Stereochemical information is defined in terms of atom and bond 'parities', but this is much less easily verified and modified than in SMILES codes. For an atom, parity can be even (2) or odd (1), depending respectively whether the order of connecting the substituents is clockwise or anticlockwise; values of 0 and 3 are used to define atoms which are non-chiral or of undefined chirality respectively. Another indicator of relative stereochemistry can also be used in the bond definition records; possible values are 0 (bond in drawing plane), 1 (bond above plane), 4 (undefined), and 6 (bond below plane). The direct modification of stereochemistry in the SDfile CT can be very complicated, for example a double bond is defined *cis* or *trans* by the 2D coordinates.

3D structure generation. The most robust and universally used program is CONCORD (Pearlman, 1987; Pearlman *et al.*, 1988). This program uses SMILES or SDfile CT files as input, and produces reasonable low energy conformations for most small organic structures; large ($>$ 12) rings and peptides are not well handled, and most organometallic compounds cannot be converted. The program is rapid (10 000–15 000 structure/hour on a VAX 4000) and gives useful structure quality information, including a close-contact ratio; many output file formats are possible. A complementary but much slower method is that of distance geometry, used in programs such as DGEOM (Blaney *et al.*, 1990) and DG-II (Havel, 1991; now commercialised ready for database conversion use by Biosym Technologies and called CONVERTER). Complex ring systems are readily handled, but the speed of structure

[4] Daylight Chemical Information Systems Inc., 18500 Von Karman Avenue, Suite 450, Irvine, California 92715, USA.

generation is much slower, and the structures generally need to be optimised by energy minimisation. Experiences with CONVERTER, which currently only accepts SDfile CT file format for input and output file formats, indicated that generated 3D structures need some verification and lack uniformity in bond geometries, but that successful structure generation was often possible for cases where CONCORD fails. Other 3D structure generation programs include COBRA/WIZARD (Dolata *et al.*, 1987; Leach *et al.*, 1988; Leach and Prout, 1990), CORINA (Gasteiger *et al.*, 1990) and the Chem-X builder (Chemical Design Ltd). 3D structure generation methods have recently been reviewed (Pearlman, 1994), and various groups have published their experiences (Rusinko *et al.*, 1989; Davies and Upton, 1990; Henry *et al.*, 1990; Mason, 1993).

Stereochemistry. If no stereochemical information is defined, then 3D structure generation programs such as CONCORD usually produce a structure with *R* stereochemistry; 'random' stereoisomers can be generated, up to a predefined limit, using programs such as CONVERTER and COBRA. Stereochemical information can be added to SMILES and SDfiles by means of parity indicators, or with readily interpretable *R,S* and *E,Z* indicators in the 'conversational' SMILES used additionally by CONCORD. The parity indicators for SDfiles were described on p. 141; with SMILES 'isomeric' codes are generated by using '@' and '@@' for atoms to indicate the clockwise and anticlockwise respectively order of connection of substituents, with '/' and '\' for the direction of bonds. Using fluorohydroxybut-2-ene as an example, for the *R cis* isomer, the CT descriptions for stereochemistry are compared in Figure 6.1.

The conversion of 'parity' information used in SDfile CT files to define stereochemistry to stereochemical identifiers for SMILES codes is not readily accomplished directly, but is possible to convert 3D structures (e.g. generated

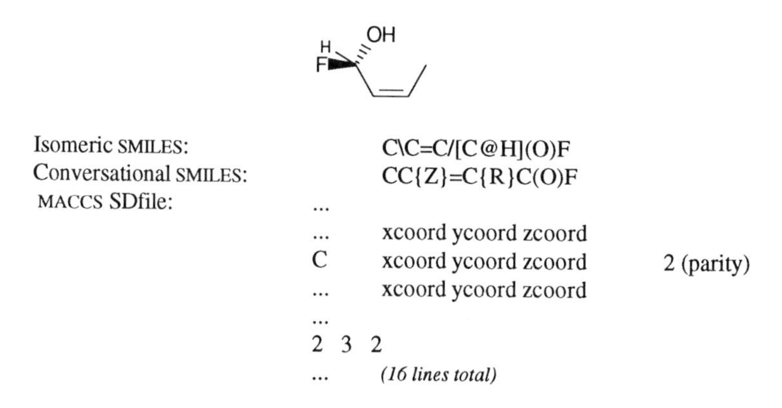

Figure 6.1 CT file descriptions of stereochemistry.

by CONCORD from SD files) to isomeric SMILES using GEMINI, and 2D and 3D structures to conversational SMILES using the SMILES interface to Chem-X.

6.2.2 Dealing with conformational flexibility

Many of the theoretical problems of handling conformational flexibility have been discussed in Chapter 3. The key to the successful use of 3D searching for many drug design applications is the ability to deal with the conformational flexibility present in most drug molecules. For example, in one of our company site databases (62 000 compounds registered for biological screening), only 5% had structures where flexibility would not play a role in defining potential pharmacophores. This was determined by whether any rotatable bonds connected the four centre types used in ChemDBS-3D: hydrogen bond donor and acceptor centres, aromatic ring centroids and positively charged centres. The first commercial 3D database searching systems searched only explicitly stored conformations of a structure; often only one conformation per structure would be present in the database, particularly when disk space was in shorter supply. This deficiency is partially compensated for by the advanced searching capabilities available in some of these systems: for example ALADDIN (Martin *et al.*, 1993; van Drie *et al.*, 1989) can automatically modify and re-evaluate hit structures, DOCK (DesJarlais *et al.*, 1988; Kuntz *et al.*, 1982) finds fits based on shape and electrostatic complementarity and CAVEAT (Lauri and Bartlett, 1994) uses bond vectors. An approach to dealing with conformationally flexible compounds by using flexible queries was proposed by Guner *et al.* (1992).

The first method used in commercial software to tackle directly the problem of conformational flexibility was rule-based conformational analysis, developed for ChemDBS-3D, a module of Chem-X (Chemical Design, 1990). The basis of the method is to store a single conformation and a set of distance keys (Murral and Davies, 1990, 1993) and to generate conformational flexibility during the search process using a rule-based approach. This principle of storing only one conformation is also used in the newly developed CFS module for the ISIS/3D (Moock, 1993) and MACCS-3D (Christie *et al.*, 1990) systems and the new UNITY/3D (Hurst, 1994) system, but the handling of conformational flexibility is different. Here a torsional minimisation technique is used—'directed-tweak' (Hurst, 1994) and 'CFS' (Moock, 1994)—which is query-directed rather than systematic; this method is also now available as an alternative in ChemDBS-3D ('flexifit'). Another query-directed conformational exploration method is distance geometry, but a recent report (Clark *et al.*, 1994) suggests that this approach is too slow to merit consideration. From a comparison of the efficiency and the effectiveness of the distance geometry, systematic search, random search, genetic-algorithm, and directed-tweak methods for conformational searching, the directed-tweak and genetic-algorithm methods were found to be most effective.

Our experiences with ChemDBS-3D, UNITY/3D and ISIS/3D-CFS also find the fastest method to be torsional minimisation, which can rapidly and thoroughly determine if a structure can fit a given set of distance constraints. The type of feature that can be used in addition to distance constraints is dependent on the implementation. The method readily handles very flexible molecules, and small tolerances on the distances, but is limited in that it is not systematic and can identify only a very limited number of possible conformational solutions. This limitation is not important where the goal of a search is to identify for biological screening all compounds in a database which can fit a particular pharmacophore, but can be restrictive for applications such as pharmacophore identification and idea evaluation. Even more restrictive for such computer-aided drug design (CADD) applications are database implementations which allow the identification of only one substructure mapping, particularly noticeable with more generalised queries; examples are found in sections 6.3.1 and 6.4.2 for the use of ideas databases and links to DOCK. Ring flexibility can be investigated in some implementations, and a steric bump check (van der Waals contacts) can be applied to avoid high energy structures; this is best done at search time, or with further constrained minimisation, to avoid the false elimination of suitable molecules.

The fast rule-based conformational analysis that can be applied during the building and searching stages, currently unique to the ChemDBS-3D system, enables a systematic evaluation of all conformational and substructure mappings of a molecule to a particular 3D query to be performed within a feasible time scale for many structures. This systematic method is not suitable for very flexible molecules (for example, >14 rotatable bonds, $\sim 5\%$ in one of our 60 000 structure databases) or for use with small tolerance distances (a minimum of 0.5 Å or 20% is recommended). A full energy calculation cannot be feasibly done, but a fast steric bump check is usually adequate. Depending on the molecules and time considerations, large bond-rotation increments are used; the default in ChemDBS-3D is three for single bonds, 6 for sp^2–sp^3 ('alpha') bonds, and two for conjugated and double bonds. An alternative method for very flexible molecules is random (rule-based) conformational analysis, although this is much slower than torsional minimisation and there may be uncertainties as to the level of conformational exploration. This method allows however for potentially fuller conformational exploration than torsional minimisation, and multiple conformation solutions can be identified; a usable implementation is available in ChemDBS-3D. An alternative approach to the problem of conformational flexibility, based on the discrete conformer storage approach, is used in the CATALYST software (Sprague, 1991). Here a subset of up to 100 conformations, selected to maximise distance space, is stored and searched for each structure.

Each method of conformational regeneration has distinct advantages and disadvantages, depending on the application. There are requirements for successful use of the different methods, such as tolerance settings and the type of feature to be searched; for example, very small tolerances are not suitable for rule-based conformational analysis and some torsional minimisation implementations have very restricted non-distance constraints. The practical use of the different methods is also strongly affected by other factors associated with the database searching system, such as the ability to easily differentiate and link different atom types and environments (discussed in section 6.2.3) and the level of integration with a molecular modelling system (see examples in sections 6.3.2 and 6.4.1).

A comparative study was made using 58 000 identical 3D structures of company registry compounds, generated by CONCORD, built into 3D databases for ChemDBS-3D, ISIS/3D, and UNITY/3D. The results of some simple 3D pharmacophoric distance searches are given in Table 6.1.

It was clear from the studies that tolerance values need to be adjusted according to the method of conformational exploration used, and that for very flexible molecules torsional minimisation techniques are the most effective (discussed further in section 6.3.2). The effect of allowing for ring flexibility can be seen with the UNITY results. The different number of hits

Table 6.1 Comparison of 3D database search results*

Aromatic nitrogen to oxygen of carboxylic acid, 8.5–10 Å (i.e. 9.25 ± 0.75 Å):

	CONCORD conformation	Conformational flexibility
ChemDBS-3D:	695	919
ISIS/3D	594	961 954 (with van der Waals bump check)
UNITY/3D	641	1016 (with ring flexibility) 990 (with van der Waals bump check) 1009 (with no ring flexibility)

Nitrogen to oxygen of carboxylic acid to aromatic ring (6-membered), N↔O 14 ± 0.5 Å; N↔AR6 14 ± 0.5 Å; O↔AR6 7 ± 0.5 Å:

	CONCORD conformation	Conformational flexibility
ISIS/3D	1	152
UNITY/3D	1	151 (with no ring flexibility) 160 (with ring flexibility) 122 (with N = basic nitrogen)
ChemDBS-3D:	1	118 of 122 hits from UNITY too flexible (> 14 rotatable bonds) for systematic rule-based analysis Random conformation analysis: → 60/122 (tol = ±0.5 Å) → 90/122 (tol = ±1.0 Å) Flexifit (torsional minimisation): → 122/122 (tol = ±0.5 Å)

*ChemDBS-3D = rule-based systematic conformational analysis (single = 3, alpha = 4, double/conjugated = 2) or random rule-based, max. time = 2 min, or flexifit (no van der Waals bump check); ISIS/3D and UNITY/3D = torsional minimisation (CFS and directed tweak).

for the CONCORD conformation search may be due to the way an aromatic nitrogen is interpreted (e.g. from Kekulé structures in ISIS database). Considerations such as atom types versus element types or speed of search versus thoroughness of conformational and substructure mapping search, together with visualisation and analysis possibilities have an influence on the effectiveness of any particular implementation or algorithm.

The yield of compounds from a 3D search can also be affected by how a potential hit compound is fitted to the query. In some implementations there is no further requirement than that the final distances are within the tolerances defined in the query, so only other factors such as unacceptable steric bumps could stop the compound being a valid hit (e.g. ISIS/3D, UNITY/3D). However, in implementations where the query is actually in 3D, rather than being a 2D sketch with distance constraints, the hit conformation is fitted to the query, using the corresponding atoms to define the superimposition (e.g. ChemDBS-3D). In this latter method, a second fitting tolerance comes into play, and the acceptance of a particular molecule may be affected by the chirality of the query and whether this final fitting uses just the atoms or centroids defining the distances, or all atoms in the query (i.e. including those defining a particular substructure such as an amide NH or an aromatic ring centroid). The most striking difference is with a non-planar four or more point query; because this is chiral in 3D the searches will only identify structures where the 'chirality' defined by the corresponding points is correct in addition to the interpoint distances. The fitting tolerance can be varied in addition to the distance tolerances to modulate these effects.

6.2.3 *Defining atom and property environments*

The ability to differentiate and link different atom types and environments can greatly affect the ease of use and efficiency of 3D searching. With many pharmacophoric searches used for lead generation it is often very important to be able to search for centres with a particular property but not limited to a single group; common examples are a basic nitrogen (both N sp^3 aliphatic amines and N sp^2 amidines and guanidines, but no other N), an acidic centre (both oxygens of a carboxylic acid and all nitrogens of a tetrazole) and aromatic ring centroids and/or hydrophobes (both five- and six-membered aromatic rings, optionally with other hydrophobic groups such as isopropyl or butyl groups). Here concepts of molecular similarity go beyond the atom model to that of associated properties, and the definition of a pharmacophore often involves only a centre with a particular property, such as hydrogen bond acceptor. Indeed, although 3D queries are currently normally centred on the defining feature, such as the oxygen of a carbonyl group for a hydrogen bond acceptor or an aromatic ring centroid, the ultimate goal of many 3D searches is the identification of another molecule with corresponding site extension points; the two defining atoms themselves do not have to be superimposable. The property-based approach requires a very flexible method of defining a feature centre. An acidic centre needs to selectively identify unsubstituted carboxylic acids and tetrazoles, and then consider equally all the oxygen and nitrogen atoms. A basic centre needs to selectively identify the sp^3 nitrogen of an amine and the sp^2 nitrogen of an amidine or guanidine, but not other sp^2 nitrogens (e.g. N–C=O, N–C=C, N–aromatic ring, N=O, aromatic N).

Two methods are currently used in pharmacophoric 3D searching: in the first, an extension of the special atom types as used in molecular modelling software are exploited; in the more widely used second method only element type is stored, and atom environment is identified at search time. Each method has advantages and disadvantages, and both need carefully predetermined atom-type assignment or template search time definitions for efficient use.

Use of atom types. In this method the atom types for each structure in the database are determined at the time of construction of the database. ChemDBS-3D is designed around this method of atom environment definition, and great flexibility is possible through a user-definable parameterisation file and database. New atom types and centres (e.g. hydrophobes) can be added by the use of additional fragments in a substructure database which is automatically used when a structure is read into Chem-X. In the definition of a query, different atom types can be readily made equivalent, or the general element type can be used. In our modified parameterisation, 16 different atom types can be assigned to a nitrogen atom, and five to an oxygen (default = 9 and 4), as outlined in Table 6.2.

Two types of parameterisation fragment are used in the ChemDBS-3D method, one which requires an exact connectivity match but disregards bond order, the other which defines a minimum connectivity count and requires an exact bond order match; the latter is also used to define 'dummy' centroid atoms such as for aromatic rings. The atom modifications are applied in the order of the fragments in a special parameterisation database, making the

Table 6.2 Possible atom types for nitrogen and oxygen (e.g. for use in ChemDBS-3D)

Nitrogen		
14A (nitrile)	14B (general sp^2)	14C (quarternary)
14D (nsp^3 amine)	14E (amide NH)	14F (pyrrole)
14G (azo)	14H (aromatic)	14J (nitro)
14S (sulphonamide)	14T (tetrazole-H)	14U (amidine, guanidine)
14V (Ar–NH_2)	14W (tertiary amine)	14Y (tautomeric, e.g. imidazole)
14Z (Nsp^2 trisubstituted, e.g. NMe amide or imidazole)		
Oxygen		
16A (ether, alcohol);	16B (carbonyl);	16C (aromatic, e.g. furan);
16D (carboxylic acid)	16E (tautomeric, as in 2- or 4-pyridone or β-diketones)	

order very important and allowing for selective modifications after a generic modification. An atom in a fragment may force a modification of the corresponding atom in the molecule, or be solely there to identify the environment of another; the different actions are determined by the name given to the atom in the fragment.

A query point can thus be fairly intuitively defined by simply allowing the drawn atom type to be equivalent to additional required atom types. By this method both oxygens of a carboxylic acid can be selectively coded as the same atom type, to simulate physiological conditions, and made equivalent during a search for all the nitrogens of an unsubstituted tetrazole, which also have a unique atom type; all amide nitrogens can be 'deactivated', and only monosubstituted ones be 'reactivated'. Solutions to the problem of dealing with tautomers can also be achieved: both nitrogens of tautomeric nitrogen heterocycles such as imidazole can be identically coded; in compounds such as 2- or 4-pyridone, the nitrogen and oxygen can be identified as both hydrogen bond donors and acceptors.

Aromatic ring centroids are added by default to five- and six-membered aromatic rings, and this can be readily modified to add new hydrophobe 'centroids' to groups such as propyl, butyl and cyclohexyl. These centres can be conformationally keyed at database build time, allowing a rapid elimination of all structures which could not possibly fit the query (see section 6.3.2).

Using the three examples of an amide nitrogen, a basic nitrogen and an acidic centre, these could be selectively searched for by simply selecting one or more of the predefined atom types stored in the database. The basic nitrogen is defined by equivalencing the atom types for aliphatic primary, secondary and tertiary amines with that for all the nitrogen atoms in an amidine or guanidine. The acidic centre is defined by equivalencing the oxygen of a carboxylic acid with the nitrogen of a tetrazole; as mentioned above, the parameterisation done at database build time enables all the oxygens of the carboxylic acid or the nitrogens of a tetrazole to be coded as equivalent and discernible with special atom types, uniquely when unsubstituted. The three atom environment definitions are illustrated in Table 6.3.

Search-time defined atom environment. In this method only the element type is stored in the database, and the query must define the desired environment of the atom. This can be simple, for example with an amide nitrogen, but can be complicated when the definition of an environment requires the exclusion of unwanted chemical groups or the linking of different groups

Table 6.3 Defining atom environments: ChemDBS-3D atom types

Amide nitrogen:	14E
Basic nitrogen:	14D or 14W (sp^3: 1°–3° amines) or 14U (sp^2: amidine/guanidine)
Acidic centre:	16D (either O of CO_2H) or 14T (N of tetrazole)

(e.g. a carboxylic acid oxygen and a tetrazole nitrogen, the centroid for five- or six-membered aromatic rings) for 3D searching. This method of defining a query atom type is used by ISIS/3D and UNITY/3D. In UNITY the atom environment is defined at search time using Sybyl Line Notation (SLN), an extended SMILES-like language; limited environments can be also defined directly via graphical sub-structures. The SLN language is powerful and flexible, but rather complicated for some atom environments. A powerful graphical definition method is used in ISIS/3D as the principal method for definition; this is an advantage for simple cases, but is more complicated for environments such as basic and acidic centres; the use of templates to predefine certain environments makes this easier. The search-time method is, however, potentially very flexible, even though complicated, as there are no predetermined restrictions as to the types of environment that can be defined and linked for a particular search. Additional complications can arise as fragments must often be used to define the required environment, and the correct attachment of the 3D constraint for the distance definition must be assured. For example, with a carboxylic acid in SLN a new feature 'AcidO' could be defined as {AcidO:OHC=O}, but this would measure the distance from the centroid of all atoms defining the query; to use only the oxygen atom would need the definition {AcidO:O&!(!OHC=O)|(O&!(!O=COH))}.

Using the three examples of amide nitrogen, basic nitrogen and acidic centre, these could be selectively searched for by defining at search time the atom environment, potential approaches are illustrated in Table 6.4 for UNITY in SLN and in Figure 6.2 for ISIS/3D. With ISIS/3D R-groups are used for the basic nitrogen definition, set at zero occurrence to eliminate unwanted environments and at less than four to eliminate quarternary nitrogens; the illustrated definition would only identify the singly-bonded nitrogen of an amidine or guanidine. The oxygen and nitrogen for the acidic centre have to be outside of an R-group definition to allow for attachment of a 3D constraint. With SLN in UNITY different atom environments can be easily made equivalent, but as mentioned above the point of attachment of the 3D constraint needs to be considered. The SLN definition can also become very complicated, particularly where tautomeric possibilities have to be taken into account. Templates of commonly used environments can be stored in both systems for future use.

Table 6.4 Defining atom environments: UNITY SLN

<table>
<tr><td>Amide nitrogen:</td><td>{AmideN:NC=O}</td></tr>
<tr><td>Basic nitrogen:</td><td>{BasicN:(N&!(!N(Grp)(Grp)Grp))&!NAny=Grp1|
(N&!(!NH=CNHAny))|(N&!(!NHC=NH))}
{Grp:H|C}{Grp1:C|O|S}</td></tr>
<tr><td>Acidic centre:</td><td>{AcidON:Carb|Tet}
{Carb:O&!(!OHC=O)|(O&!(!O=COH))}
{Tet:N&!(!N[1]H:N:N:N:C:@1)|(N&!(!N[2]H:N:N:C:N:@2))}</td></tr>
</table>

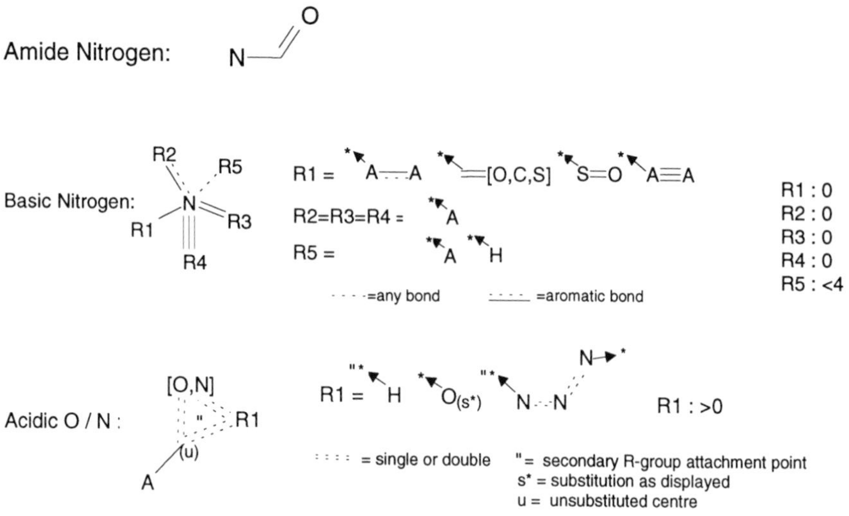

Amide Nitrogen:

Basic Nitrogen:

R1 = A—A =[O,C,S] S=O A≡A

R2=R3=R4 = A

R5 = A H

- - - =any bond - - - =aromatic bond

R1 : 0
R2 : 0
R3 : 0
R4 : 0
R5 : <4

Acidic O / N :

R1 = H O(s*) N--N

R1 : >0

: : : : = single or double

" = secondary R-group attachment point
s* = substitution as displayed
u = unsubstituted centre

Figure 6.2 Defining atom environments: ISIS/3D.

Other features. In addition to distance constraints, other geometric constraints can be readily defined and used in the analysis. Additional features such as exclusion spheres can be defined, but these are not so readily used in conformationally flexible 3D searching; currently ISIS/3D (MACCS-3D) allows the full use of additional features in 3D searching by torsional minimisation.

6.2.4 *Database keys and transferability*

The 3D database is created by reading a file of 3D structures; depending on the database system different actions occur at this stage. If atom types are used, then a parameterisation database of special fragments is used to assign the desired atom types, ring centroids, and hydrophobe centres; this method is used in ChemDBS-3D the user-definable process allowing extensive tailoring to suit particular needs or structures. Various element, substructure or distance range keys are used in 3D searching to rapidly screen out unsuitable structures. These can be set during or after the registration of structures in the database, or generated at search time. A relatively rapid distance key calculation based on a maximum and minimum distance defined by the connectivity is generally used, which is suitable for torsional fitting. An alternative method is the use of conformationally coded 3D keys; this option is available in ChemDBS-3D. The conformational analysis (normally rule-based systematic) is applied to each suitable database structure. During this analysis 10 3D distance bit keys for all the combinations of four centres (normally hydrogen bond donors, hydrogen bond acceptors, positively charged centres and aromatic ring centroids) are calculated and the composite result for all conformations stored (Davies and Upton, 1990); the conformations generated are discarded later. These keys define which distance ranges are accessible, and are used at the start of a search to eliminate all structures which could not possibly fit the query; the conformational analysis is then repeated to try and identify one or more conformations which simultaneously satisfies all the query requirements. These keys, which represent accessible conformational space, can also be used for other applications such as the generation of potential pharmacophores (see section 6.3.1).

The databases are generally hardware platform system dependent, except for the ChemDBS-3D databases which are binary compatible between the various systems supported (VAX/VMS, Silicon Graphics/Unix, PC/DOS and Macintosh), allowing the convenience of use on a platform different to the one used for the construction; this includes the results database, which is written directly as a 3D database of superimposed hit structures, usable for further 3D searches.

6.3 Identification of new leads by 3D database pharmacophoric searching

Lead generation through selective screening sets is an important application of 3D database searching in pharmaceutical and agrochemical research. Pharmacophoric models can be proposed from the structures of molecules that interact with a desired biological target. Using these 3D requirements for bioactivity, 3D database searching of large company registry compounds or external databases allows the identification of structures that match the hypothesis to varying degrees. The use of this method to produce screening sets of 50–1000 compounds from our corporate databases of about 150 000 compounds has been of great value in lead generation for new biological screens. As the compounds are already synthesised and should be available in stock bottles, results can be rapidly obtained and the time lag between starting a new biological screen and finding a lead suitable for a medicinal chemistry programme can be greatly reduced. This method requires sufficient data about the target or active ligands to define a 3D model and effective pharmacophore identification and definition are vital to a successful application. Without this specific data, general 'rational' screening sets produced by different methods of database and structural analysis, such as a clustering or partition of molecular properties can be used.

6.3.1 Defining the pharmacophoric query: pharmacophore identification

A pharmacophoric model can be proposed from the analysis of a set of biologically active ligands or from the analysis of a receptor or enzyme active site structure, preferably determined with a ligand bound. New methods are becoming available to provide automated analyses of sets of active ligands to produce hypothetical pharmacophoric models. Enzyme and receptor structures can be surveyed energetically by a program such as GRID (Goodford, 1985) to identify favourable interaction sites for a large variety of different functional group probes, such as a carbonyl group, an amine NH and an aromatic CH. These regions for favourable interactions can then be used to define sets of pharmacophoric queries for 3D searches based on different combinations of inter-region distances.

A more common situation is where only the structures of several active but flexible ligands are known. Several automated methods that try to identify potential pharmacophores are now available. The most common approach is to analyse a predefined set of conformations for each ligand, such as in DISCO (Martin *et al.*, 1993; Bures *et al.*, 1994), APEX (Golender and Vorpagel, 1993) and CATALYST (Sprague, 1991). An alternative approach is to use the distance bin keys from conformationally coding a database of the active ligands to rapidly suggest possible common features. These proposed models are then validated by using them as queries for a 3D search on the active-ligand database. This method is used in Chem-X/ChemDBS-3D; clustering and similarity methods are available to aid the selection of a suitable structure set for analysis.

Thus, using the defined centre types (hydrogen bond acceptors and donors, aromatic ring/hydrophobe centres, positive charge centres) potential pharmacophore models of combinations of three or four of these centres are rapidly identified by an analysis of the distance keys. These models then need to be evaluated and potentially validated by using them as 3D queries to analyse, with conformer regeneration, the initial database of active compounds. This step identifies any conformations of compounds in the ligand training set which can fit a particular proposed pharmacophore model; where multiple centres exist in a structure in close proximity to a proposed model point, the centre type in the model can be modified (e.g. an aromatic ring centroid to a hydrogen bond acceptor for an imidazole ring) and the new model rapidly evaluated.

An investigation of this method was made using eight non-peptidic structurally distinct angiotensin II antagonists (see Figure 6.3). All possible hydrophobe centres (e.g. ethyl and butyl groups) were explicitly coded together with the aromatic ring centres, and the 3D keys were generated through a systematic conformational analysis of all the compounds with three (single bonds) and six (conjugated and sp^2–sp^3 bonds) rotamers per bond being calculated. Depending on the tolerances defined, 3–44 potential four-point pharmacophores were proposed; different combinations of the four possible centre types occurred in these four-point models, mainly hydrogen bond acceptors and aromatic ring/hydrophobe centres (carboxyl oxygens and tetrazole nitrogens were all coded in the databases equivalently

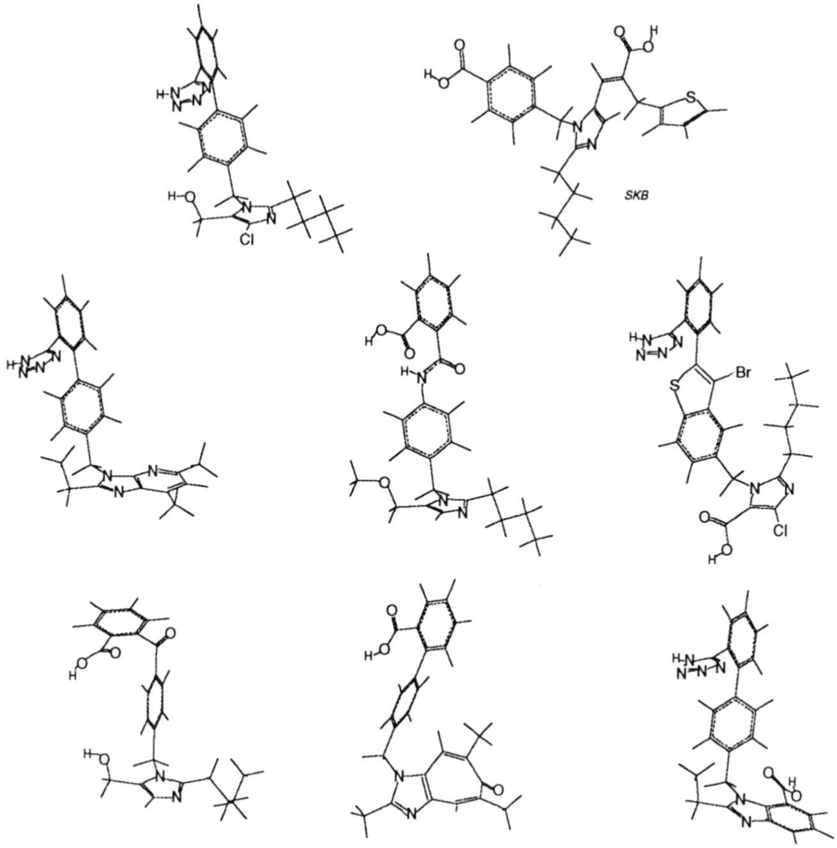

Figure 6.3 Eight angiotensin II antagonists used for the automated pharmacophore identification.

and as hydrogen bond acceptors only). There were many false positives, which failed to validate for all ligands when used as 3D pharmacophore queries to analyse the initial eight-compound database, but some interesting ideas were rapidly obtained, including models close to those which had been derived from time-consuming classical molecular modelling studies.

The easy and rapid visualisation of fitted multiple matches (different conformations and substructure mappings) per structure is a key feature of this method, from which new ideas are suggested and easily evaluated. Figure 6.4 illustrates this multiple mapping and conformation matching using an example pharmacophore model proposed (three hydrogen bond acceptor points and one aromatic ring centroid/hydrophobe point) and four conformations/mappings of the SKB structure (identified in Figure 6.3).

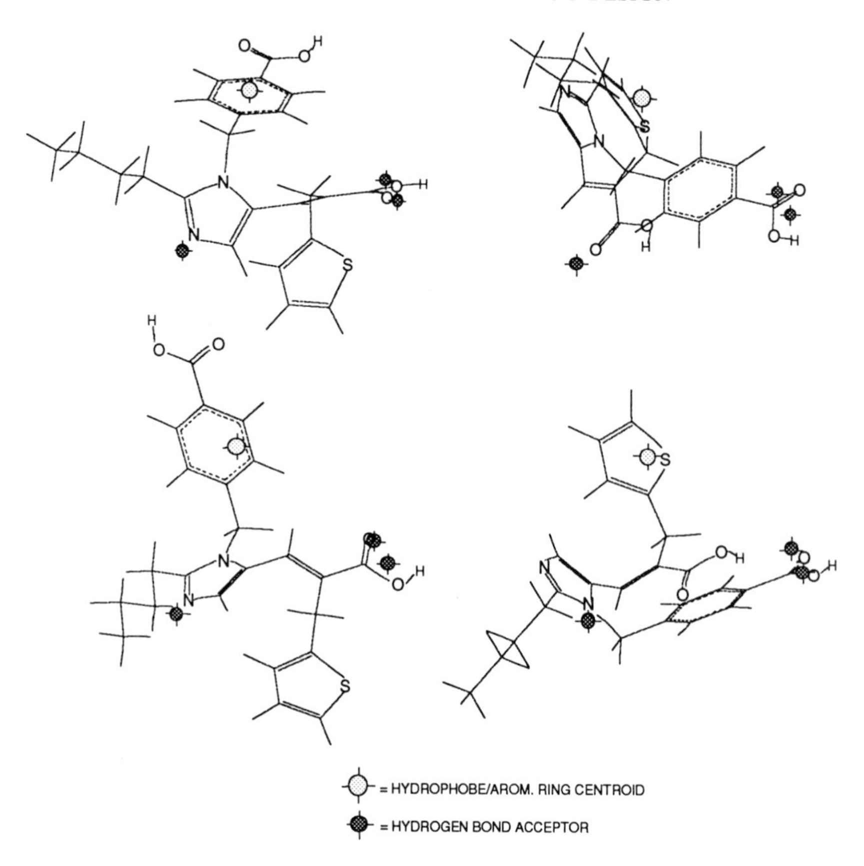

Figure 6.4 Different fits of one of the angiotensin II antagonists (SKB) on a proposed model from the Chem-X-pharmacophore identification analysis.

6.3.2 *Identification of matching structures: conformational, query definition and search strategies*

Depending on the database searching system, information about the environment of an atom and the effects of conformational flexibility is stored in the database or generated at search time. In ISIS/3D, MACCS-3D, and UNITY the environment (e.g. amide N) is defined at search time and flexibility is considered through torsional fitting, also only at search time. The creation of a ChemDBS-3D database involves definition of particular atom types (see section 6.2.3) with automated recognition of certain important features (hydrogen bond acceptors, hydrogen bond donors, aromatic ring centroids/hydrophobes, positively charged atoms) and conformational sampling, normally by rule-based conformational analysis, to generate composite 3D distance keys between all possible combinations of these centres; the keys encode all the allowed distances between each pair of centres. These keys enable a very rapid elimination of structures which for none of the generated conformations could fit a particular 3D query; up to 95% of all structures can be eliminated at this first pass.

We have found that structures with up to 14 rotatable bonds are readily handled using the rule-based conformational analysis; two (conjugated and double bonds), three (single bonds) and four (sp^2–sp^3 bonds) rotamers per bond were calculated in our databases for such structures. The default for 'alpha' (sp^2–sp^3) bonds is a six-point analysis, but four points can offer a large gain in time without necessarily a reduction in yield. For a database of 60 000 company compounds, a total of 29 million conformations were accepted by the rules from a total of 912 million considered (on average, there were ~500 conformations accepted per compound, and about 3 seconds of central processor unit time for the calculation on a Silicon Graphics workstation; a maximum time of 60 seconds of central processor unit time was allowed). About 6000 compounds had no suitable bonds for rotation, and 1600 were too flexible (>14 bonds for rotation); for the latter class of compounds, other analysis strategies such as torsional fitting or random conformation generation are more effective. Indeed for a particular query with a basic nitrogen at 15 Å to an acidic centre, most of the eligible compounds were too flexible for a systematic rule-based conformational analysis, but the use of torsional fitting in any of the searching programs rapidly identified 122 hits from a 55 000 company compound database, as described in section 6.2.2; a random conformational analysis search yielded about half this number.

Pharmacophoric searches using rule-based conformational analysis in ChemDBS-3D normally proceed at acceptable rates. This is, however, generally slower than torsional fitting methods, although less refined distance keys are then used, usually leading to much less elimination or screening-out of compounds from the database prior to searching. The yield of hits is also generally higher with torsional fitting, but not all the compounds identified from the conformational search are necessarily found. Indeed, for handling *cis–trans* double-bond isomerism/mixtures a conformational method may be necessary, whereas ring flexibility is better handled by other methods.

Different query definition strategies are needed according to the method used to handle conformational flexibility. If torsional fitting is used, then small tolerances (<0.5 Å) can be used; using a conformational analysis the limited number of rotamers generated per bond (e.g. as low as 2–4) makes the use of too small tolerances (e.g. <0.5Å or 20% for non-bonded atoms) inappropriate, and much reduced yields may be obtained with suitable compounds being missed. Another factor which may affect both the yield and quality of hits is substructure mapping. It is often important that many more than one mapping of the query points onto suitable target molecule points are both considered in order to identify that molecule as a hit, and used in producing a results set of conformations/orientations (mappings) for later evaluation and analysis.

As an example for rule-based analysis, a three-point model consisting of a basic nitrogen, a doubly-bonded oxygen and an aromatic ring centroid, all at 7 Å to each other was considered, using a tolerance of ± 1 Å. The 3D key search on a 55 000 company compound database screened out about 93% of the entries in less than 1 second, a geometry search on the single stored conformer proceeded at greater than 300 compounds per minute to yield 56 matches, and the geometry search with full conformer regeneration proceeded at greater than 50 compounds per minute to yield a further 63 matches (timings on a Silicon Graphics 4D/35 workstation). The single- and multi-conformer searches both produce new databases of structures and conformations which match the 3D query, automatically reoriented to a best fit to the query atoms. The searches may be limited to one conformation or substructure match, but have the very useful ability to find multiple fits (mappings) for a given structure on a 3D query; such a facility was found to be essential for uses such as idea evaluation described later.

Pharmacophoric searches on the company databases have given very encouraging and interesting results. Many of the most interesting hits found have been in conformations other than the nominal low energy conformation stored in the database, illustrating the extra power of 3D searching with conformational flexibility. The power of such 3D database searches was shown using a model developed for a potential protein kinase C (PKC) pharmacophore by molecular modelling studies on the natural product staurosporine and a potent PKC inhibitor from Hoffmann–la Roche (see Figure 6.5).

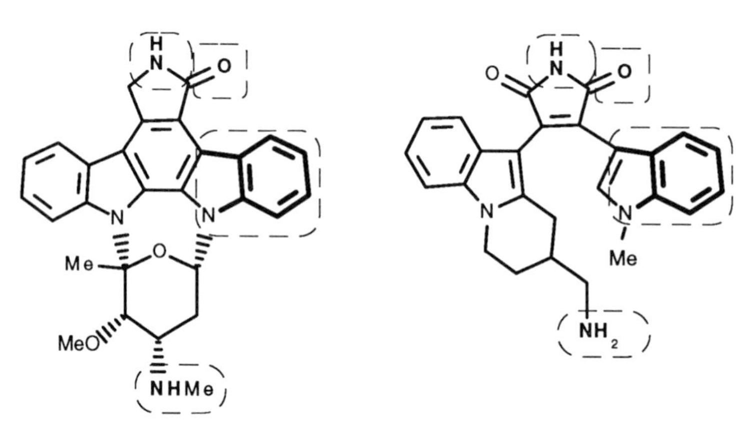

Figure 6.5 PKC inhibitors used to derive the pharmacophoric model.

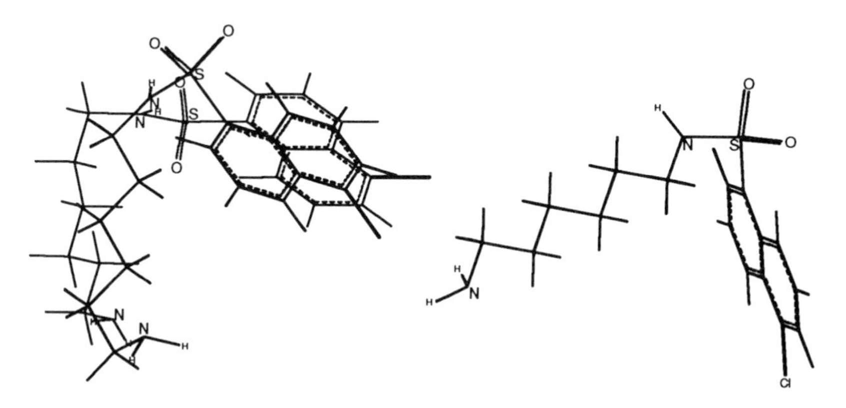

Figure 6.6 Two conformations of W7 from the 3D search, and the CONCORD generated conformation stored in the database.

This search identified many structurally diverse compounds that fit this (partial) PKC pharmacophore, some of which have shown PKC inhibitory activity. One of the compounds identified was a naphthyl sulphonamide (see Figure 6.6); the results database of superimposed hits contained many different conformations which could fit the model, with quite different sulphonamide orientations (220 conformers from search, 1192 from 18 941 conformations keyed during database build). This compound turned out to be W7, a known non-specific inhibitor of PKC, an early result which encouraged us as to the effectiveness of the method. Figure 6.6 illustrates two of the hit conformations found, and the CONCORD generated low-energy conformation for the molecule, which clearly could not fit the model.

6.3.3 *Analysis of matching structures: visualisation and further database analysis*

A 3D database search can produce large numbers of hit structures. If the sole purpose of the search is to identify compounds for a large capacity screen then no further analysis may be necessary. However, in many cases post-search analysis is needed, either to evaluate the quality of the hit or to enable a representative solution to be chosen. The quality of the hit can be defined by many different criteria, including divergence from the desired constraint values or the presence or absence of an additional feature; the concept of a 'pharmacophoric volume' can be envisaged here, defined by the amount of the molecule that is involved in the query relative to the total size of the molecule. The selection of representative sets can be achieved through clustering or partitioning of the results based either on the 3D structures or characteristic 2D substructures. The need to prioritise or organise hits lists produced by 3D searching has recently been addressed by Pearlman (1991, 1994).

Visualisation. The visualisation of hits is an important stage in most 3D searches. In its simplest form this may be a 2D representation of the structure, but is most powerful in 3D with the hit conformation reoriented to a best fit on the query centres. This type of visualisation is best achieved when the 3D searching is well integrated into molecular modelling software. The most useful mode of operation found is where the results are automatically fitted to the query centres, with centres corresponding to the query renamed, and written to a new 3D database. This new database is thus safely stored to disk, and can be readily visualised with simultaneous display of the query, or can be used for new 3D searches. This mode is only used by ChemDBS-3D, although now in SYBYL/3DB-UNITY the results can be visualised fitted to the query and used as the basis for a new search, but are not stored directly to disk; ISIS/3D presents a 3D projection of the hit structures with the query centre distance labelled, but any further analysis other than manipulation in ISIS/Draw requires a SDfile of results to be generated.

The results of a 3D database search can be used as input for the refinement of a query; additional constraints which may slow an initial search such as included and excluded volumes can be added, or distance tolerances refined. For additional analysis using the new conformations of the hit structures, the automatic creation of a new, fully searchable, 3D database is particularly convenient; this is the default in ChemDBS-3D.

6.4 Generation and optimisation of novel leads using 3D databases

6.4.1 *Using ideas databases*

Another use of conformationally flexible 3D database searching is to classify and optimise a set of 'idea' structures for a particular target. Given a potential pharmacophoric model, thorough systematic analysis, including all conformational and substructural possibilities, of a 3D database of designed structures enables the rapid selection of a small set of the most favourable compounds for synthesis. The automation achievable in both the construction of the 3D structures and in their evaluation in different pharmacophoric models means that all substitution and isomeric possibilities can be easily evaluated, an otherwise time-consuming effort.

This capability was investigated in the design of angiotensin II antagonists. Using pharmacophore models already proposed from molecular modelling studies of various non-peptidic angiotensin II antagonists, 3D searches using ChemDBS-3D were used to classify a database of about 1000 idea structures. Simple manipulation of the SMILES codes before 3D structure generation with CONCORD enabled the investigation of structural variations for the idea structures; different chain length, substitution and isomeric possibilities could thus be evaluated in a systematic and automated way. From the 3D searches, with full conformer regeneration, superimposed best fit conformations for each structure were obtained; this facilitated fine tuning of the idea and the incorporation of other features not readily defined in the 3D query. The need to be able to view more than just the first conformation or substructure mapping found during the search became apparent during this work; the systematic method used for conformer regeneration identified many different matches. One of the compounds identified as being of interest was the secondary amide shown in Figure 6.7. This particular compound fitted both the pharmacophoric models used for the 3D searches, where it used different low energy conformations, neither of which were the stored conformation. Shortly after this analysis was carried out, a compound very similar to this amide appeared in a patent as a potent angiotensin II antagonist, another encouraging result for the effectiveness of the method to identify interesting flexible molecules from a database.

6.4.2 *Molecular scaffolds and synthons*

Another use of 3D database searching in drug design is the identification of molecular scaffolds. This can be achieved using the distance-based 3D searching programs already described, or using specialist programs such as CAVEAT (Lauri and Bartlett, 1994) which focuses on the orientation of bonds rather than the location of atoms in the 3D searching; using this approach structures can be identified which match the position and direction of bonds. A major application of these types of 3D searches is in the design of peptidomimetics; potential replacements for secondary structures such as β-turns can be identified from both corporate and public databases, now including massive resources such as the CAS database. The identification of suitable 3D scaffolds or linking groups is of general application in small molecule design and combinatorial screening library design.

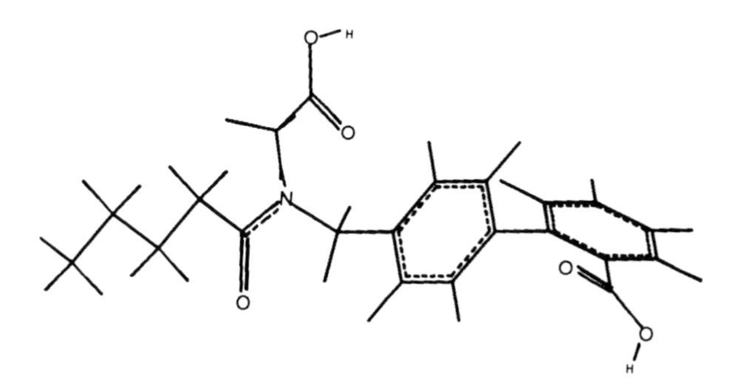

Figure 6.7 Secondary amide found as hit in angiotensin II antagonist 3D searches of ideas database.

6.4.3 *Complementarity-based searching*

Where the 3D structure of an enzyme or receptor is known, or via the definition of a site by projecting complementary properties from a ligand, complementarity-based 3D searching can be applied. The major method used for this is DOCK; the structure-based design of inhibitors of thymidylate synthase (Shoichet *et al.*, 1993) illustrates its use to identify new complementary structures, dissimilar to the enzyme substrate. The method is based on first identifying shape complementarity, using a negative image of the site, and from Version 3 an additional scoring for electrostatic complementarity is also applied. Conformational flexibility is handled by searching multiple conformations of a molecule stored in the database. New methods to generate suitable subsets of conformations for searching are being developed. One such method we are developing is to identify some key pharmacophoric points in the site, and use these to perform a 3D pharmacophoric type search using rule-based conformational analysis and the retention of multiple conformational hits (Lewis, 1993). The resulting 3D database of structures can then be used as input for a DOCK search, to identify those structures and conformers that fit the active site of the receptor. The structures are superposed onto the active site, making the evaluation of the results much simpler. The hit structures may also be used to provide seed ideas for *de novo* design. The advantage of using ChemDBS-3D is that alternate conformers of equal or higher energy that match the query are generated. The advantage of DOCK is that structures that cannot fit into the site (due to bulky substituents) but still fit the query are eliminated rapidly. Each hit also receives a score, including electrostatic complementarity, allowing screening lists to be prioritised. Thus by linking ChemDBS-3D to DOCK, a set of conformations for DOCK analysis can be generated where at least one key interaction distance is fulfilled.

6.5 Conclusion

3D database searching is a powerful tool in the lead generation and optimisation stages of drug design. The ability to search conformationally flexible molecules is a key development in this area, and has been found to yield interesting and useful results, with reasonable search times. 3D database searching has already added many new capabilities to computer-aided drug design. The future promises yet more exciting new possibilities as conformationally flexible searching is further developed; the possibilities include more emphasis on non atom-centred features such as site-interaction points or molecular fields (e.g. electrostatics) in distance-based searching, the incorporation of flexibility into complementarity-based searching, and new tools for hit-list prioritisation including an assessment of free-energy requirements for a fit to the query.

Acknowledgements

The author thanks Dr I.M. McLay of the Dagenham Research Centre (DRC) for his work on the angiotensin II modelling project and in contributing to our development of the use of ChemDBS-3D database searching. The linking of DOCK and Chem-X/ChemDBS-3D is a project with Dr R.A. Lewis (DRC).

References

Blaney, J.M., Crippen, G.M., Dearing, A. and Dixon, J. S. (1990) *DGEOM, QCPE Catalog*, **10**, no. 590.

Boyd, D.B. (1994) Compendium of software for molecular modeling. In *Reviews in Computational Chemistry*, Vol. 5 (eds K.B. Lipowitz and D.B. Boyd), pp.381–428. VCH, New York.

Bures, M.G., Danaher, E., DeLazzer, J. and Martin, Y.C. (1994) New molecular modeling tools using three-dimensional chemical substructures. *Journal of Chemical Information and Computer Sciences*, **34**(1), 218–223.

Chemical Design (1990) ChemDBS-3D. *The Chem-X User*, **4**, 1, 5, 9.

Christie, B.D., Henry, D.R., Wipke, W.T. and Moock, T.E. (1990) Database structure and searching in MACCS-3D. *Tetrahedron Computer Methodology*, **3-6C**, 653–664.

Clark, D.E., Jones, G., Willet, P. *et al.* (1994) Pharmacophoric pattern matching in files of three-dimensional chemical structures: comparison of conformational-searching algorithms for flexible searching. *Journal of Chemical Information and Computer Sciences*, **34**(1), 197–206.

Dalby, A., Nourse, J.G., Hounshell, W.D. *et al.* (1992). Description of several chemical structure file formats used by computer programs developed at Molecular Design Limited. *Journal of Chemical Information and Computer Sciences*, **32**(3), 244–255.

Davies, K. and Upton, R. (1990) Experiences building and searching the Chapman & Hall Dictionary of Drugs. *Tetrahedron Computer Methodology*, **3**, 665–671.

DesJarlais, R.L., Sheridan, R.P., Seibel, G.L. *et al.* (1988) Using shape complementarity as an initial screen in designing ligands for a receptor binding site of known three-dimensional structure. *Journal of Medicinal Chemistry*, **31**, 722–729.

Dolata, D.P., Leach, A.R. and Prout, K. (1987) WIZARD: AI in conformational analysis. *Journal of Computer-Aided Molecular Design*, **1**, 73–85.

Gasteiger, J., Rudolph, C. and Sadowski, J. (1990). Automatic generation of 3D atomic coordinates for organic molecules. *Tetrahedron Computer Methodology*, **3**, 537–547.

Golender, V.E. and Vorpagel, E.R. (1993) Computer-assisted pharmacophore identification. In *3D QSAR in Drug Design: Theory, Methods and Applications* (ed. H. Kubinyi), pp. 137–149, ESCOM, Leiden.

Goodford, P.J. (1985) A computational procedure for determining energetically favourable binding sites on biologically important macromolecules. *Journal of Medicinal Chemistry*, **28**, 849–857.

Guner, O.F., Henry, D.R. and Pearlman, R.S. (1992) Use of flexible queries for searching conformationally flexible molecules in databases of three-dimensional structures. *Journal of Chemical Information and Computer Sciences*, **32**, 101–109.

Haraki, K.S., Sheridan, R.P., Venkataraghavan, R. *et al.* (1990) Looking for pharmacophores in 3D databases: does conformational searching improve the yield of actives? *Tetrahedron Computer Methodology*, **3**, 565–573.

Havel, T.F. (1991) An evaluation of computational strategies for use in the determination of protein structures from distance constraints obtained by NMR. *Progress in Biophysics and Molecular Biology*, **56**, 43–55.

Henry, D.R., McHale, P.J., Christie, B.D. and Hillman, D. (1990) Building 3D structural databases: experiences with MDDR-3D and FCD-3D. *Tetrahedron Computer Methodology*, **3**, 531–536.

Hurst, T. (1994) The directed tweak technique in 3D searching. *Journal of Chemical Information and Computer Sciences*, **34**(1), 190–196.

Jorgenson, W.L. (1991) Rusting of the lock and key model for protein–ligand binding, *Science*, **254**, 954–955.

Kuntz, I.D., Blaney, J.M., Oatley, S.J. *et al.* (1982) A geometric approach to macromolecule–ligand interactions. *Journal of Molecular Biology*, **161**, 269–288.

Lauri, G. and Bartlett, P.A. (1994) CAVEAT: a program to facilitate the design of organic molecules. *Journal of Computer-Aided Molecular Design*, **8**(1), 51–66.

Leach, A.R., Prout, K. and Dolata, D.P. (1988) An investigation into the construction of molecular models by the template joining method. *Journal of Computer-Aided Molecular Design*, **2**, 107–123.

Leach, A.R. and Prout, K. (1990) Automated conformational analysis: directed conformational search using the A* algorithm. *Journal of Computational Chemistry*, **11**, 1193–1205.

Lewis, R.A. (1993) unpublished results.

Martin, Y.C. (1992) 3D database searching in drug design. *Journal of Medicinal Chemistry*, **35**, 2145–2154.

Martin, Y.C., Bures, M.G., Danaher, E.A. *et al.* (1993) A fast new approach to pharmacophore mapping and its application to dopaminergic and benzodiazepine agonists. *Journal of Computer-Aided Molecular Design*, **7**, 83–102.

Mason, J.S. (1992) Experiences with conformationally flexible databases. In *Trends in QSAR and Molecular Modelling* 92: *Proceedings of the* 9th *European Symposium on Structure–Activity Relationships: QSAR and Molecular Modelling, September* 7–11, 1992, *Strasbourg, France* (ed. C.G. Wermuth), pp.252–255, Escom, Leiden.

Mason, J.S. (1993) Drug design using conformationally flexible 3D databases. In *Trends in Drug Research* (ed. V. Claassen), pp.147–156. Elsevier, Amsterdam.

Mason, J.S., McLay, I.M. and Lewis, R.A. (1994) Applications of computer-aided drug design techniques to lead generation. In *New Perspectives in Drug Design* ed. P.M. Dean, C.G.N. Newton and G. Jolles). Academic Press, London. In press.

Moock, T. (1993) Conformational searching in ISIS 3D databases. Presented at the *Third International Conference on Chemical Structures, Leewenhorst Congress Centre, Noordwijkerhout, The Netherlands, June* 6–10.

Moock, T. (1994) Conformational searching in ISIS/3D databases. *Journal of Chemical Information and Computer Sciences*, **34**(1), 184–189.

Murrall, N.W. and Davies, E.K. (1990) Conformational freedom in 3-D databases. I. Techniques. *Journal of Chemical Information and Computer Sciences*, **30**, 312–316.

Murrall, N.W. and Davies, E.K. (1993) Conformational freedom in 3-D databases. In *Chemical Structures II*. Springer-Verlag, Berlin.

Pearlman, R.S. (1987) Rapid generation of high quality approximate 3D molecular structures. *Chemical Design Association News*, **2**, 1–7.

Pearlman, R.S., Balducci, R., Rusinko, A. *et al.* (1988) CONCORD User's Manual, Tripos Associates, St Louis, Missouri 63144.

Pearlman, R.S. (1991) Prioritizing the hits from a 3D search. *Abstracts of the 202nd National Meeting of the American Chemical Society*, CINF20, New York.

Pearlman, R.S. (1994) 3D molecular structures: generation and use in 3D searching. In *3D QSAR in Drug Design: Theory, Methods and Applications* (ed. H. Kubinyi), pp. 41–70. ESCOM, Leiden.

Rusinko, A., Sheridan, R.P., Nilakantan, R. *et al.* (1989) Using CONCORD to generate a large database of three-dimensional coordinates from connection tables. *Journal of Chemical Information and Computer Sciences*, **29**, 251–255.

Sheridan, R.P., Rusinko III, A., Nilakantan, R. *et al.* (1989) Searching for pharmacophores in large coordinate data bases and its use in drug design. *Proceedings of the National Academy of Sciences USA*, **86**, 8165–9.

Shoichet, B.K., Stroud, R.M., Santi, D.V. *et al.* (1993) Structure-based discovery of inhibitors of thymidylate synthase. *Science*, **259**, 1445–1450.

Sprague, P.W. (1991) CATALYST: a computer aided drug design system specifically designed for medicinal chemists. In *Recent Advances in Chemical Information — Proceedings of the* 1991 *Chemical Informational Conference* (ed. H. Collier), pp.107–111. Royal Society of Chemistry, Cambridge.

van Drie, J.H., Weininger, D. and Martin, Y.C. (1989) ALADDIN: an integrated tool for computer-assisted molecular design and pharmacophore recognition from geometric, steric and substructure searching of three-dimensional molecular structures. *Journal of Computer-Aided Molecular Design*, **3**, 225–251.

Weininger, D., Weininger, A. and Weininger, J.L. (1989) SMILES 2. Algorithm for generation of unique SMILES notation. *Journal of Chemical Information and Computer Sciences*, 29, 97–101.

7 Molecular surface comparisons

B.B. MASEK

7.1 Introduction to surface similarity and complementarity

This chapter focuses on the use of molecular surfaces for the comparison of molecules. These molecular surfaces provide a description of the spatial or three-dimensional (3D) characteristics of a molecule. This is important since molecular recognition, the binding of a drug to its target receptor, is inherently a 3D phenomenon. Hence, the comparison of molecular surfaces is a search for the similarity (or complementarity) of the 3D properties of molecules. Because of their 3D nature, these comparisons will be sensitive to changes in the geometry or conformation, and to the relative orientation of the molecules being compared. While simple steric or geometrical surfaces are often used in the search for surface similarity or surface complementarity, other chemical properties such as electrostatic or hydrophobic properties are important and have also been used in surface comparisons.

A number of computational methods for molecular surface comparison have been developed recently and will be reviewed. Many of these methods have been successful in reproducing the alignment of two molecules expected from experimental studies. With a number of well-validated methods now in hand, it is likely that predictions for experimentally unknown systems will be close at hand.

The comparison of molecular surfaces can be conceptually divided into two applications, *molecular surface similarity* and *molecular surface complementarity* (Figure 7.1) Both approaches seek to identify molecular surfaces which closely match and to align molecules to superpose or mesh their surfaces. Molecular surface complementarity provides an approach to the molecular docking problem: given the structures of a ligand and its receptor, can the geometry of the ligand–receptor complex be predicted? Molecular docking is currently a very active area of research. Docking methods have been based on a variety of approaches including molecular surface complementarity, potential energy functions, and geometrical criteria (for a recent review, see Cherfils and Janin, 1993). The latter two are beyond the scope of this chapter and emphasis will be placed on methods which employ molecular surface complementarity as the basis for molecular docking. The complementarity of protein–protein interfaces has been well documented (for examples, see Chothia and Janin, 1975; Chothia *et al.*, 1985; Janin and Chothia, 1990).

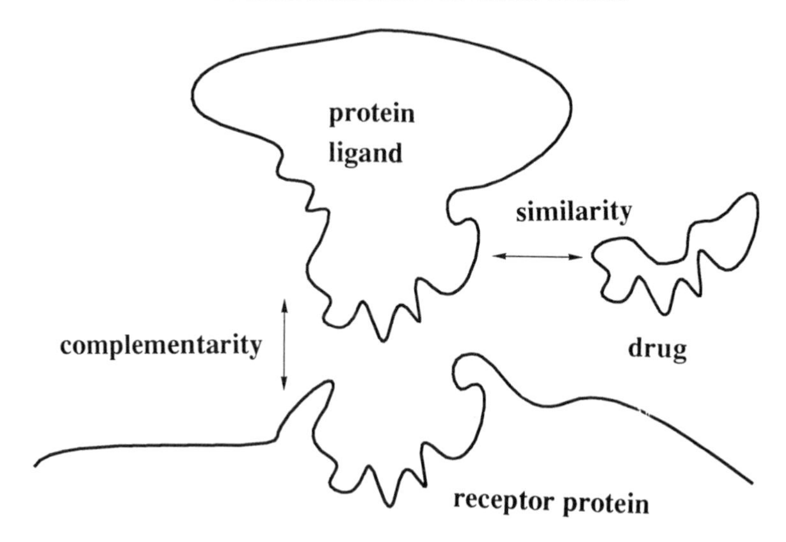

Figure 7.1 Molecular surface similarity versus molecular surface complementarity.

Molecular similarity is directed at the inverse of the molecular docking problem. Medicinal chemists are routinely faced with the task of designing drug candidates in the absence of a 3D structure of the target receptor. By comparison of a set of ligands for the same binding site, the active analog approach (Marshall *et al.*, 1979) attempts to determine the common spatial arrangement of chemical features which may be responsible for recognition and binding. Molecular surface and molecular shape similarity methods can be used as a tool to identify and spatially overlay the similar chemical features of a set of active analogs.

Molecular surface comparison methods can measure either the similarity or the dissimilarity of the surfaces. The two approaches are not necessarily equivalent. Good surface matching may be limited to the binding faces and encompasses only part of the molecular surfaces. Least-squares based approaches to surface matching attempt to minimize the differences between the two surfaces being compared. Areas of surface which mismatch will contribute most to the comparison measurement. A window of interest can be chosen to focus the comparison on a particular face of the molecules (Dean and Chau, 1987; Bacon and Moult, 1992), but this decision may be somewhat arbitrary. Conversely, methods which gauge the extent two molecular surfaces overlap focus on areas which closely correspond and ignore mismatched areas of surface which do not add to the overlap. These methods seek to maximize the similarity rather than minimizing the differences.

7.2 Defining molecular surfaces

Molecular surfaces can be defined in a variety of ways, depending on the property of the molecule which is represented. Rigorous definitions, based on the quantum mechanical wavefunction of a molecule, have been derived (see, for example Mezey, 1990 and Chapter 10). However, different physical properties, such as the molecular electron density or the electrostatic potential, can be derived from the wavefunction leading to different, but equally rigorous, definitions of molecular surfaces. For many molecules of interest, especially large molecules, obtaining the wavefunction may not be practical. Other surface definitions, such as the van der Waals surface, have proven to be useful and practical.

Three types of molecular surfaces are commonly used to represent the steric properties or space occupied by a molecule. These will be denoted as the van der Waals surface, the solvent-accessible (SA) surface, and the accessible molecular (AM) surface. For the van der Waals surface, the molecule is represented by a set of overlapping spherical atoms with atomic radii set to the van der Waals radii. The exposed surface of these spheres is taken as the van der Waals surface of the molecule. For large molecules, such as proteins or nucleic acids, much of the van der Waals surface may be buried within the interior of the molecule. To define the outer, exposed surface, a solvent-accessible surface (SA surface) was originally proposed by Lee and Richards (1971). The SA surface is generated by the center of a probe sphere as it rolls over the van der Waals surface. This is equivalent to increasing the atomic radii of a van der Waals surface by the probe radius. A second type of solvent-accessible surface, denoted here as the AM surface for clarity, was later proposed (Richards, 1977; Connolly, 1983a,b). Unlike the original SA surface, the AM surface is not displaced from the van der Waals surface. The AM surface consists of two parts, the contact surface and the re-entrant surface, and is again generated by rolling a probe sphere over the van der Waals surface. The contact surface is that part of the van der Waals surface which comes in direct contact with the probe. The re-entrant surface is generated by the face of the probe when it is in simultaneous contact with two or more atoms, smoothing over the crevices between atoms. All of these surfaces, van der Waals, SA, and AM, have been represented as either a set of discrete points (dot surface) or as a continuous surface (Figure 7.2).

In addition to the steric properties, it is often important to consider other physical or chemical properties to understand the chemical behavior of the atoms and groups which make up a molecule. For purposes of molecular surface comparison, patches on a steric surface have been assigned to classes based on chemical properties, such as hydrophobicity, or hydrogen-bonding ability (Jiang and Kim, 1991; Masek *et al.*, 1993). Electrostatic potential values have been mapped onto steric surfaces, such as an AM surface, to represent and compare electrostatic properties (Namasivayam and Dean,

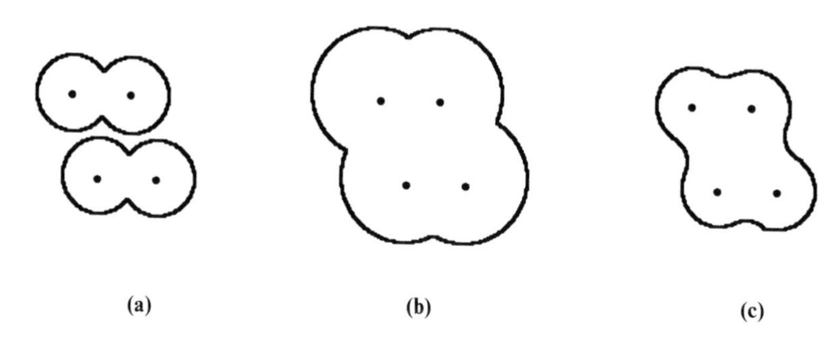

Figure 7.2 Schematic of (a) van der Waals surface, (b) SA surface of Lee and Richards (1971) and (c) AM surface. Redrawn from Connolly (1983b).

1986; Chau and Dean, 1987; Dean and Chau, 1987; Bladon, 1989; Payne and Glen, 1993; Blaney *et al.*, 1993).

7.3 Surface comparison: alignment and conformation

In comparing molecular surfaces for either similarity or complementarity, alignments of the two molecules which maximize the comparison are very often sought. Several different alignments may exist, many of which have high-quality matching of the two molecular surfaces. To specify the relative alignment of two rigid molecules, six degrees of freedom, three translational and three rotational, are required. A means to search this six-dimensional space for optimal alignments is required and a variety of approaches have been developed. Several comparison methods which are independent of molecular alignment have recently been developed (Mezey, 1990; Zachmann *et al.*, 1992, 1993; see also Chapter 6). However, these methods do not give information on the relative alignments of the two molecules that are compared.

Surface comparison is also complicated by the fact that most molecules of interest are flexible and adopt multiple conformations. In molecular docking, conformational changes (induced fit) may occur at the surface of molecules upon complex formation. These changes may be substantial and difficult to predict (for example, see Jorgensen, 1991). When searching for the 3D similarity of a set of flexible analogs, a collection of conformations for each analog need to be considered. Conformational flexibility remains a challenging issue in the application of moelcular surface comparison methods.

7.4 Methods

In the remainder of this chapter, a number of surface comparison methods will be discussed in detail. For each method, a number of issues will be discussed: the intended application of the method, molecular docking (surface complementarity) or molecular surface similarity; the computational algorithm used to represent the surface—is it discrete or continuous?; the algorithm used to compute or quantitate the surface comparison; the methods used to search for and optimize the molecular alignments; and the computational resource requirements. Finally, results obtained with each method will be briefly reviewed.

7.4.1 DOCK and other geometrical methods

Kuntz *et al.* (1982) introduced a geometrical approach to molecular docking based on creating a cast, or reverse image of the binding pockets on a protein. These reverse images consist of a set of overlapping spheres which complement the molecular surface of the binding site. Docking is accomplished by geometrical matching of the spheres in the cast to atoms in the ligand. Many orientations are produced and various scoring functions have been used to prioritize and screen them. These methods have been very successfully applied to protein–small molecule docking (Kuntz, 1992), protein–protein docking (Shoichet and Kuntz, 1991), and docking of a database of ligands to a protein active site (DesJarlais *et al.*, 1988, 1990). Methods for incorporating conformational flexibility into the ligand as it is docked have been presented (DesJarlais *et al.*, 1986; Leach and Kuntz, 1992).

Casts of a protein's indentations were produced by the program SPHGEN (Kuntz *et al.*, 1982) by filling pockets on the surface of a protein with a set of spheres. Starting with an AM surface, spheres were placed so as to touch the surface at two points, with the center of the sphere sitting on a normal to the surface at one of the points. For each surface atom, the largest sphere which does not intersect the surface is generally retained.

Dockings are generated by matching four ligand atoms to four corresponding sphere centers. This has been accomplished efficiently by presorting the distances into bins (Shoichet and Kuntz, 1991; Shoichet *et al.*, 1992). An alternative geometric hashing method has also been found to be computationally very efficient (Fischer *et al.*, 1993; Norel *et al.*, 1994). Many docking orientations, from tens of thousands to millions, were typically generated. These orientations were subjected to a number of filters or screens, including a contact score, an electrostatic score, and a force field score. Grid-based evaluations have been introduced to speed this process (Meng *et al.*, 1992).

These methods have been applied to numerous systems; highlights will be given here to provide a flavor for the scope of this work. These methods have successfully docked the separated components from an X-ray structure of complexes; systems studied include thyroxine docking to prealbumin (Kuntz *et al.*, 1982), methotrexate docking to dihydrofolate reductase, NAD-lactate docking to lactate dehydrogenase, uridine vanadate docking to ribonuclease, and protein–protein systems such as pancreatic trypsin inhibitor docking to trypsin (Shoichet *et al.*, 1992). Other protein–protein systems successfully docked from both bound and unbound conformations include chymotrypsin inhibitor docking to subtilisin, pancreatic trypsin inhibitor docking to trypsin, and ovomucoid inhibitor third domain docking to chymotrypsin (Shoichet and Kuntz, 1991). Finally, application of the methods to screening of databases of molecules for pharmacological testing has produced weak leads (Kuntz, 1992) in several assays.

7.4.2 *Matching knobs and holes*

These methods seek to identify convex or concave features, knobs and holes, of the molecular surface and to match complementary features of two molecular surfaces; knobs of one surface are matched to holes of similar size and shape on a second surface and vice versa. By concentrating on the matching of a small number of prominent features, these methods reduce the search for complementarity from an exhaustive six-dimensional grid (three rotations + three translations) down to a small number of discrete possible matchings. While these methods could be applied to assessing molecular similarity, in practice they have been applied principally to assessing geometrical complementarity as a means to predicting molecular docking.

An initial approach to define surface features was based on locating the features on two-dimensional slices of a molecular surface (Lee and Rose, 1985). The algorithm was applied to two complementary α-helices of flavodoxin protein; for helix 63–75, 14 bumps and 16 holes were identified; for helix 94–106, 13 bumps and 12 holes were identified. It was noted that in the protein, the helix–helix interaction was defined primarily by the matching of a single knob–hole pair and it was proposed that docking the two helices would require considering only $(14 \times 12) + (13 \times 16) = 376$ possible matches. The method did not consider the relative size or shape of the knobs and holes identified.

The concepts of solid angle and local volume were introduced by Connolly (1986a) to describe the size and shape of the knob–hole and were used for the docking of proteins in protein–protein complexes (Connolly, 1986b). A number of criteria were derived by which the size and shape complementarity of knob–hole pairs could be judged. The local volume was defined by placing a sphere (6 Å radius) at points distributed on the molecular surface of the protein and calculating the volume of the sphere which was buried by the protein at that point. A point with a local volume greater (less) than all neighbors was identified as a hole (knob). For complementary knob–hole pairs, the sum of the local volumes of the pair were required to be close to the total volume of the sphere (905 Å^3). Furthermore, vectors were defined from the centroid of the knob's or hole's local volume to the sphere center. For a matched knob–hole pair, these vectors were required to point in roughly the opposite direction (see Figure 7.3).

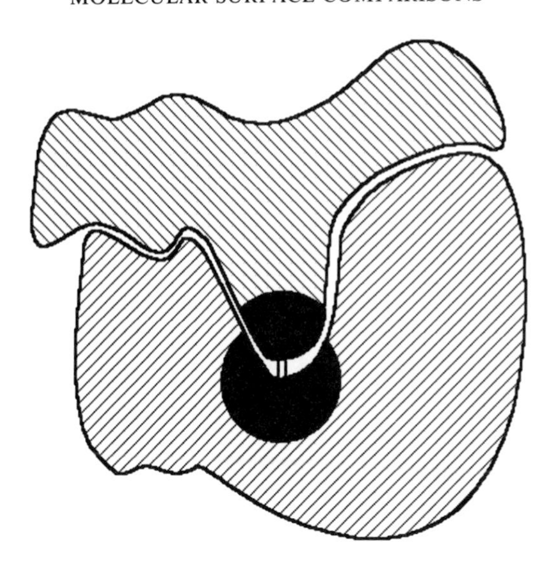

Figure 7.3 Complementarity of a knob–hole pair. The local volumes are shown unshaded. The sum of the local volumes are approximately equal to the total volume of the sphere. Complementary antiparallel normals were also used to discern complementarity. Redrawn from Connolly, 1986b.

The solid angle was also used to define a criterion for assessing the size and shape complementarity of a knob–hole pair. The solid angle was defined by placing a sphere at points distributed on the molecular surface of the protein and calculating the surface area of the sphere that was buried by the protein; this area was normalized to a range of 0 to 4π by dividing the buried area by the sphere radius squared. To get a better sampling of the shape, five concentric spheres (radii of 4, 5, 6, 7, and 8 Å) were used for each feature to give a set of solid angles for each knob or hole. For a complementary knob–hole pair, the sum of the solid angles at given radii should be approximately 4π, and the values should be negatively correlated.

Once complementary knob–hole pairs were identified, docking was accomplished by matching, via rigid body overlap, the four knob–holes of one surface with four hole–knobs of the other surface. A tree pruning algorithm was used to avoid comparing all possible quartets and enhance efficiency. An rms separation of < 1.0 Å was required for the matching of the four pairs of points. Finally, all atoms in the docked complex are checked for collisions (atoms separated by < 2 Å). If more than 15 collisions are found, the docking was rejected.

The method was successfully applied to the docking of the α and β subunits of horse methemoglobin. The α subunit had 168 knobs and holes, the β subunit had 165. This yielded 23 dockings before collision checking. Only three dockings met all the criteria. The surface area buried in each of these complexes was calculated and the correct docking was found to have the highest buried area. A total of about 17 central processor unit hours on a VAX-11/750 were required for the docking. The method was unable to dock trypsin and trypsin inhibitor in a reasonable amount of computer time. Finally, a number of refinements to the method have been recently described (Connolly, 1992).

Wang (1991) has modified this approach by requiring only a single knob–hole pair to be matched. The matching of knobs with holes was used to generate a set of initial docking alignments which were then refined using a procedure based on a double-layered surface to score and optimize the alignment of the docked proteins. A detailed description of the method and results are presented in section 7.4.8.

7.4.3 Least-squares methods

The methods grouped under this heading work by constructing a pattern of points on the molecular surfaces to two molecules to be compared, and superimposing the surface patterns based on least-squares best fit criteria. They are computationally very efficient. These methods have been successfully applied to both protein–protein and protein–small molecule docking problems.

Bacon and Moult (1992) developed a method based on a least-squares comparison of web or net-like patterns of points interpolated on the AM surfaces of the two molecules to be docked. These webs preserve the essential notion of distance and map surface regions consistently. Web points were arranged on a pattern of rings which were separated by a radial distance of approximately 1 Å and typically extended to 10–15 Å in radius. For ring k, the number of points in the ring was the nearest integer to $2\pi k$. A complex heuristic was used to decide the position of the points. For the protein, the user must decide the location of the binding site and the position of the center of the pattern. For the ligand, a large number of patterns were generated, each centered at a different point on the surface of the ligand and each was compared with the active site pattern. A list of the best matches, based on rms fit, was kept. A round of coarse sampling at 1 ligand pattern center/Å^2 was followed by fine sampling at 4 patterns/Å^2 in the neighborhood of the coarse solutions. The relative twisting of the two patterns defined the single remaining degree of freedom which was searched by generating active site patterns that differed by a twist angle increment.

The pattern matchings were evaluated by least-squares fitting. A Gaussian weighting scheme was used such that the outer points in the pattern contribute roughly 10% relative to the center. Every other ring was used for the fitting. In addition, not all parts of the patterns were expected to match, because the shape of the interface was not necessarily circular. To overcome this problem, pairs of points which are most distant from each other, after superimposition, were discarded. In the coarse pass, the 40% were discarded and the remaining subset were rematched. In the fine pass, four rounds of pruning were performed, with 25% of the points removed at each pass. Final solutions were filtered with a clash test and a test of the electrostatic interaction energy.

A number of systems were studied by docking the separated components from an X-ray structure of complexes. The method correctly docked methotrexate to dihydrofolate reductase, antibody fragment HyHel5 with hen egg-white lysozyme, lysozyme with a trisaccharide inhibitor, protease B from *Streptomyces griseus* with the third domain of turkey ovomucoid inhibitor, and bovine pancreatic trypsin inhibitor to trypsin. Dockings with the lowest electrostatic score were within 1–2 Å rms of the crystallographically observed geometries. When alternate geometries, obtained from sources other than X-ray structure of the complex, were used results were again generally good. However, one failure, the docking of uncomplexed trypsin to BPTI from PDB entry 6PTI, was noted. Ampicillin was properly docked to β-lactamase only after inclusion of a crucial active site water. Dockings were generally completed in 30–90 minutes of central processor unit time on an SGI Personal Iris workstation.

Lin *et al.* (1994) used a geometric hashing paradigm, adapted from computer vision techniques, to find the geometrical correspondence of critical points on two AM surfaces. Connolly's (1983b) smooth representation of the AM surface was used. Each convex, concave, or saddle-shaped face of the surface was characterized by a point and a normal. Some pruning was used to remove points that were too close together or covered a face that was very small or very large. This gave rise to 40–60 critical points on a small molecule ligand or about 200–600 critical points on the surface of a protein. A modified geometric hashing paradigm (Nussinov and Wolfson, 1991; Bachar *et al.*, 1993; Lin *et al.*, to be submitted) was used to rapidly identify the matching of these points and normals on two surfaces. The location of these points on the surface was found to give good correspondence in experimentally determined structures. Nine crystallographic complexes were surveyed; at the interfaces, 261 to 471 pairs of critical points were observed to match within 1 Å. The normals of these pairs were antiparallel, at angles of $156.1 \pm 15.0°$. The docking of several complexes were studied using the separated components from an X-ray structure of complexes. The systems studied were heme docking to myoglobin, methotrexate docking to dihydrofolate reductase, hemoglobin α subunit docking to the β subunit, pancreatic trypsin inhibitor docking to trypsin, and ovomucoid third domain docking to chymotrypsin. Solutions with good fits (rms deviation of 0.25–1.0 Å) were found; however, the total number of docking solutions was not stated and no method was given for separating the correct dockings from the incorrect dockings produced by the method. Dockings were computationally very efficient; typical central processor unit times on an SGI-Indigo workstation totaled 2–5 minutes.

7.4.4 *Surface comparison by projection onto a plane*

Topographical maps of molecular surfaces, obtained by projecting a molecular surface onto a plane, have been used to find molecular surface complementarity. Early graphical methods (Greer and Bush, 1978) were used to visualize the complementarity at protein–protein interfaces. Zielenkiewicz and Rabczenko (1984) projected the chemical properties (hydrophobic, positively charged, negatively charged, etc.) of a surface layer of atoms onto a two-dimensional grid; however, topographical information was not represented by the grid. By counting the number of grid points with coincident properties, insulin dimer was correctly docked. Finally, Walls and Sternberg (1992) proposed a multistage docking method; the initial stage involved the matching of topographical surface maps. However, the surface matching algorithm generated more than 200 000 possible dockings for lysozyme complexed with its antibody; these required clustering and further screening.

7.4.5 *Surface comparison by gnomonic projection*

The gnomonic projection method for the comparison of molecular surfaces was introduced by Dean and coworkers (Chau and Dean, 1987; Dean and Chau, 1987; Dean and Callow, 1987; Dean *et al.*, 1988; Dean, 1990). These methods have been applied to shape comparison of dissimilar 'small' drug molecules and to the comparison of two proteins.

Gnomonic projection involves mapping the properties of the molecular surface onto a set of nearly equally spaced points on a sphere. These points are generated by tessellation of a regular polyhedron; an icosahedron is typically used. The sphere, which is large enough to contain the molecule, is placed at the center of interest (COI) which is typically the centroid of the molecule. A ray from the COI is projected to each point on the sphere and these rays meet the molecular surface at the pierce points (Figure 7.4). The properties at the pierce points are then projected onto the corresponding points on the sphere. Typically, a van der Waals surface is used to calculate the position of the pierce points. A potential problem arises if the ray passed through the molecular surface more than once. Because only the outermost pierce point is used, information about portions of the surface can, in theory, be lost.

To represent the steric properties of a molecule, the distance along the ray from the COI to the pierce points is mapped as the property which represents the surface (Chau and Dean, 1987). The value of the electrostatic potential at the pierce points is used to represent the electrostatic properties of molecules. Alternative representations of these properties have also been examined (Bladon, 1989; Payne and Glen, 1993; Blaney *et al.*, 1993).

Several functions have been employed to quantitate the similarity (or dissimilarity) of the gnomonic projections of two surfaces, including Spearman's

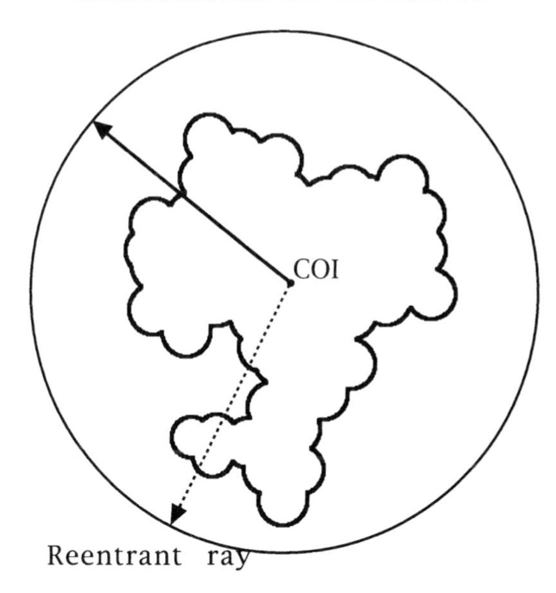

Figure 7.4 Gnomonic projection onto the surface of a sphere. Rays projected from the center of interest (COI) meet the molecular surface at the pierce point. Re-entrant rays pierce the surface in multiple places; typically only the outermost is used.

rank correlation coefficient, Pearson's product moment correlation coefficient, and most commonly, the sum of the squares of residual errors. The latter two have the advantage that they are continuous. The sum of squares of the residual errors is given by:

$$f(x) = \sum_{i=1}^{n} (P_{iA} - P_{iB})^2 \tag{7.1}$$

where P_{iA} is the value of the property for point i of molecule A, with a corresponding P_{iB} for molecule B. The sum can be over any subset of the entire set of points; for example, a hemisphere of points has been used. This will focus the comparison to one face of the molecule. The matches produced, however, will be dependent on the sizes and number of the faces over which the comparison is made.

The relative orientation of the two molecules must be searched in order to identify alignments which maximize the surface similarity or minimize the surface dissimilarity. The simplest approach involves graphically displaying a map of the residual errors and allowing the user to adjust the molecular orientations in realtime. This approach has been applied to an analysis of β-lactam antibiotics and to the comparison of phosphodiesterase inhibitors (Blaney *et al.*, 1993). Both electrostatic and steric properties were used in the comparisons. Differences in the details of the comparisons were noted upon changing the COI.

Searching procedures allow the comparison to be automated and ensure a more systematic exploration of the molecular alignments. A procedure examined by Dean *et al.* (1988) starts with a large number of random orientations of the two molecules to be compared. These orientations are subjected to short bursts of minimization (40 evaluations of the similarity); the molecular alignment is adjusted to minimize the residual error defined above. Similar orientations are grouped together via a cluster analysis of the Eulerian angles. From each cluster, the orientation with the minimum residual error is subjected to prolonged minimization. Typically, 20 clusters were found to be adequate. This method was applied to the comparison of two neurotoxins, tetrodotoxin and saxitoxin. Both electrostatic and steric properties were used in the comparisons. A hemispherical window, encompassing either 25 or 90 points, was used to define a face over which the dissimilarity was measured. The orientation of both molecules, with respect to the window, was searched. No indication of the central processor unit time required was given, however.

Bladon (1989) explored a systematic procedure for searching molecular alignments. The procedure makes use of the symmetry of the icosahedron and dodecahedron to enhance the efficiency of the search. For example, there are 60 equivalent orientations of an icosahedron. Furthermore, these can be thought of as simple permutations of the indices of the points to be compared. This eliminates the need for recalculation of pierce points or properties for these orientations and effectively reduces the rotational space which must be systematically searched. The limits on the rotations were -90 to $+90$, -30 to $+30$, and -36 to $+36°$ about the x-, y-, and z-axes, respectively. Two grid resolutions were employed; the first involves 10° increments about the x- and y-axes and 9° increments about the z-axis. A finer 2° grid was also used. Both steric and electrostatic properties were used for the comparisons; 32 points (icosahedron + dodecahedron) were projected over the entire surface of the sphere. The systems studied include the comparison of tetrodotoxin and saxitoxin (as above), and also the comparison of two proteins, chymotrypsin and trypsin. Central processor unit times on a VAX 8650 varied from 30 seconds to 2 minutes for the small molecule comparisons to 15 minutes for the comparison of the proteins.

Finally, the genetic algorithm has been used as a method to search for optimal alignments based on gnomonic projection of surface properties (Payne and Glen, 1993). Conformational flexibility and relative orientation were varied simultaneously in the search procedure. Rigid body rotation and translation, rigid rotation about flexible bonds, and corner flipping in rings were all coded in a binary 'genome'. A population of genomes corresponds to a set of different conformation/alignments. At each generation, genetic operations such as crossover and mutation were used on these genomes to generate new orientations/conformations. The residual errors of electrostatic and steric properties were combined into a fitness function. Selection, based on this fitness, was used to evolve the population of high-quality alignments. The effects of changing population size, mutation probability, and number of generations on the convergence of the surface comparison were examined by self-comparison (comparing a molecule with itself). For example, for the self-comparison of 2-deoxyribose, 129 projection points were used for the gnomonic projection over the surface of a sphere. Four rotatable bonds, plus rigid body rotation and translation were combined to give ten degrees of freedom which were searched. Under optimal conditions (500 individual genes, 300 generations, mutation rate of 0.005) 2.5 central processor unit hours were required on an IBM RS6000/320 workstation. By the end of the simulation, the moving molecule was overlaid on the static molecule as expected. A number of other comparisons were also examined, including the comparison of benzodiazepine receptor ligands, the fitting of Leu-enkephalin to morphine, and the comparison of GABA analogs.

7.4.6 *Surface complementarity via projection onto a cylinder*

The PUZZLE method seeks to identify molecular surface complementarity by comparison of molecular surfaces projected onto a cylinder (Helmer-Citterich and Tramontano, 1994). The cylindrical projection of a molecular surface is represented by a two-dimensional matrix. The method is conceptually similar to the spherical gnomonic projection discussed previously. The difference is that the cylindrical projection uses an axis as the COI while the gnomonic projection uses a single point as the COI. The method is directed at finding large regions (e.g. 15 Å × 15 Å) of geometrical or steric complementarity and has been successfully applied to reproduce the docking of several protein–protein complexes. The results were reasonably insensitive to the source of geometries for the protein components.

The process of projecting the molecular surface onto a cylinder begins by placing the geometrical center of the molecule on the cylinder axis, which was arbitrarily chosen to coincide with the z-axis. A dot representation of the AM surface, generated at a density of 2 dots/Å^2, was chosen to represent the molecular surface. The cylinder and surface were then cut into slices orthogonal to the z-axis; 2.0 Å thick slices were used. Each slice was mapped to a row of the two-dimensional matrix. For each slice, rays were projected from the center of the slice, dividing the molecular surface into segments of constant area (4.0 Å^2). Note, the number of segments varied from slice to slice. Each segment corresponds to an element in the row representing the slice. The angular positions of each segment were stored for every slice. For each segment i, the distance from the surface points to the center of the slice were computed and the largest distance, R_i, is chosen. While this distance could be used to represent the surface, it is dependent on the choice of the

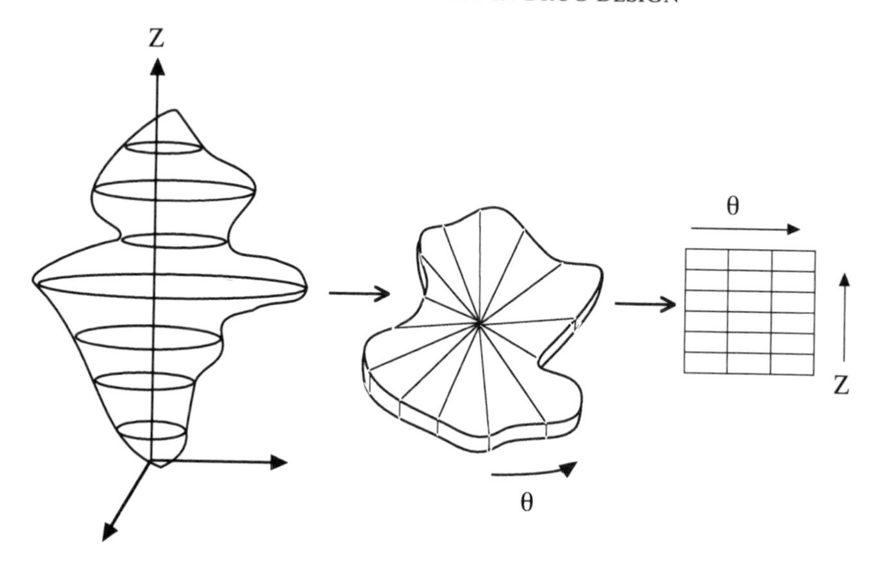

Figure 7.5 Projection of a molecular surface onto a cylinder. Surface is divided into slices. Slices are divided into segments. Segments are mapped to a two-dimensional matrix. Redrawn from Helmer-Citterich and Tramontano (1994).

COI. Instead, for each segment, the difference $(R_i - R_{i-1})$, which measures the radial change of the surface from segment to segment is stored in the matrix and used to represent the molecular surface (Figure 7.5).

Two molecular surfaces were compared by searching the 2D matrices, obtained by cylindrical projection of each molecular surface, for complementary regions. In the first step of this comparison, each row of the matrix for one molecule (target) is compared with every row of the matrix for the second molecule (probe). For this row comparison, complementarity was computed over a sliding window of 8–12 segments. Allowing a sliding window is equivalent to rotating the molecules about the cylinder (z) axis. Windows were judged to be complementary if the distance, as measured by the difference of the matrix elements, fell below a threshold value. The region of complementarity could be extended horizontally by combining or joining sets of windows judged to be complementary. Finally, complementarity was extended vertically to other slices by joining complementary regions corresponding to the same angular positions. Regions of complementarity which extend over 15 Å in each direction were retained as possible solutions to the protein docking problem.

Because the molecular surface is poorly described at the poles of the cylinder, the probe protein is rotated about the x- and y-axes to generate multiple orientations so each region of its surface will be projected optimally at least once. A coarse 10° grid search was performed, followed by a focused

1° search over a 20° × 20° region around the solutions from the coarse grid search. Solutions from the fine run were examined and those with van der Waals collisions involving the main chain atoms were discarded. Solutions were ranked based on the surface area buried upon formation of the complexes. For the target protein, information on the location of the active site was used to determine optimal placement relative to the cylinder used for projection. Two orthogonal cylinders could be used to ensure proper projection of the target protein (see below).

Two example problems were examined with the method, the docking of bovine pancreatic trypsin inhibitor (BPTI) to trypsin, and the docking of hen egg white lysozyme with HyHel5 antibody. Each problem was studied with several different geometries for the proteins used in the docking calculations. Geometries were taken from the X-ray structure of the co-crystallized complexes, from X-ray structures of unbound components, or from modelled structures of at least one component. In each case, three or four solutions to the docking problem were typically generated. In all but one case (model of HyHel10 model docked lysozyme) a solution was identified near the correct crystallographically determined geometry for the complex. This solution always ranked either first or second in terms of surface area buried upon complex formation. For these computationally docked complexes, the rms deviation for all atoms of the probe molecule were in the range of 1.7–5.4 Å from the X-ray structure. Central processor unit times of 3–24 hours on an SGI-4D320 were required for a single docking run.

7.4.7 *Fourier analysis of surfaces*

Three-dimensional Fourier analysis has been proposed as the basis for describing and comparing molecular surfaces (Leicester *et al.*, 1988). The surface model used was the AM surface calculated using Connolly's (1981) MS program. An interpolation scheme was used to convert the surface from a Cartesian to a polar coordinate system to give a function $r(\theta,\phi)$. In order to properly represent the molecular surface, this function must be single valued. That is, a ray from the origin in the polar direction θ,ϕ should cross the surface only once. A large probe radius (6.4 Å) was used to circumvent this problem on the test molecule hemerythrin B. The function $r(\theta,\phi)$ is then represented by a series of spherical harmonics $Y_l^m(\theta,\phi)$ which is given by:

$$r(\theta,\phi) = \sum_{l=0}^{+\infty} \sum_{m=-l}^{+l} a_{lm} Y_l^m(\theta,\phi) \tag{7.2}$$

The authors found for hemerythrin B that values of $l \geqslant 20$ give reasonable representations of the surface (rms) error $< 1.0\%$). Solving for the coefficients with $l = 50$ took 9 hours on a VAX8600. This process grows as $(l + 1)(l + 2)/2$.

The expansion coefficients, a_{lm}, represent a description of the molecular shape. The magnitude of the difference in these coefficients was proposed as a method to quantitate the comparison of the shapes of two molecular surfaces; however, no examples were given. The use of discrete Fourier transforms to compare a grid-based surface representation (Katchalski-Katzir *et al.*, 1992) is detailed in the next section.

7.5 Methods utilizing surfaces with thickness

In attempting to compare two molecular surfaces, there may be small mismatchings which are acceptable. Defining molecular surfaces with thickness allows a means of including these tolerances in the comparison process, the common conceptual link to the methods grouped under this heading. Perhaps because of this tolerance, these methods have been successful in docking a number of protein–ligand complexes, are tolerant of changes in the geometry of the docking partners, and have also been successfully used to identify similarities between a protein and its small molecule mimic.

A number of methods are detailed in this section. A double-layer discrete representation of a protein surface was first introduced by Wang (1991). Two concentric dot surfaces, an inner and an outer surface, were used to quantitate and optimize the complementarity of two protein surfaces. A cube representation of the molecular surface, introduced by Jiang and Kim (1991), was used to identify molecular surface complementarity and enable the 'soft' docking of both small ligands and larger proteins to their binding sites. Katchalski-Katzir *et al.* (1992) also made use of a discrete grid to represent protein surfaces with thickness and accelerated the comparison by using discrete Fourier transforms. This method was successful in docking both small ligands and larger proteins to their binding sites. Masek *et al.* (1993) derived an analytical method for comparing molecular surfaces with thickness, which were dubbed 'molecular skins'. Two continuous surfaces were used to define the inner and outer boundaries of the molecular skin. The method was applied to identifying surface similarities for molecules of differing sizes such as matching proteins with their small molecule mimics. The details of each of these methods are given below.

7.5.1 *Using a discrete double surface for surface complementarity*

This method (Wang, 1991) is based on finding the geometrical complementarity of two steric surfaces as a means of docking two proteins. The method starts by calculating two dot surfaces for the probe molecule, which is to be docked to the target molecule. A SA dot surface was used for the outer surface. By moving each point in the SA surface inward toward the center of the atom that generated it, a van der Waals inner surface was generated. This resulted in a dot density of about 6 dots/$Å^2$ for the inner surface. The object is to find dockings in which the outer surface of the probe overlaps with the target molecule, while avoiding overlaps of the inner surface of the probe with the target molecule. The complementarity is calculated as Score $= 2N_i - N_o$, where $N_i(N_o)$ is the number of points on the inner (outer) surface which are buried by the target molecule. Initial alignments of the two molecules were optimized with respect to rigid body rotation and translation using 1000 steps of a Monte Carlo random walk procedure. Because of the discrete nature of the scoring function, more efficient gradient optimization procedures cannot be used.

A clever procedure, based on matching topological features of the molecular surface, 'knobs with holes' (see section 7.4.2), was used to generate a set of initial docking alignments for refinement. A knob (hole) was defined by placing a sphere (6 Å radius) at each point on the van der Waals dot surface of the probe or target protein and calculating the local volume, V_r, the volume of the sphere which is buried by the protein at that point. A point with a local volume greater (less) than all neighbors within 4 Å was identified as a hole (knob). By matching a single complementary knob–hole pair, three translational degrees of freedom are fully determined. Two rotational degrees of freedom can be removed by requiring that the vector from the knob to its centroid be antiparallel to the vector from the hole to its centroid. The remaining degree of freedom, the twist of the probe protein around the centroid/centroid axis, was searched in 18° steps to generate 20 initial dockings for optimization for each knob–hole pair.

A knob–hole pair are complementary if the local volume of the hole plus the local volume of the knob is similar to the total volume of the sphere used to measure local volume; that is:

$$0.85 < (V_r(\mathrm{A}) + V_r(\mathrm{B}))/(\tfrac{4}{3}\pi r^3) < 1.05$$

for $r = 4, 5, 6, 7, 8, 9$ Å where A and B are the probe and target molecules, respectively. A similar criterion, based on local surface area, was also used.

The method successfully docked the α_1/β_1 dimer of hemoglobin as well as docking trypsin to bovine pancreatic trypsin inhibitor. For each docking, a combination of approximately 20 hours' central processor unit time on a MicroVax and 60 hours' time on an ST100 was required.

7.5.2 *Matching molecular surface cubes*

The 'Soft Docking' method (Jiang and Kim, 1991) finds the complementarity of two molecular surfaces, based on both geometrical and chemical properties, as a means of docking two molecules. The method is complex and will only be broadly outlined here. To calculate the complementarity, the AM dot surface was used along with surface normals attached to each point on the surface (1.0 or 0.3 dots/Å^2 for fine or coarse matching, respectively). The probe and target molecules were each placed in a cubic grid (1.4 or 2.8 Å step size). Grid cubes which contain one or more points from the dot surface were designated as surface cubes, all other cubes on the interior of the molecule were volume cubes. This cube representation of the surfaces acts to soften the matching critera, allowing for some degree of mismatching. For a given orientation of the probe and target molecules, overlaps of surface and volume cubes were checked. In the overlapping surface cubes, each probe point was paired with every target point. The angle between the associated pair of normals was calculated and must fall within a tolerance (30° or 50°) of antiparallel to be scored. In this way, local surface complementarity was ensured. The number of overlapping volume cubes was used to exclude those alignments which have large sterically overlapping regions. If the number of overlapping volume cubes exceeds a tolerance (50–220), the orientation was discarded.

To find the optimal docking alignment, a rotational/translational search procedure was used. Between 8280 and 28 043 rotational orientations were sampled. For each orientation, a procedure was used to quickly and efficiently identify optimal translations, taking advantage of the grid nature of the surface representation and allowing all translations to be recorded in a single pass over pairs of cubes.

The method, as described so far, involves only geometric complementarity of the surfaces. Chemical complementarity was used as a final step to enhance the signal of the correct solutions. Six categories of surface points were assigned, positively charged, negatively charged, hydrogen-bond donor, hydrogen-bond acceptor, polar, and hydrophobic, depending on the type of atom which generated the surface point. Interactions between a pair of points in an overlapping surface cube will be either favorable (added to score) or unfavorable (subtracted) depending on the properties of the points. The authors found that most incorrect dockings were eliminated when chemical complementarity was included in the scoring.

A number of systems were studied by docking the separated components from an X-ray structure of complexes. Based on purely geometrical matching, the method correctly docked NADPH to DHFR (dihydrofolate reductase), methotrexate (MTX) to NADPH–DHFR complex, and docked trypsin inhibitor to trypsin. Approximately 5 central processor unit days on a VAX/785 were required for this docking. Understandably, the method failed to dock MTX to DHFR itself. The binding of MTX and NADPH to DHFR is cooperative. When the structures of isolated trypsin and trypsin inhibitor were used, the correct binding orientation was still located, but was one of many geometrical solutions and one of nine solutions with a consistently high score based on chemical complementarity. A similar result was obtained for the docking of the isolated structure of lysozyme with lysozyme antibody.

To find the geometrical complementarity of two molecular surfaces, Katchalski-Katzir *et al.* (1992) also use a discrete grid-based representation of the molecular surface and interior. Each of the two molecules to be docked was placed in a three-dimensional grid. For the moving probe molecule's grid, any grid point on the interior (within 1.8 Å of an atom in the probe molecule) was assigned a value of +1; all other grid points were assigned a value of zero. For the stationary target molecule's grid, grid points on the interior were assigned a value of −15. For a layer of grid points within 1.5–2.5 Å of the surface of the target molecule, the value is reset to +1. All other grid points were assigned a value of zero. The product of these grids is positive where the probe molecule overlaps the surface of the target molecule, but negative when the probe overlaps too deeply with the target. A surface layer with thickness allows for some tolerance to small conformational changes, and lack of explicit hydrogen's in the structures.

For a given orientation of the probe and target molecules, the matching of the surfaces was calculated using a correlation function, $\bar{c}_{\alpha,\beta,\gamma}$, between the target molecule's grid, $\bar{a}_{l,m,n}$ and the probe molecule's grid, $\bar{b}_{l,m,n}$, as:

$$\bar{c}_{\alpha,\beta,\gamma} = \sum_{l=1}^{N} \sum_{m=1}^{N} \sum_{n=1}^{N} a_{l,m,n} \times \bar{b}_{l+\alpha,m+\beta,n+\gamma} \tag{7.3}$$

where α, β, γ are the number of steps the probe molecule is shifted with respect to the target molecule. A discrete Fourier transform was used to accelerate the calculation of the correlation function. Positive peaks in $\bar{c}_{\alpha,\beta,\gamma}$ correspond to translational shifts where the surfaces of the target and probe molecule are well matched. To complete a general search, all orientations about the three Euler angles of the probe molecule were scanned in 20° steps; this resulted in 2916 orientations for the correlation function being evaluated using a coarse grid step size. Typically the 10 orientations which yielded the highest peaks in the correlation function using the coarse grid (1.0–1.2 Å grid steps) were recalculated with a finer grid (0.7–0.8 Å grid steps). For matching two molecules (1100 atoms each), a total of about 7.5 central processor unit hours on a Convex C-220 computer was required.

The method was successful in docking the separated components from X-ray structures of complexes. Systems successfully docked include the α and β subunits of human deoxyhemoglobin, docking the α and β subunits of horse methemoglobin, docking tyrosinyl adenylate to tRNA synthetase, docking a peptide inhibitor to aspartic proteinase, and docking trypsin to trypsin inhibitor. When the native structure of aspartic proteinase was used, the correct docking was again obtained. However, when the native structure of trypsin inhibitor was used, the method was unable to find the correct docking with trypsin.

7.5.3 Comparison of molecular skins

Masek *et al.* (1993) derived an analytical method for comparing molecular surfaces with thickness, which were dubbed 'molecular skins'. The method was developed to identify surface similarities for molecules of differing sizes such as matching proteins with their small molecular mimics. Two continuous surfaces were used to define the inner and outer boundaries of the molecular skin. For the inner surface, both van der Waals and SA surfaces were explored. The outer surface of the skin was obtained by increasing all atomic radii by 0.4 Å. An analytical expression was used to calculate the intersection volume of two molecular skins, $V(\alpha \cap \beta)$, and is given by:

$$V(\alpha \cap \beta) = V(A^* \cup B) + V(A \cup B^*) - V(A \cup B) - V(A^* \cup B^*) \quad (7.4)$$

where A represents the inner surface of one molecule's skin, and A* represents the outer surface and similarly for the second (B) molecule, and where, for example $V(A^* \cup B)$ is the volume of A* union B. These union volumes were calculated analytically using the method of Connolly (1985).

The method was tested by comparing two inhibitors of the enzyme elastase, turkey ovomucoid inhibitor (TOMI), a 56-residue peptide, and a difluoroketone inhibitor (DFKi). The X-ray structures of each of these inhibitors, complexed with elastase, were used to generate an alignment of the complexes based purely on aligning the enzyme atoms. This generated an 'experimental' alignment of the two inhibitors. Each inhibitor was removed from the complex, in the bound geometry, and the elastase coordinates were discarded. The two inhibitors were then subjected to skin comparison to see if the experimental alignment could be regenerated. While a method for incorporating chemical properties into the skin comparison was proposed, only geometric similarity was examined (Figure 7.6).

Two situations were examined with the skin comparison method. First, a TOMI fragment, the binding loop Ala-15–Tyr-20, was excised from the TOMI inhibitor and compared to the DFKi inhibitor. In the comparison, the relative orientation of the two molecules was searched in order to identify alignments which maximized the skin overlap. The centers of geometry for DFKi and TOMI fragment were superposed and random rotations were used to generate starting alignments. Each starting alignment was then optimized by maximizing the overlap to the two molecular skins with respect to both rotation and translation. Analytical first and second derivatives of the skin intersection with respect to rotation and translation were derived

	P_4	P_3	P_2	P_1	P_1'	P_2'
TOMI	Ala15	Cys16	Thr17	Leu18	Glu19	Tyr20
DFKi	Ace	Ala	Pro	Val	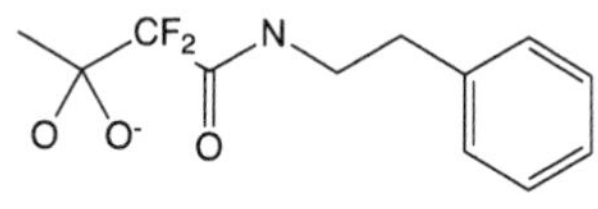 CF_2 N O O^- O	

Figure 7.6 Schematic of the 'experimental' aligment of two elastase inhibitors. Only the binding loop of TOMI is given.

and used with a BFGS variable metric optimization algorithm. The experimental alignment was found and had the highest molecular skin intersection among the many local maxima located.

Next, a comparison of DFKi with the entire TOMI protein was carried out. DFKi was placed at random points on the surface of the TOMI protein and random rotations were used to generate starting geometries. Optimization of 172 starting points, focused within 9.5 Å of the binding loop, yielded the experimental alignment three times. This alignment had the highest score (skin intersection) among the local maxima identified. These calculations required 12 central processor unit days on an SGI-R4000 processor. Further searching of 192 starting points over the entire TOMI surface failed to generate any additional high-scoring matches. The method was successful in discriminating the proper matching of DFKi to the TOMI protein.

Molecular dynamics was used to relax the TOMI structure away from the crystallographically determined structure. A focused comparison over 118 starting points within 9.5 Å of the binding loop was performed. The second best match was found to correspond to the experimental alignment in this case.

Finally, a molecular volume comparison method was also examined. For comparison of similar sized molecules, such as DFKi and the TOMI fragment, molecular volume comparison was successful. However, volume comparison was unsuccessful in matching DFKi to the entire TOMI protein. To obtain the correct result for matching DFKi to a large molecule, such as the entire TOMI protein, it was necessary to focus the comparison on the surface of the TOMI protein by using skin matching.

7.6 Conclusion

Molecular recognition, the binding of a drug to its target receptor, demonstrates remarkable complementarity in the three-dimensional arrangement of interactions between ligand and receptor. Molecular surfaces provide a three-dimensional representation of the properties of a molecule which can aid understanding and prediction of these interactions. Methods for molecular surface similarity identify the common spatial features of a set of active, but dissimilar, molecules and provide a hypothesis for interaction with a common binding site or pharmacophore. Molecular complementarity methods allow prediction of the binding of a ligand to a target protein. A variety of promising molecular surface comparison methods have been reviewed here. Many of these methods are capable of identifying 'experimentally verified' similarities or complementarities. However, more experience with these methods will be needed to identify the most efficient approaches, and to compare the reliability and results of the various methods over a range of problems. Challenges remain. Conformational flexibility is an important property of both drug and protein. Methods to deal with this issue have begun to emerge, but more work is needed to fully and efficiently include the effects of conformational flexibility.

References

Bachar, O., Fischer, D., Nussinov, R. and Wolfson, H. (1993) A computer vision based technique for 3D sequence independent structural comparison of proteins. *Protein Engineering*, **6**, 279–288.

Bacon, D.J. and Moult, J. (1992) Docking by least-squares fitting of molecular surface patterns. *Journal of Molecular Biology*, **225**, 849–858.

Bladon, P. (1989) A rapid method for comparing and matching the spherical parameter surfaces of molecules and other irregular objects. *Journal of Molecular Graphics*, **7**, 130–137.

Blaney, F., Finn, P., Phippen, R. and Wyatt, M. (1993) Molecular surface comparison: application to drug design. *Journal of Molecular Graphics*, **11**, 98–105.

Chau, P.L. and Dean, P.M. (1987) Molecular recognition: 3D surface structure comparison by gnomonic projection. *Journal of Molecular Graphics*, **5**, 97–100.

Cherfils, J. and Janin, J. (1993) Protein docking algorithms: simulated molecular recognition. *Current Opinion in Structural Biology*, **3**, 265–269.

Chothia, C. and Janin, J. (1975) Principles of protein–protein recognition. *Nature*, **256**, 705–708.

Chothia, C., Novotny, J., Bruccoleri, R. and Karplus, M. (1985) Domain association in immunoglobin molecules. *Journal of Molecular Biology*, **186**, 651–663.

Connolly, M.L. (1981) MS. *Quantum Chemistry Program Exchange Bulletin*, **1**, 75.

Connolly, M.L. (1983a) Solvent-accessible surfaces of proteins and nucleic acids. *Science*, **221**, 709–713.

Connolly, M.L. (1983b) Analytical molecular surface calculation. *Journal of Applied Crystallography*, **16**, 548–558.

Connolly, M.L. (1985) Computation of molecular volume. *Journal of the American Chemical Society*, **107**, 1118–1124.

Connolly, M.L. (1986a) Measurement of protein surface shape by solid angles. *Journal of Molecular Graphics*, **4**, 3–6.

Connolly, M.L. (1986b) Shape complementarity at the hemoglobin $\alpha_1\beta_1$ subunit interface. *Biopolymers*, **25**, 1229–1247.

Connolly, M.L. (1992) Shape distributions of protein topography. *Biopolymers*, **32**, 1215–1236.

Dean, P.M. (1990) Molecular recognition: the measurement and search for molecular similarity in ligand–receptor interaction. In *Concepts and Applications of Molecular Similarity* (eds M.A. Johnson and G.M. Maggiora), Chapter 8. John Wiley, New York.

Dean, P.M. and Chau, P.L. (1987) Molecular recognition: optimized searching through rotational 3-space for pattern matches on molecular surfaces. *Journal of Molecular Graphics*, **5**, 152–158.

Dean, P.M. and Callow, P. (1987) Molecular recognition: identification of local minima for matching in rotational 3-space by cluster analysis. *Journal of Molecular Graphics*, **5**, 159–164.

Dean, P.M., Callow, P. and Chau, P.L. (1988) Molecular recognition: blind searching for regions of strong structural match on the surfaces of two dissimilar molecules. *Journal of Molecular Graphics*, **6**, 28–34.

DesJarlais, R.L., Sheridan, R.P., Dixon, J.S., Kuntz, I.D. and Venkataraghavan, R. (1986) Docking flexible ligands to macromolecular receptors by molecular shape. *Journal of Medicinal Chemistry*, **29**, 2149–2153.

DesJarlais, R.L., Sheridan, R.P., Seibel, G.L., Dixon, K.S., Kuntz, I.D. and Venkataraghavan, R. (1988) Using shape complementarity as an initial screen in designing ligands for a receptor binding site of known three-dimensional structure. *Journal of Medicinal Chemistry*, **31**, 722–729.

DesJarlais, RL., Seibel, G.L., Kuntz, I.D., Furth, P.S., Alverez, J.C., Ortiz de Montellano, P.R., DeCamp, D.L., Babe, L.M. and Craik, C.S. (1990) Structure-based design of nonpeptide inhibitors specific for the human immunodeficiency virus 1 protease. *Proceedings of the National Academy of Sciences USA*, **87**, 6644–6648.

Fischer, D., Norel, R., Wolfson, H. and Nussinov, R. (1993) Surface motifs by a computer vision technique: searches, detection, and implications for protein–ligand recognition. *Proteins: Structure, Function and Genetics*, **16**, 278–292.

Greer, J. and Bush, B.L. (1978) Macromolecular shape and surface maps by solvent exclusion. *Proceedings of the National Academy of Sciences USA*, **75**, 303–307.

Helmer-Citterich, M. and Tramontano, A. (1994) PUZZLE: a new method for automated protein docking based on surface shape complementarity. *Journal of Molecular Biology*, **235**, 1021–1031.

Janin, J. and Chothia, C. (1990) The structure of protein–protein recognition sites. *Journal of Biological Chemistry*, **265**, 16027–16030.

Jiang, F. and Kim, S.-H. (1991) 'Soft Docking': matching of molecular surface cubes. *Journal of Molecular Biology*, **219**, 79–102.

Jorgensen, W.L. (1991) Rusting of the lock and key model for protein–ligand binding. *Science*, **254**, 954–955.

Katchalski-Katzir, E., Shariv, L., Eisenstein, M., Friesem, A.A., Aflalo, C. and Vakser, I.A. (1992) Molecular surface recognition: determination of geometric fit between proteins and their ligands by correlation techniques. *Proceedings of the National Academy of Sciences USA*, **89**, 2195–2199.

Kuntz, I.D. (1992) Structure-based strategies for drug design and discovery. *Science*, **257**, 1078–1082.

Kuntz, I.D., Blaney, J.M., Oatley, S.J., Langridge, R. and Ferrin, T.E. (1982) A geometric approach to macromolecule–ligand interactions. *Journal of Molecular Biology*, **161**, 269–288.

Leach, A.R. and Kuntz, I.D. (1992) Conformational analysis of flexible ligands in macromolecular receptor sites. *Journal of Computational Chemistry*, **13**, 730–748.

Lee, B. and Richards, F.M. (1971) The interpretation of protein structures: estimation of static accessibility. *Journal of Molecular Biology*, **55**, 379–400.

Lee, R.H. and Rose, G.D. (1985) Molecular recognition. I. Automatic identification of topographic surface features. *Biopolymers*, **24**, 1613–1627.

Leicester, S.E., Finney, J.L. and Bywater, R.P. (1988) Description of molecular surface shape using Fourier descriptors. *Journal of Molecular Graphics*, **6**, 104–108.

Lin, S.L., Nussinov, R., Fischer, D. and Wolfson, H.J. (1994) Molecular surface representations by sparse critical points. *Proteins: Structure, Function and Genetics*, **18**, 94–101.

Marshall, G.R., Barry, C.D., Bosshard, H.E., Dammkoehler, R.A. and Dunn, D.A. (1979) The conformational parameter in drug design: the active analog approach. In *Computer Assisted Drug Design, ACS Symposium 112* (eds E.C. Olson and R.E. Christoffersen), pp. 205–226. American Chemical Society, Washington, DC.

Masek, B.B., Merchant, A. and Matthew, J.B. (1993) Molecular skins: A new concept for quantitative shape matching of a protein with its small molecule mimics. *Proteins: Structure, Function and Genetics*, **17**, 193–202.

Meng, E.C., Shoichet, B.K. and Kuntz, I.D. (1992) Automated docking with grid-based energy evaluation. *Journal of Computational Chemistry*, **13**, 505–524.

Mezey, P.G. (1990) Three-dimensional topological aspects of molecular similarity. In *Concepts and Applications of Molecular Similarity* (eds M.A. Johnson and G.M. Maggiora), Chapter 11. John Wiley, New York.

Namasivayam, S. and Dean, P.M. (1986) Statistical method for surface pattern-matching between dissimilar molecules: electrostatic potentials and accessible surfaces. *Journal of Molecular Graphics*, **4**, 46–50.

Norel, R., Fischer, D., Wolfson, H. and Nussinvo, R. (1994) Molecular surface recognition by a computer vision-based technique. *Protein Engineering*, **7**, 39–46.

Nussinov, R. and Wolfson, H.J. (1991) Efficient detection of three-dimensional structural motifs in biological macromolecules by computer vision techniques. *Proceedings of the National Academy of Sciences USA*, **88**, 10495–10499.

Payne, A.W.R. and Glen, R.C. (1993) Molecular recognition using a binary genetic search algorithm. *Journal of Molecular Graphics*, **11**, 74–91.

Richards, F.M. (1977) Areas, volumes, packing and protein structure. *Annual Review of Biophysics and Bioengineering*, **6**, 151–176.

Shoichet, B.K. and Kuntz, I.D. (1991) Protein docking and complementarity. *Journal of Molecular Biology*, **221**, 327–346.

Shoichet, B.K., Bodian, D.L. and Kuntz, I.D. (1992) Molecular docking using shape descriptors. *Journal of Computational Chemistry*, **13**, 380–397.

Walls, P.H. and Sternberg, M.J.E. (1992) New algorithm to model protein–protein recognition based on surface complementarity. *Journal of Molecular Biology*, **228**, 277–297.

Wang, H. (1991) Grid-search molecular accessible surface algorithm for solving the protein docking problem. *Journal of Computational Chemistry*, **12**, 746–750.

Zachmann, C.-D., Heiden, W., Schlenkrich, M. and Brickmann, J. (1992) Topological analysis of complex molecular surfaces. *Journal of Computational Chemistry*, **13**, 76–84.

Zachmann, C.-D., Kast, S.M., Sariban, A. and Brickmann, J. (1993) Self-similarity of solvent-accessible surfaces of biological and synthetical macromolecules. *Journal of Computational Chemistry*, **14**, 1290–1300.

Zielenkiewicz, P. and Rabczenko, A. (1984) Protein–protein recognition: method for finding complementary surfaces of interacting proteins. *Journal of Theoretical Biology*, **111**, 17–30.

8 Neural networks in the search for similarity and structure–activity

D.J. LIVINGSTONE and D.W. SALT

8.1 Introduction

This book is concerned with molecular similarity, what it is, how it may be quantified and how it can be used in the design of new drugs. Perhaps one of the most important aspects of similarity, however it is defined, is the question of how it may be perceived. One of the skills that we humans can lay particular claims to is our ability to recognize patterns, in other words to perceive similarity, and thus many of the similarity tools used in drug design are intended to express similarity, leaving the task of perception to the human 'expert'. Unfortunately, our ability to recognize patterns is generally restricted to relatively low dimensional data sets (i.e. 2, 3 or 4D) and thus we need help when similarity is described by hundreds or thousands of variables. This help can take the form of a variety of statistical methods, some of the most useful being pattern recognition techniques (Livingstone, 1991a) which literally set out to identify any underlying patterns in sets of data. An alternative approach in trying to tackle the problem is by the use of comparative molecular field analysis (CoMFA) (see Chapter 12).

One of the sources of pattern recognition methodology is artificial intelligence (AI) research which has aimed to reproduce human intelligence by imitating 'what we do'. The linear learning machine, for example, is an algorithm which learns to distinguish between different classes of samples (Nilsson, 1965) and expert systems are excellent examples of computer programs which are designed to mimic decision making by human experts (Ayscough *et al.*, 1987; Jakus, 1992; Cartwright, 1993). A more recent AI approach to the simulation of intelligence is an imitation of 'the way that we do it', in other words an attempt to construct artificial brains. The rationale underlying this approach is that since we do not know precisely how a brain works then perhaps a model of a brain, physical or otherwise, may exhibit some of the intelligent properties of a brain. How can this be achieved? Figure 8.1 shows a simplified picture of the basic building block of biological brains, the neuron, along with its connections by means of which it forms networks of neurons. The figure also shows an illustration of an artificial neuron which, like its biological counterpart, forms the basic building block of systems designed to simulate such biological networks. These systems are known as

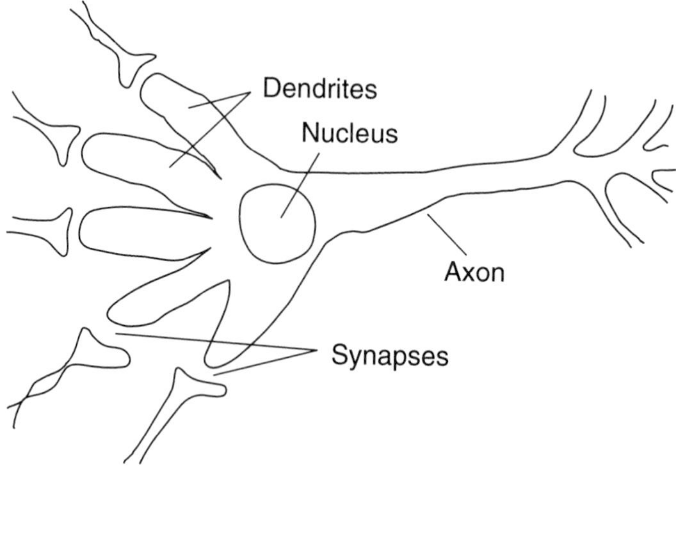

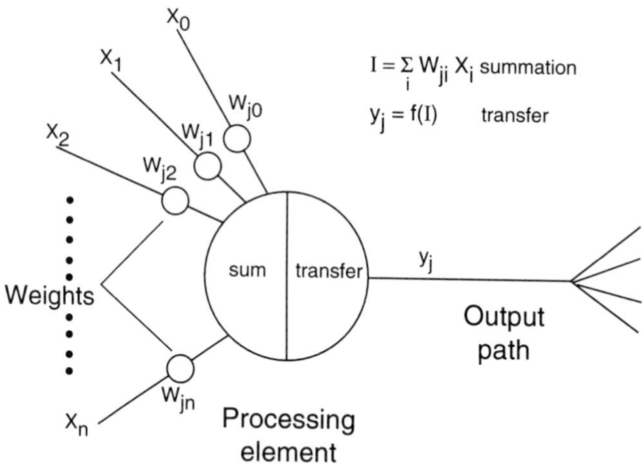

Figure 8.1 Simplified representation of a biological neuron and an artificial equivalent. (Reproduced with permission from Syu *et al.*, 1993.)

artificial neural networks (ANN) and the whole area has received much 'hype' as the following extract from a recent book[1] shows (Eberhart and Dobbins, 1990):

[1] These authors are at pains to distinguish between real substance and hype and present a very useful set of case studies along with source code listings (in C) for neural network programs.

In the past few years, neural networks have received a great deal of attention and are being touted as one of the greatest computational tools ever developed. Much of the excitement is due to the apparent ability of neural networks to imitate the brain's ability to make decisions and draw conclusions when presented with complex, noisy, irrelevant, and/or partial information. Furthermore, at some primitive level, neural networks appear able to imitate the brain's 'creative' processes to generate new data or patterns.

Implementations of ANN may be a recent development but in fact the field has a longer history than electronic computers. Turing made a theoretical examination of the possibility of imitating the function of the human brain in 1936 and McCulloch and Pitts set up an electronic brain cell in the 1940s (McCulloch and Pitts, 1943). Progress in ANN research has been irregular since those early studies (see chapter 1 of Eberhart and Dobbins, 1990, for an illuminating history) but has blossomed in the last half dozen years or so to such an extent that ANN are being used in applications as diverse as stock market forecasting, running nuclear reactors and drug detection. ANN have also found their way into many areas of chemical research as a number of recent reviews (Lacy, 1990; Zupan and Gasteiger, 1991; Katz *et al.*, 1992; Burns and Whitesides, 1993; Gasteiger and Zupan, 1993) and a book (Zupan and Gasteiger, 1993) will testify. This chapter is concerned with the way that ANN may be useful in the search for similarity as an aid to the drug-design process. The next section gives a basic introduction to the way that the neural networks operate so that the reader will be able to understand how ANN are used in the applications described in section 8.3.

8.2 What are neural networks?

Artificial neural networks are constructed by assembling a set of artificial 'neurons' or units (as shown in Figure 8.1) and connecting them together in some particular way with respect to their input and output connections. The units may be created using dedicated computer hardware, a collection of neurons on a silicon chip for example, or a single microprocessor for each neuron, or they may be simulated using software. This latter approach is more flexible since networks can be constructed on virtually any kind of computer and the organization (number of units and their 'wiring') of the networks may be readily changed. Hardware implementations have the advantage that they are usually very quick to train and, where individual processors are assigned to individual units in the network, the whole system becomes a parallel computing device (much like biological neural networks) as opposed to 'everyday' serial computers. All of the examples described in this chapter have been carried out using software implementations of the ANN.

Having connected up the artificial neurons in a particular way, referred to as the architecture of the network (see later), what happens next? ANN, like their biological counterparts, require training. As biological neural networks develop, they increase in size and establish connections between their component neurons. These connections modify the way that signals are passed between different parts of the network since they amplify signals to different extents and may have an inhibitory or stimulatory effect. The strength and number of connections between neurons are a result of learning following input from the senses as organisms interact with their environment (and other organisms). This combination of neurons, the connections between them and the strength (and sign) of the connections is, to a large extent, responsible for the processing power and reasoning ability of biological neural networks. Training of ANN involves a similar process, they are given sets of input patterns and associated target patterns and, through an iterative process, the internal representation of the data is modified until the predicted results are as close as desired to the targets. Patterns in this case means input signals which can be binary, e.g. present or absent, or continuous such as a physicochemical property (e.g. log P). The target patterns can similarly take either binary or continuous values. Networks trained in this way are often referred to as supervised networks (Salt *et al.*, 1992); the target patterns in some applications are the input patterns themselves (see section 8.3.4) and the network is essentially performing an identity mapping. Such networks are referred to as self-supervised networks (Sanger, 1989).

The architecture of a neural network is determined by the way in which the outputs of the neurons are connected to the other neurons. While there are a number of different ANN architectures (McClelland and Rumelhart, 1986), the most popular network used in similarity and structure–activity applications is the multilayer feed forward network. These networks are also known as the back-propagation multilayer perceptron or the feedforward back-propagation network. In this chapter they will be referred to simply as back-propagation (BP) networks. In this type of network the neurons are arranged into groups called layers — an input layer, an output layer and a number of hidden layers. The number of neurons in a layer and the number of layers depends directly on the number of variables in the data set, the number of samples and the type of output required. For example, in a network which was carrying out the equivalent of regression analysis there would be the same number of neurons on the input layer as there are independent variables in the data set and, with one response variable, there would be just one neuron making up the output layer. Such a network with two hidden layers of neurons is illustrated in Figure 8.2.

How is a network trained? Training involves alteration of the connection strengths (weights) between each pair of connected neurons and of biases to each neuron so that the network produces some desired output. Computation of an output from this type of ANN proceeds as follows. An input pattern is presented to the neurons on the input layer. The outputs from this layer are propagated to neurons in the next layer and eventually to the neurons

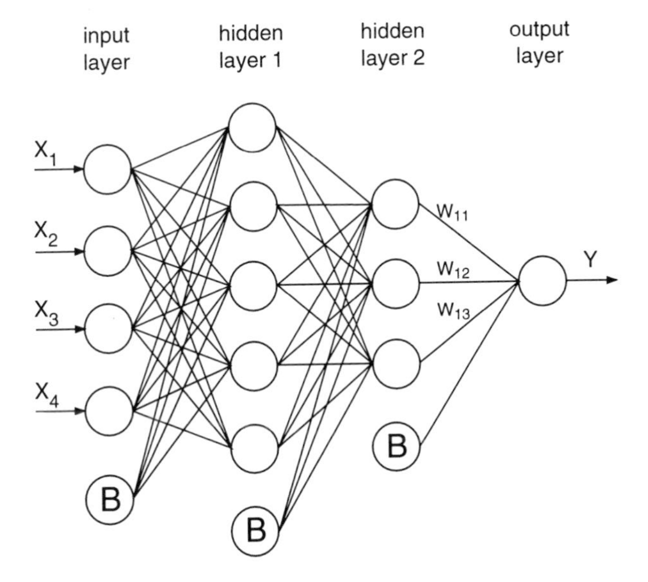

Figure 8.2 A back-propagation artificial neural network with four inputs (X_i, $i = 1, 2, 3, 4$), two hidden layers and one output (Y). B denotes the bias neurons. (Reproduced with permission from Salt *et al.*, 1992.)

of the output layer. The neurons in the input layer perform no processing and act essentially as 'distributors' since their function is to transmit the value of an input signal to the neurons in the hidden layer to which they are connected. Each of the other neurons (j) in the network computes the weighted sum of all outputs, out_i, from neurons in the preceding layer connected to it. With the addition of a bias term the sum, denoted inp_j, may be expressed as:

$$\text{inp}_j = \sum w_{ji}\,\text{out}_i + \text{bias}_j \tag{8.1}$$

where w_{ji} is the connection weight to neuron j from neuron i. The bias can be treated as the connection weight from a neuron with a constant output of 1.0, bias neurons act as shift operators for the subsequent processing of input signals to a layer. The output from neuron j, out_j, is calculated by processing inp_j through a non-linear transfer function. The form of this transfer function (sometimes referred to as a 'squash function') is quite arbitrary, but in applications of neural networks the most commonly used form is the sigmoidal function $f(x) = 1/[1 + \exp(-x)]$ (Figure 8.3). Consequently the output from neuron i is given by equation (8.2). The transfer function is a very important element of the operation of ANN. Its non-linear nature allows a network to build up complex, non-linear relationships

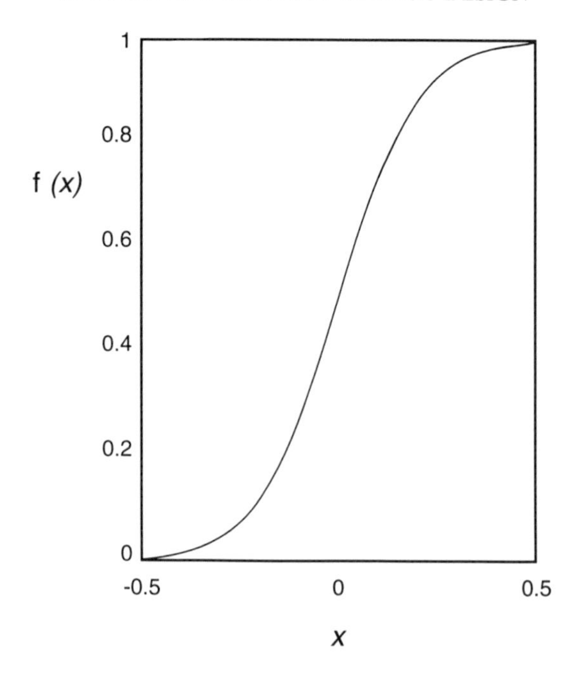

Figure 8.3 The logistic transfer function $f(x) = 1/(1 + e^{-x})$, which processes the input to each neuron on the hidden and output layers. (Reproduced with permission from Salt *et al.*, 1992.)

between the input data and the target output.

$$\text{out}_j = f(\text{inp}_j) = 1/[1 + \exp(-\text{inp}_j)] \tag{8.2}$$

The flow of information as described here shows how the term 'feed forward' arises. Since the outputs from one layer pass through the network from input layer to the output layer, without ever returning to a previous layer, this network design is known as a 'feed-forward network'.

Initially, all of the weights in a network are set to random values (usually between -0.3 and 0.3). The output from neurons in the outer layer will, therefore, differ initially from the desired target values. It is this difference, or error, between calculated output and target output that forms the basis for updating the connection weights in back-propagation learning. The general principle behind most commonly used back-propagation algorithms is the delta rule (Aoyama and Ichikawa, 1991) whereby connection weights are iteratively changed in order to minimize an error function, E, as shown in equation (8.3).

$$E = \sum_p \sum_m (\text{t}_{pj} - \text{out}_{pj})^2 \tag{8.3}$$

The summation is over m output-neurons; t_{pj} and out_{pj} are respectively the desired target output and calculated output from neuron j on the last layer of the network following the input of the pth pattern. In the delta rule the change, $\Delta_k w_{ji}$, in the weight for the connection between neuron j and neuron i at the kth iteration is given by:

$$\Delta_k w_{ji} = \beta \delta_{pj} \text{out}_{pi} + \alpha \Delta_k - w_{ji} \tag{8.4}$$

where β is the learning rate and α is a momentum term which prevents divergent oscillations. The error signal δ_{pj} of the jth neuron on presentation of the pth pattern in equation (8.4) is determined as follows:

If neuron j belongs to the output layer

$$\delta_{pj} = (\text{t}_{pj} - \text{out}_{pj}) f'(\text{inp}_{pj}) \tag{8.5}$$

If neuron j belongs to the hidden layer(s)

$$\delta_{pj} = f'(\text{inp}_{pj}) \sum_z \delta_{pz} \text{w}_{zj} \tag{8.6}$$

z represents the neurons to which neuron j sends its output and where $f'(\text{inp}_{pj})$ is the derivative with respect to in p_{pj} of the transfer function given in equation (8.2).

From equation (8.6) it can be seen that the error δ_{pj} for neurons on hidden layers is proportional to the errors δ_{pz} in neurons on subsequent layers, so the error given by equation (8.5) must be calculated for neurons on the output layer first and then passed back to neurons on earlier layers to allow them to alter their connection weights. It is this process of the passing back of the error value that gives rise to the name 'back propagation' for this method of network training. The training phase of a network is continued for either a fixed number of iterations or until the error given by equation (8.3) reaches some acceptable level.

The learning rate and momentum terms are two adjustable parameters which can be used to control how quickly a network will train and, to some extent at least, how closely it will approach a global minimum in the error surface. There is no guarantee, of course, that this minimization process will achieve a global minimum, should one exist, but there are strategies which may help to ensure adequate network training (Tetko *et al.*, 1993; Manallack *et al.*, 1994). These parameters affect the way a given network will train with a particular data set but perhaps the most important factors which control network training and subsquent performance are the number of layers and the number of neurons in each layer. Too few layers or too few neurons in each layer and the network will not contain sufficient connections to construct a satisfactory relationship between the input patterns and the target patterns. Too many layers or, more importantly, too many neurons and the network will simply 'memorize' the input and target patterns and will produce a network that appears to fit the data very well but which will be unable to generalize (Andrea and Kalayeh, 1991; Maggiora *et al.*, 1992; Manallack and Livingstone, 1992; Manallack *et al.*, 1994).

8.3 Applications

The following sections give examples of the use of artificial neural networks in protein structure prediction, relationships between chemical and biological properties and chemical structure and the display of multivariate data (as an aid to the establishment of structure–activity relationships). In addition to the review articles on general applications of ANN to chemistry quoted in the introduction, there are recent reviews on the prediction of protein structure (Holley and Karplus, 1991; Holbrook, 1993), the prediction of biochemical sequences (Presnell and Cohen, 1993), the application of ANN to quantitative structure–activity relationships (Manallack and Livingstone, 1994a) and a provocative comment on the use of ANN in pharmacodynamic modelling (Veng-Pedersen and Modi, 1993; see Siegel, 1993 for a 'commentary' and Veng-Pedersen, 1993 for a reply).

8.3.1 Protein structure prediction

Knowledge of the structure of a macromolecular target such as an enzyme, drug receptor, ion channel or antigen binding site can be of considerable use in drug design. Ligands (inhibitors, agonists, antagonists, etc.) can be docked into binding sites and their interaction energies calculated, new potential binding interactions may be identified and existing ones enhanced. Unfortunately, experimental 3D structures are available for only a small number of such targets whereas the amino acid sequences of many more are known. Techniques of experimental structure determination will undoubtedly continue to improve but it is likely to be some time before the number of known structures begins to match the number of known sequences if, indeed, it ever does. In the meantime, can use be made of the knowledge of protein sequence to construct models of interesting protein targets? The answer is a qualified yes. Where the structure of an homologous protein is known then all is well; it is possible to construct models by substitution of amino acid residues followed by minimization. Deletions and insertions into the sequence can cause problems as can the question of the degree of homology between two sequences. In other words, how similar is the target protein to the protein of known structure? This seems like a good problem for artificial neural networks. Since biological neural networks are good at recognizing patterns and identifying similarity then if ANN are good models of biological networks it should be possible to train a network to associate the sequence of amino acids in a protein with its 3D structure.

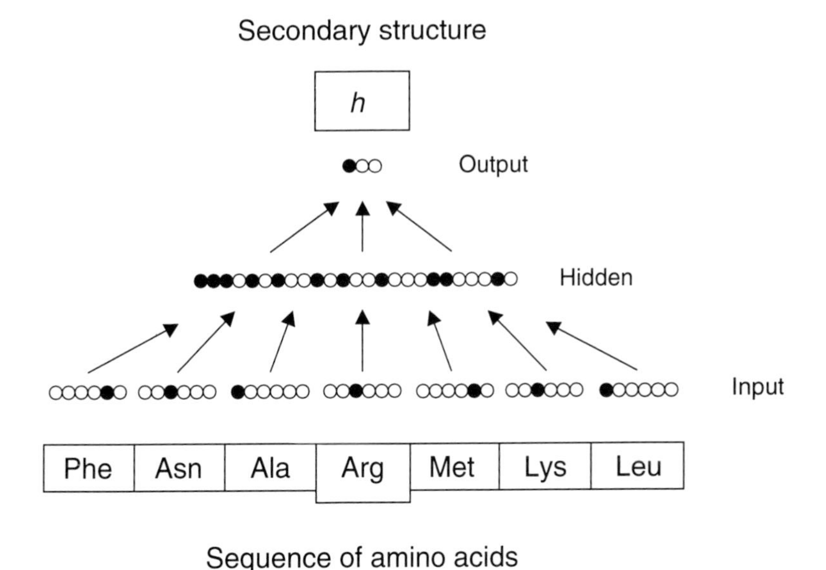

Figure 8.4 Diagram of the network architecture used for protein structure prediction. (Reproduced with permission from Qian and Sejnowski, 1988.)

One of the earliest attempts to do this was reported by Qian and Sejnowski (1988) following their success in the use of ANN to generate speech from text. These workers employed back-propagation networks consisting of an input layer of 13 groups of neurons (each group containing 21 units), a hidden layer and an output layer of three neurons, as shown in Figure 8.4. Each group of neurons on the input layer represents an amino acid position in the protein sequence, 20 of the 21 units in each group represent the 20 naturally occurring amino acids and one unit represents a chain spacer (to signify the end or beginning of a sequence). Thus, protein sequences could be fed into the input layer, one amino acid at a time, and the appropriate unit activated in each of the input groups according to the type of amino acid presented or chain spacer. The target output for this kind of network was the secondary structure of the central amino acid, no. 7 in the set of 13 input groups, described as α-helix, β-sheet or random coil. Each of these possible secondary structures was represented by one of the three output units and thus training needed to be continued until only one output unit was activated as each new amino acid reached position 7 in the input layer.

Experiments were carried out with artificial data, based on the statistical distribution of residues in known sequences, and with training and testing sets which contained either homologous or non-homologous proteins. As might be expected, the prediction success rate (test set predictions) was higher for the set of homologous proteins than for the non-homologous set; Figure

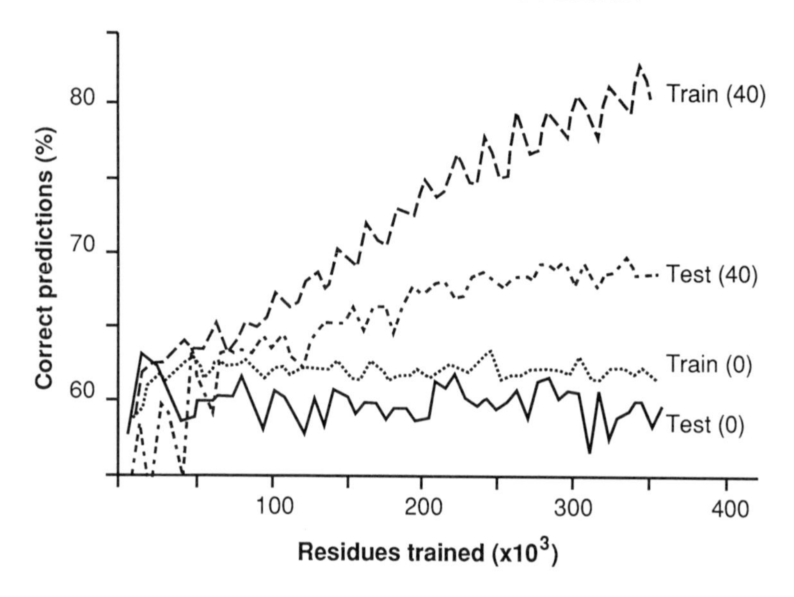

Figure 8.5 Plot of the percentage of correct predictions against the number of residues trained for a set of homologous proteins. (Reproduced with permission from Qian and Sejnowski, 1988.)

8.5 shows the percentage of correct predictions as a function of the number of residues trained (i.e. training cycles) for networks containing 40 and 0 units in the hidden layer. The networks without a hidden layer performed at about the same success rate for both training and test sets, around twice as well as the chance expectation of 33%. Performance on the training set by the network with a hidden layer of 40 neurons continued to improve as training was continued, probably indicating that the network was 'memorizing' the data. Test set performance for this network levelled off at around about 70%, which compares favourably with the results obtained using other well established methods. Table 8.1 shows a comparison of network performance for pairs of homologous proteins in terms of their sequence homology, structural homology and a measure of prediction success, Q_3. This statistic is based on the percentage of correct predictions for all three structural types as shown in equation (8.7):

$$Q_3 = \frac{P_\alpha + P_\beta + P_{\text{coil}}}{N} \tag{8.7}$$

For this comparison, each network was trained on the training protein of the pair and predictions were made for the test protein. The prediction success rate generally tended to be at some intermediate value between the sequence homology and structural homology. It was noted, however, that this was

Table 8.1 Results of overall prediction success for homologous pairs of proteins (from Qian and Sejnowski, 1988)[†]

Homologous pairs*		Residues (n)	Sequence homology (%)[1]	Structural homology (%)[2]	Q_3 (%)[3]
Test	Train				
1azu	1aza	125	69	84	78
1lzt	1lzl	129	65	96	83
1pfc	1fc2	111	66	62	63
1ppd	2act	212	54	93	83
2gch	1tgs	237	46	87	70
1gfl	1fr2	70	71	94	99
1p2p	1bp2	124	83	91	90
2ape	2app	318	67	80	61
2rhe	1ig2	114	77	92	77
2sga	3sgb	181	65	91	76
3hhb	2dhb	287	85	91	89
5ldh	1ldx	333	71	86	68

[†] Weighted averages: 1, 68; 2, 87; 3, 76.
* The abbreviations refer to the codes used in the Brookhaven protein databank.

less than that obtained by alignment of the two proteins and assignment of structural class to the test protein amino acids based on the class of the training amino acids. Judgement of the use of ANN for protein-structure prediction should not be made just on the basis of their success compared with other methods. Once a network has been trained on a set of proteins, it contains information, coded in the weights, which may not be evident when using other techniques. The weights in a network may be visualized by use of a diagram called a Hinton diagram (Figure 8.6). This shows the sign and the magnitude of the weight (see legend for details) for each amino acid when it is in any one of the 12 positions on either side of the central amino acid residue. The three parts of Figure 8.6 give the different effects of each amino acid in terms of their influence in causing the central amino acid to adopt α-helix, β-sheet or random coil structures.

Several other workers have adopted a similar approach to the use of ANN to predict protein structure (Bohr *et al.*, 1988; Holley and Karplus, 1989; Kneller *et al.*, 1990; Andreassen *et al.*, 1990; Fariselli *et al.*, 1993). Certain improvements were obtained by the division of the sequence data into structural classes and it is encouraging to see that there are similarities between the Hinton diagrams produced by networks trained on different data sets (Kneller *et al.*, 1990). Other improvements have involved the addition of tertiary structure information by the inclusion of contacts with the central residue (Vieth and Kolinski, 1991) and the use of multiple sequence alignment information (Rost and Sander, 1993a,b). This latter technique relies on the fact that multiple sequence alignments contain more information than single sequences. Figure 8.7 compares the prediction accuracy of five different networks; interestingly a 'jury' of 12 different networks (predicting structure

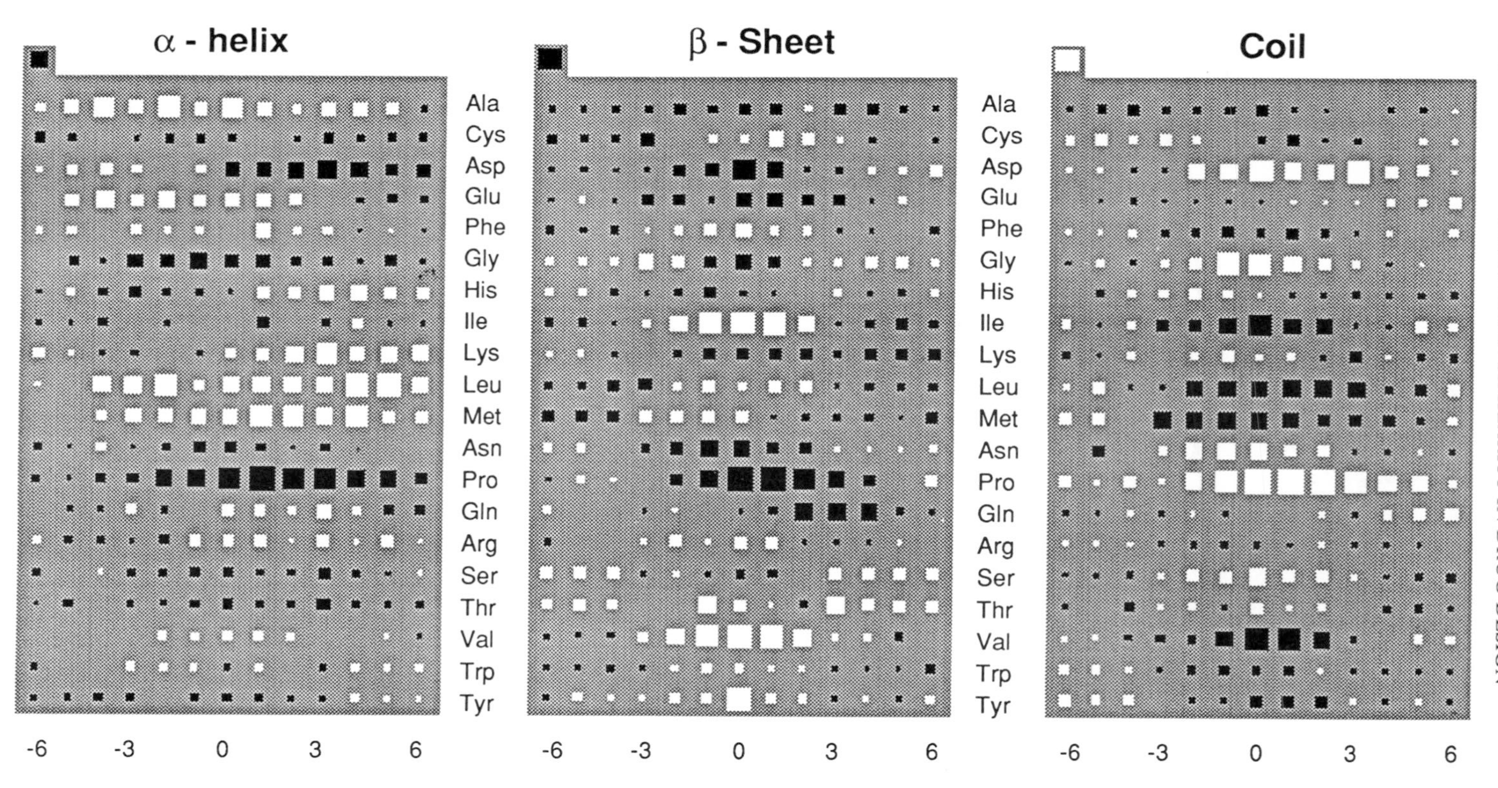

Figure 8.6 Hinton diagram showing the weights from the input units to three output units. A weight is represented by □ if it is positive (excitatory) and ■ if it is negative (inhibitory); the size of the square represents the magnitude of the weight. The three rectangular blocks represent the three output units for α-helix, β-sheet and random coil. (Reproduced with permission from Qian and Sejnowski, 1988.)

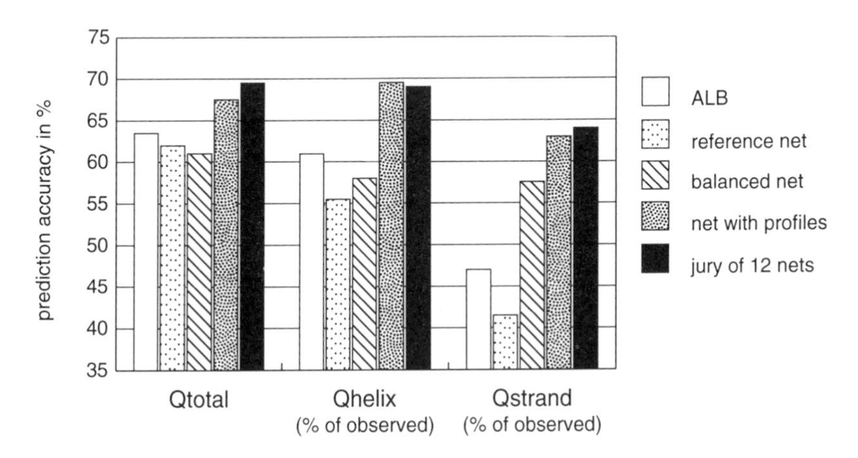

Figure 8.7 Comparison of the prediction accuracy of five different neural networks. See reference for details of the different types of network but note that the 'jury' of 12 networks had the highest prediction accuracy. (Reproduced with permission from Rost and Sander, 1993b.)

by a majority vote) gave the highest prediction accuracy. Prediction of protein structure is not restricted to simple yes/no decisions concerning membership of one of three structural classes. Bohr and coworkers (1990), for example, have trained a neural network to reproduce the distance matrix between C_α backbone carbons.

Other structural features of proteins have been successfully predicted using ANN. McGregor and coworkers (1989), for example, have used networks to predict β-turns and a network has been used to assign the disulphide bonding state of cysteine residues (Muskal *et al.*, 1990). This latter study was based on 128 protein structures which contained 689 non-identical cysteine containing sequences (379 disulphide bonded, 310 non-bonded). Table 8.2 shows the quite impressive results obtained with seven different sets of training and test sequences randomly extracted from the overall set. Information other than simple lists of the amino acids in a sequence can also be used as input to a network. Metfessel and coworkers (1993), for example, used amino acid composition and hydrophobic pattern frequency to predict structural classes, while Wade *et al.* (1992) used seven physicochemical properties and an assignment of structural class to predict water-binding sites on proteins. Schneider and Wrede (1993) used physicochemical descriptors of residues to achieve a 97% success rate for the prediction of *E. coli* signal peptidase cleavage sites using a three-layer network. Finally, it should be pointed out that all of these ANN approaches to protein-structure prediction, and other techniques for that matter, are critically dependent on the choice of training set structures (Ruggiero *et al.*, 1993).

Table 8.2 Results of the prediction of the disulphide bonding state of cysteine residues (from Muskal *et al.*, 1990)

Set*	% Train		% Test	
	S–S	S–H	S–S	S–H
1	89.7	83.3	80.0	80.0
2	89.4	82.3	80.0	80.0
3	89.7	83.3	90.0	70.0
4	90.2	83.0	70.0	90.0
5	90.5	83.0	70.0	100.0
6	90.5	84.3	90.0	70.0
7	90.0	82.7	90.0	70.0
Average:	90.0	83.1	81.4	80.0

* Each set consisted of a randomly selected set of 20 sequences, the network was trained on 669 sequences and test set predictions made for the 20.

8.3.2 *Prediction of chemical properties*

Molecular similarity is clearly of great importance when attempts are made to predict (or 'explain') chemical properties. Many chemical properties may be predicted simply by consideration of chemical structure and thus must depend on the similarity that is coded by the structural information. The handbook by Lyman and coworkers (1990), for example, lists a total of 33 properties which may be calculated from molecular fragment values.[2] Physical organic chemistry similarily provides examples of the prediction of chemical properties, in this case rate and equilibrium constants, by consideration of molecular similarity. Having shown earlier in this chapter how ANN can make use of similarity information it is not surprising that they have been used to predict a variety of chemical properties ranging from water solubility to the foaming and emulsifying properties of proteins.

Bodor and coworkers (1991) have reported the estimation of water solubility using a three-layer BP network. The training set for this example consisted of 331 compounds and the input descriptors for the network included geometrical (surface area, volume, ovality), quantum chemical (dipole, charges, etc.) and structural (number of NH bonds, etc.) parameters. A total of 17 input variables were used and this gave a fitted standard deviation of 0.23 (log units) which compared favourably with the results obtained for the same data set using regression analysis (0.3). At first sight, the ANN appears to be doing well but this particular example employed a hidden layer of 18 neurons; thus the network had a total of 324 connections, ignoring bias neurons. With a training set of only 331 compounds there are

[2] In some cases the calculation technique also requires an estimate of another property, which itself can often be derived from structure.

almost as many connections as compounds and thus the network may be memorizing the data (Andrea and Kalayeh, 1991; Maggiora *et al.*, 1992; Manallack and Livingstone, 1992; Manallack *et al.*, 1994a). It seems, in fact, that this is not likely to be the case as predictions worked well for a test set of 19 compounds as shown in Table 8.3. This does, however, highlight a potential danger in the use of ANN in data analysis generally and warns against using them as 'black boxes'.

Boiling point is another property which has been estimated using a three-layer BP network and a similar set of descriptors to those used for the prediction of water solubility (Egolf and Jurs, 1993). This report also compared the performance of the ANN with regression analysis and concluded that they were comparable. An interesting extension to the use of the trained networks was made by an examination of the connection weights between the input and hidden layer neurons. Figure 8.8 shows a graphical representation of these weights from which the most (and least) important decriptors may be identified. Lohninger (1993) has examined a variant of BP networks based on radial basis functions for the prediction of boiling points. These networks are similar in layout to BP networks but use a more complex transfer function on the hidden layer neurons and a linear function on the output; effectively the output neuron(s) simply sums the hidden layer signals. Prediction errors for both training and test set compounds were reduced using these networks compared with the results from regression analysis.

Table 8.3 Predicted results for log W (water solubility) using a neural network and regression analysis (reproduced with permission from Bodor *et al.*, 1991)

Compound*	Experimental	ANN (estimated)	Regression (estimated)
4-heptanol	−1.40	−1.40	−1.61
menthone	−2.35	−2.72	−2.03
1,1-diphenylethylene	−4.52	−5.23	−5.28
p-cresol	−0.81	−0.53	−0.44
testosterone	−4.08	−4.66	−4.49
2,4,4′-PCB	−6.24	−5.96	−5.66
dexamethasone	−3.59	−3.47	−3.58
4-chloronitrobenzene	−2.85	−1.83	−2.66
2,5-PCB	−5.06	−5.75	−5.45
2,6-PCB	−5.21	−5.52	−5.35
2,4,6-PCB	−6.06	−6.24	−5.88
fluorene	−4.92	−4.43	−4.78
pyrene	−6.17	−6.04	−6.39
indan	−3.04	−3.10	−3.27
2-methylpyridine	0.04	−0.01	−0.17
isoquinoline	−1.45	−1.11	−1.24
tetrahydrofuran	0.48	0.59	0.74
cortisone	−3.27	−2.95	−3.55
2-naphthol	−2.25	−2.08	−1.61

* The abbreviation PCB is not defined in the reference.

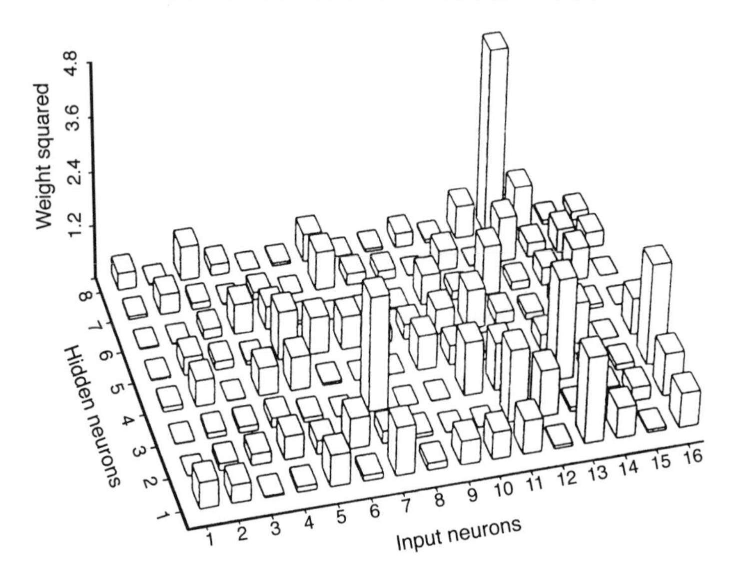

Figure 8.8 Identification of the important descriptors in a neural network study of the prediction of boiling points. (Reproduced with permission from Egolf and Jurs, 1993.)

Spectroscopic properties and chemical structure may also be linked by ANN. The ^{13}C chemical shifts of norbornene derivatives (*exo* and *endo*) have served as a useful test set in demonstrating the application of pattern recognition techniques such as the linear learning machine to chemical data (Kowalski, 1977). A three-layer BP network was trained with a set of 25 norbornene derivatives and was shown to make successful predictions for 12 out of a set of 13 test compounds (Aoyama *et al*., 1989). This performance was better (by one compound) than either cluster analysis or the linear learning machine applied to the same set, although the network contained a large number of connections and might have been demonstrating the ability of networks to memorize data. Weigel and Herges (1992) have examined the ability of neural networks to assign aromatic substitution patterns on the basis of infrared spectra. Input to the networks consisted of intensities from infrared spectra divided up into 12.5 or 6.25 cm^{-1} intervals; peak lists were also used as input but their performance was inferior. Training targets were the substitution pattern of the molecules for which the spectra were measured. Three different types of neural network were evaluated (two were of the self-supervised kind) but a BP network without a hidden layer was found to give 93% and 87% prediction accuracy for di- and tri-substituted compounds, respectively. Interestingly, this report showed that networks could be 'over-trained': prediction performance was shown to peak at a particular number of training cycles and would then be reduced if the network was trained further. Prediction in the opposite direction, that is spectra from structure, has been shown for ^{13}C NMR shifts for a set of alkanes using topological descriptors and a BP network (Doucet *et al.*, 1993). The network result for a training set of 208 different chemical environments was a standard error of prediction of 0.32 ppm, which compares very favourably with the application of a widely used model (Linderman and Adams, 1971) to the same data set giving a standard error of 0.79 ppm.

Somewhat more complex chemical properties may also be predicted using neural networks. For example, Kvasnicka and coworkers (1993) have reported a classification of inductive and resonance substituent constants using a three-layer BP network. Input to the network consisted of 14 structural descriptors and good correlations (0.957 for σ_I and 0.994 for σ_R) were obtained for the training sets. Broughton *et al.* (1992) were able to predict the pK_a value of a histidine residue in the enzyme triose phosphate isomerase using a three-layer BP network which had been trained on a set of substituted imidazoles. The training set compounds were described by a grid of electrostatic potential values and thus the trained network could be used to predict the pK_a value of an imidazole in any particular chemical environment, e.g. an enzyme. The predicted value of 4.5 for this histidine residue is in good agreement with the value estimated from experimental data. Finally, ANN have been used to predict some of the functional properties of proteins using a combination of experimental and theoretical descriptors (Arteaga and Nakai, 1993). Figure 8.9 shows a plot of emulsion activity index (EAI) predicted using three-layer BP networks. As a test of predictive ability, cross-validation was used by training a network with one protein omitted and making a prediction for the omitted protein, this being repeated for each member of the set. As can be seen from the figure, both the networks trained on all the proteins and the cross-validated networks performed well and these networks gave higher correlation coefficients than principal components regression carried out on the same data sets.

8.3.3 Biological properties and chemical structure

Attempts to explain or predict the biological properties of compounds in terms of their chemical structure have their origins in the middle of the last century. Since then a considerable effort has been expended in the characterization of chemical structure in quantitative terms and in the exploration of the statistical techniques which may be used to establish quantitative relationships (Livingstone, 1991b). An early application of the use of ANN to construct quantitative structure–activity relationships (QSAR) involved the discriminant analysis of a set of anticarcinogenic mitomycin derivatives using a three-layer BP network (Aoyama *et al.*, 1990a). This was soon followed by an example of the construction of a regression type QSAR model, again using a three-layer BP network, but in this case the output layer neuron used a linear transfer function rather than a non-linear function

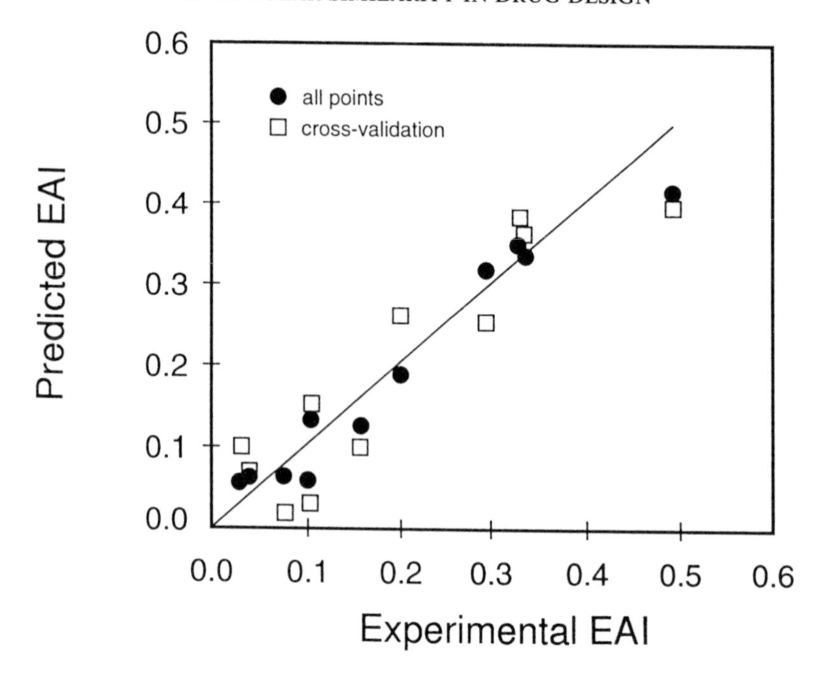

Figure 8.9 Plot of predicted versus experimental emulsion activity index (EAI). ● represent predictions from a network trained on the complete set, □ from cross-validated networks. (Reproduced with permission from Arteaga and Nakai, 1993.)

such as that shown in Figure 8.3 (Aoyama *et al.*, 1990b). The linear transfer function was used so that the network would simulate the way that a linear regression model combines terms but Livingstone and Salt (1992) have demonstrated that BP networks using a non-linear transfer function on the output neuron give a better fit with fewer connections (Table 8.4). The question of the number of connections in a network, and problems of networks

Table 8.4 Fitting errors* with different numbers of units in the hidden layer (reproduced with permission from Livingstone and Salt, 1992)

Units	MSE ($\times 10^3$)	RV ($\times 10^3$)	Units	MSE	RV
4	6.09	6.25	9	2.64	2.71
5	2.64	2.71	10	2.64	2.71
6	2.63	2.71	11	2.64	2.72
7	2.63	2.71	12	2.64	2.71
8	2.63	2.71	12†	42.9	44.1

* Fitting errors are mean square error (MSE) and residual variance (RV) as described by Aoyama *et al.*, 1990b.

† Results from Aoyama *et al.*, 1990b.

memorizing data, has been mentioned earlier in this chapter. Andrea and Kalayeh (1991), in a report of the use of ANN to construct QSAR for dihydrofolate reductase inhibitors, suggested that a parameter, ρ, could be used to characterize networks in terms of their connections and the number of samples used to train them:

$$\rho = \frac{\text{Number of datapoints}}{\text{Number of connections}} \tag{8.8}$$

For the examples considered by them, it was found that optimal network models were produced when $1.8 < \rho < 2.2$. Networks with $\rho > 2.2$ contained too few connections to allow the construction of sufficiently complex models to give good predictions whereas networks with $\rho < 1.8$ began to overfit the data (good models but poor predictions).

Neural networks with a large number of connections may not only overfit data but may also suffer from chance correlations as described by Topliss and Edwards (1979) for QSAR models fitted by regression. Experiments have been reported in which discriminant analysis networks (Manallack and Livingstone, 1992) and regression networks (Livingstone and Manallack, 1993) have been fitted to data sets made up from random numbers. At low values of ρ, these networks were able to give a perfect fit for the random data (which contained no 'real' correlations) as shown in Figure 8.10. It was suggested that ρ should be kept above 2.0 for discriminant networks and above 3.0 for regression networks in order to keep chance correlations below $R^2 = 0.5$. Lower ρ values than this might be acceptable if some form of

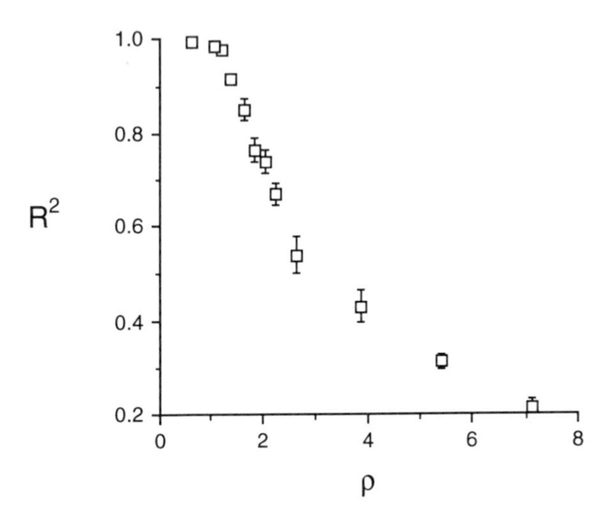

Figure 8.10 Plot of R^2 versus ρ for regression networks trained on random number data sets. (Reproduced with permission from Livingstone and Manallack, 1993.)

cross-validation scheme, preferably a test set/training set procedure, was employed to examine the predictive ability of the trained networks.

These problems raise the question of whether ANN offer any advantages over traditional statistical methods for fitting models to QSAR data sets. The results reported by Andrea and Kalayeh (1991) suggest that the networks do indeed do a better job than regression analysis; they quote a correlation coefficient (R^2) of 0.83 for network fitting to one data set compared with 0.38 for regression and 0.7 for regression using indicator variables. The network input did not include any indicator variables and thus the ANN is taking care of a number of complexities in the data set without the need to explicitly state them. Manallack and coworkers (1994), on the other hand, have examined the predictive performance of BP models for a number of different types of 'typical' QSAR data sets and they conclude that the predictive performance of the network models is poorer than regression fitting. Figure 8.11 shows the performance of a three-layer BP network used to analyse a 'linear' QSAR data set; ρ values were altered by changing the number of hidden layer neurons. It can be seen from this figure that the network gives a higher R^2 than regression analysis for fitting, but lower correlation coefficients for the cross-validated predictions.

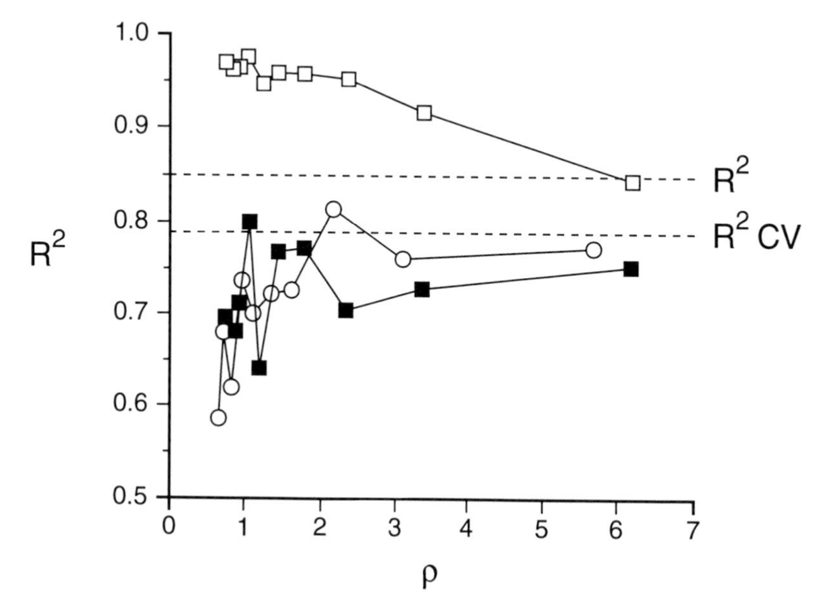

Figure 8.11 Plot comparing multiple linear regression modelling to a neural network. The top curve represents training of the data using networks employing a 3-n-1 architecture with the data scaled between 0.2 and 0.8 (□). Testing the predictive performance of the networks used cross-validation employing both leave-one-out, LOO (■) and leave-N-out (○) procedures (N was approximately 10%). The horizontal lines represent the multiple regression correlation coefficients. (Reproduced with permission from Manallack *et al.*, 1994.)

Since the early reports of the application of ANN to QSAR modelling they appear to have become quite popular, being applied to dihydropteridine reductase inhibitors (Song and Yu, 1993), 5-lipoxygenase inhibitors (Ghoshal *et al.*, 1993a), mutagenicity predictions (Villemin *et al.*, 1993; Brinn *et al.*, 1993; Ghoshal *et al.*, 1993b), structure–odour relationships (Chastrette and de Saint Laumer, 1991) and structure–biodegradability relationships (Cambon and Devillers, 1993). Wikel and Dow (1993) have demonstrated how ANN may be used for variable selection and Tetko and coworkers (1993) have suggested how the performance of neural networks might be assessed statistically. Most of these studies have made use of a BP-type network; little attention has been paid to other architectures or other ways of coding chemical similarity information. An exception is an approach called the functional link net (FUNCLINK) which makes use of various mathematical transformations of the input parameters to the network (Liu *et al.*, 1992). The FUNCLINK network is reported to perform better at prediction than standard BP networks although the approach has been criticized recently (Manallack and Livingstone, 1994b).

8.3.4 *Similarity through data display*

Computational chemistry techniques are increasingly being used to characterize small molecules in the search for quantitative structure–activity relationships (Hyde and Livingstone, 1988). Unfortunately, these methods may present the drug designer with a problem since they can often result in data matrices which contain more descriptors than there are compounds in the set, particularly when using methods based on interaction fields around molecules (see Chapter 12). A powerful approach to the analysis and interpretation of such data sets can be found in the multivariate techniques for data display such as principal components analysis and non-linear mapping (Hudson *et al.*, 1989; Domine *et al.*, 1993; Devillers, 1994). Self-supervised neural networks may also be used as a means for the display of a multivariate data set, in other words as a dimension reduction device. One such network, known as ReNDeR (reversible nonlinear dimensionality reduction), is shown in Figure 8.12 (Livingstone *et al.*, 1991). The input layer contains as many neurons as there are variables in the data set, the first hidden layer (encoding layer) contains a smaller number of neurons than the input layer and the central hidden layer (parameter layer) contains two (or three) neurons. The network is symmetrical with the decoding and output layers the same size as the encoding layer and the input layer. Input patterns are presented to a ReNDeR network one at a time and the connection weights are adjusted until the output values match the input values. Once network training has been carried out, a multivariate data set has been 'mapped' from the input layer onto the output layer by being passed through a bottleneck of two neurons. As each pattern (sample) in the data set is presented to the network it will generate

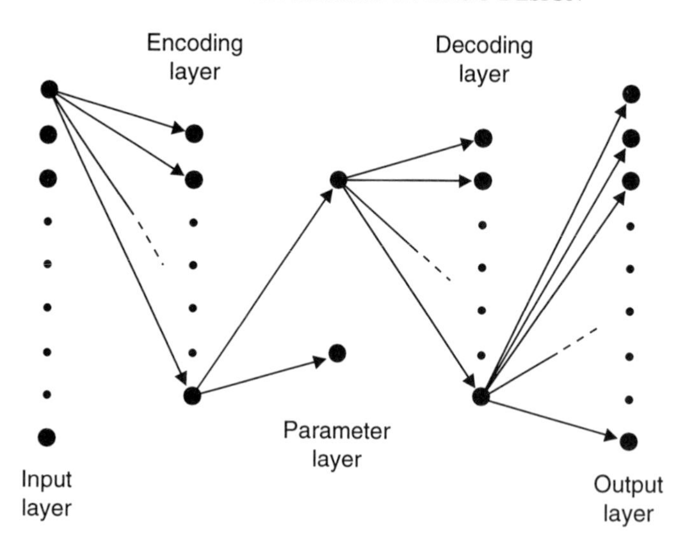

Figure 8.12 Diagram of the ReNDeR back-propagation network. The network is fully connected; only a few connections are shown for clarity. (Reproduced with permission from Livingstone *et al.*, 1991.)

output signals at the two parameter layer neurons; these output signals may be used as the *X*- and *Y*-coordinates of a two-dimensional plot to produce a 2D display of the N-D data set. The original report of this work showed that the ReNDeR method was able to generate 2D displays of data sets that were comparable to the results obtained using both principal components analysis and non-linear mapping.

Good and coworkers (1993) have examined the structure–activity relationships of the testosterone-binding globulin (TBG) and corticosteroid-binding globulin (CBG) affinities of a set of steroids. The compounds were characterized by similarity matrices, in which each compound was compared to every other, based on shape similarity, electrostatic potential similarity and a combination of the two. Two-dimensional plots produced from 31 or 62 dimensional matrices using a ReNDeR-type network showed quite marked clustering of the high-affinity compounds for both TBG and CBG. Figure 8.13 shows a plot of these compounds, described by combined shape and electrostatic potential similarity, coded for their binding affinity to CBG. The high-affinity compounds are mostly clustered together in the bottom left-hand corner of the plot, low-affinity compounds at the right-hand side and top with intermediate affinity compounds in between. Reibnegger and coworkers (1993) have also demonstrated the use of a ReNDeR-type plot to display molecular similarity information. Atomic charges of five common atoms in a set of pterin and guanine derivatives were calculated by an *ab initio* method. Figure 8.14 shows the results of a cluster analysis carried out on this set

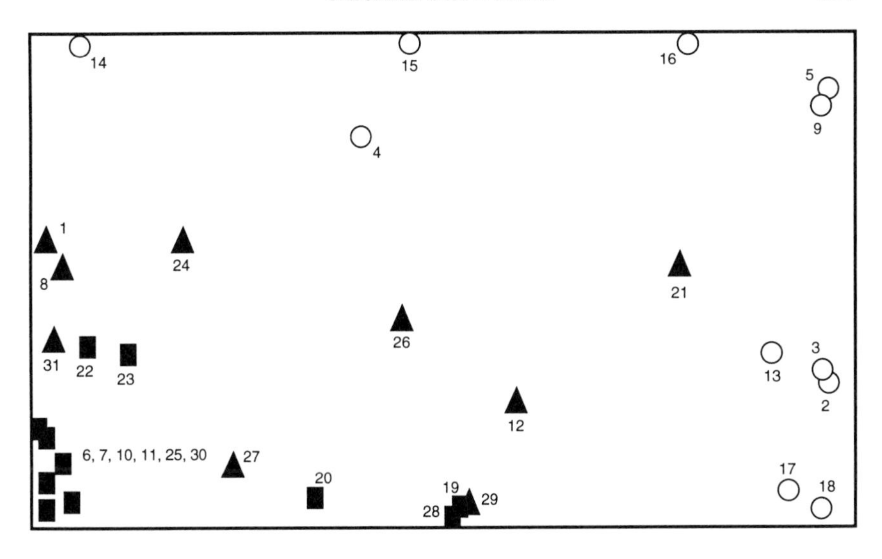

Figure 8.13 ReNDeR type plot of 31 steroids described by combined shape and electrostatic potential similarity. The compounds are coded for their CBG binding affinity: high ■; intermediate ▲; low ○. (Reproduced with permission from Good *et al.*, 1993.)

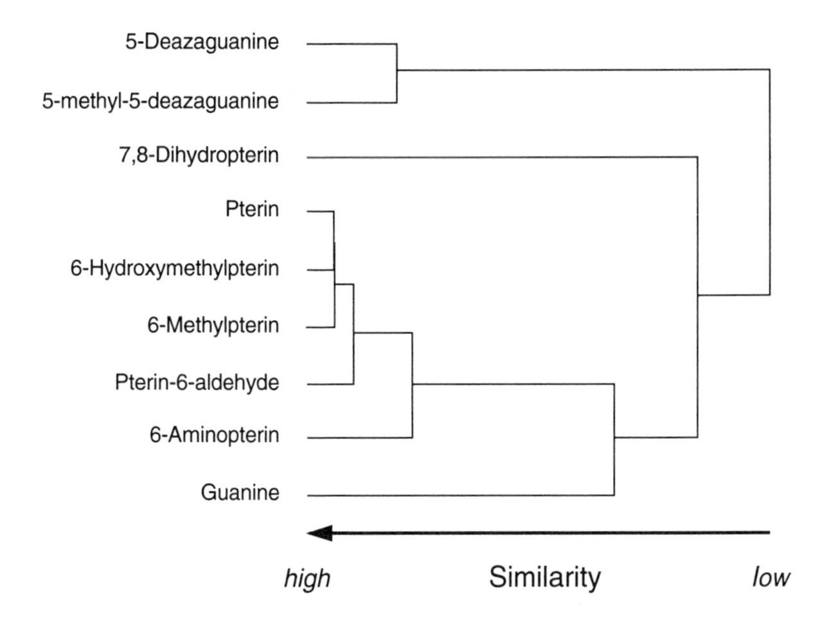

Figure 8.14 Dendrogram of the associations between a set of pterins and guanines described by five calculated atomic charges. (Reproduced with permission from Reibnegger *et al.*, 1993.)

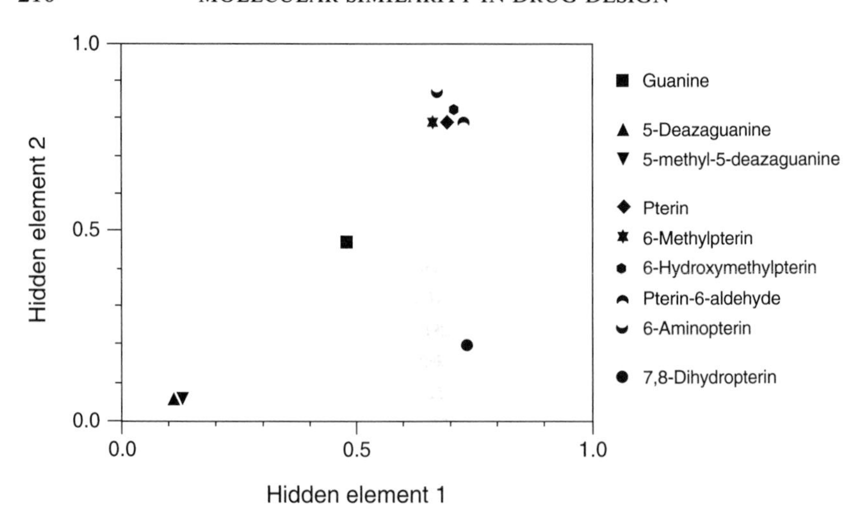

Figure 8.15 ReNDeR-type plot of a set of pterins and guanines described by five calculated atomic charges. (Reproduced with permission from Reibnegger *et al.*, 1993.)

where it can be seen that the two deazaguanines are grouped together, the pterins fall into one cluster and the dihydropterin and guanine are in separate groups to the rest. A two-dimensional ReNDeR type[3] plot is shown in Figure 8.15 where, again, the pterins are grouped together, the two deazaguanines are grouped together and guanine and dihydropterin fall into separate areas of the plot.

These examples show that self-supervised networks are able to produce similar results to more well-established techniques such as principal components analysis and non-linear mapping, but do they offer any advantages over these methods? One disadvantage of the low-dimensional display of multivariate data is the problem of moving from two- (or three-)dimensional space back to the original variables. Since the connection weights in a trained network are known it should be possible to project a point from two-space back to N-space using a ReNDeR network and the original report claimed that this was the case. Since then, Reibnegger and coworkers (1993) have shown that in certain circumstances, where data points are linear combinations of others, this mapping will fail. This, and other aspects of the use of ReNDeR-type networks for data display, is under investigation.

[3] This was actually produced using a 5-2-5 BP network, i.e. without the encoding and decoding layers of the 'true' ReNDeR network.

8.4 Summary

Artificial neural networks are an exciting new development in artificial intelligence research which, in the space of just a few years, have found many applications in the search for molecular similarity. The use of ANN in protein-structure prediction has given very encouraging results based simply on known protein sequences. As more structural information becomes available, and with the inclusion of structure determining physicochemical properties, it is expected that these methods may be greatly improved. The prediction of chemical properties and biological properties from chemical structure has also given interesting results, as has the use of ANN as data-display devices. Most of the reported applications, however, have involved only one type of network architecture — feed-forward back propagation. Other network architectures, or different ways of coding molecular similarity, may well prove of more use in the drug-design process.

References

Andrea, T.A. and Kalayeh, H. (1991) Applications of neural networks in quantitative structure–activity relationships of dihydrofolate reductase inhibitors. *Journal of Medicinal Chemistry*, **34**, 2824–2836.

Andreassen, H., Bohr, H., Bohr, J., Brunak, S., Bugge, T., Cotterill, R.M.J., Jacobsen, C., Kusk, P., Lautrup, B., Petersen, S.B., Saermark, T. and Ulrich, K. (1990) Analysis of the secondary structure of the human immunodeficiency virus (HIV) proteins p17, hp120, and gp41 by computer modeling based on neural network methods. *Journal of Acquired Immune Deficiency Syndromes*, **3**, 615–622.

Aoyama, T. and Ichikawa, H. (1991) Basic operating characteristics of neural networks when applied to structure–activity studies. *Chemical and Pharmaceutical Bulletin*, **39**, 358–366.

Aoyama, T., Suzuki, Y. and Ichikawa, H. (1989) Neural networks applied to pharmaceutical problems. I. Method and application to decision making. *Chemical and Pharmaceutical Bulletin*, **37**, 2558–2560.

Aoyama, T., Suzuki, Y. and Ichikawa, H. (1990a) Neural networks applied to structure–activity relationships. *Journal of Medicinal Chemistry*, **33**, 905–908.

Aoyama, T., Suzuki, Y. and Ichikawa, H. (1990b) Neural networks applied to quantitative structure–activity relationship analysis. *Journal of Medicinal Chemistry*, **33**, 2583–2590.

Arteaga, G.E. and Nakai, S. (1993) Predicting protein functionality with artificial neural networks: foaming and emulsifying properties. *Journal of Food Science*, **58**, 1152–1156.

Ayscough, P.B., Chinnick, S.J., Dybowski, R. and Edwards, P. (1987) Some developments in expert systems in chemistry. *Chemistry and Industry*, 515–520.

Bodor, N., Harget, A. and Huang, M-J. (1991) Neural network studies. 1. Estimation of the aqueous solubility of organic compounds. *Journal of the American Chemical Society*, **113**, 9480–9483.

Bohr, H., Bohr, J., Brunak, S., Cotterill, R.M.J., Lautrup, B., Norskov, L., Olsen, O. and Petersen, S. (1988) Protein secondary structure and homology by neural networks. *FEBS Letters*, **241**, 223–228.

Bohr, H., Bohr, J., Brunak, S., Cotterill, R.M.J., Fredholm, H., Lautrup, B. and Petersen, S. (1990) A novel approach to prediction of the 3-dimensional structures of protein backbones by neural networks. *FEBS Letters*, **261**, 43–46.

Brinn, M.W., Payne, M.P. and Walsh, P.T. (1993) Neural network prediction of mutagenicity using structure–property relationships. *Transactions of IChemE*, **71**, 337–339.

Broughton, H.B., Green, S.M. and Rzepa, H.S. (1992) Prediction of the histidine-95 pK_a perturbation in triosephosphate isomerase using an electrostatically trained neural network (SONNIC). *Journal of the Chemical Society. Chemical Communications*, 1178–1180.

Burns, J.A. and Whitesides, G.M. (1993) Feed-forward neural networks in chemistry: mathematical systems for classification and pattern recognition. *Chemical Reviews*, **93**, 2583–2601.

Cambon, B. and Devillers, J. (1993) New trends in structure–biodegradability relationships. *Quantitative Structure–Activity Relationships*, **12**, 49–56.

Cartwright, H.M. (1993) *Applications of Artificial Intelligence in Chemistry*. Oxford University Press, Oxford.

Chastrette, M. and de Saint Laumer, J.Y. (1991) Structure–odor relationships using neural networks. *European Journal of Medicinal Chemistry*, **26**, 829–833.

Devillers, J. (1994) Nonlinear mapping. In *Chemometric Methods in Molecular Design* (ed. H. Van de Waterbeemd). VCH, Weinheim. In press.

Domine, D., Devillers, J., Chastrette, M. and Karcher, W. (1993) Non-linear mapping for structure–activity and structure–property modelling. *Journal of Chemometrics*, **7**, 227–242.

Doucet, J.P., Panaye, A., Feuilleaubois, E. and Ladd, P. (1993) Neural networks and ^{13}C shift prediction. *Journal of Chemical Information and Computer Sciences*, **33**, 320–324.

Eberhart, R.C. and Dobbins, R.W. (eds) (1990) *Neural Network PC Tools. A Practical Guide*. Academic Press, Cambridge, Massachusetts.

Egolf, L.M. and Jurs, P.C. (1993) Prediction of boiling points of organic heterocyclic compounds using regression and neural network techniques. *Journal of Chemical Information and Computer Science*, **33**, 616–625.

Fariselli, P., Compiani, M. and Casadio, R. (1993) Predicting secondary structures of membrane proteins with neural networks. *European Biophysics Journal*, **22**, 41–51.

Gasteiger, J. and Zupan, J. (1993) Neural networks in chemistry. *Angewande Chemie. International Edition*, **32**, 503–527.

Ghoshal, N., Mukhopadhayay, S.N., Ghoshal, T.K. and Achari, B. (1993a) Quantitative structure–activity relationship studies using artificial neural networks. *Indian Journal of Chemistry*, **32B**, 1045–1050.

Ghoshal, N., Mukhopadhayay, S.N., Ghoshal, T.K. and Achari, B. (1993b) Quantitative structure–activity relationship studies of aromatic and heteroaromatic compounds using neural network. *Bioorganic Medicinal Chemistry Letters*, **3**, 329–332.

Good, A.C., So, S-S. and Richards, W.G. (1993) Structure–activity relationships from molecular similarity matrices. *Journal of Medicinal Chemistry*, **36**, 433–438.

Holbrook, S.R. (1993) Application of computational neural networks to the prediction of protein structural features. *Genetic Engineering*, **15**, 1–19.

Holly, L.H. and Karplus, M. (1989) Protein secondary structure prediction with a neural network. *Proceedings of the National Academy of Sciences USA*, **86**, 151–156.

Holly, L.H. and Karplus, M. (1991) Neural networks for protein structure prediction. *Methods in Enzymology*, **202**, pp. 204–224. Academic Press, San Diego.

Hudson, B., Livingstone, D.J. and Rahr, E. (1989) *Journal of Computer-Aided Molecular Design*, **3**, 55–65.

Hyde, R.M. and Livingstone, D.J. (1988) Perspectives in QSAR: computer chemistry and pattern recognition. *Journal of Computer-Aided Molecular Design*, **2**, 145–155.

Jakus, V. (1992) Artificial intelligence in chemistry. *Collection of Czechoslovak Chemical Communications*, **57**, 2413–2451.

Katz, W.T., Snell, J.W. and Merickel, M.B. (1992) Artificial neural networks. *Methods in Enzymology*, **210**, pp. 610–636. Academic Press, San Diego.

Kneller, D.G., Cohen, F.E. and Langridge, R. (1990) Improvements in protein secondary structure prediction by an enhanced neural network. *Journal of Molecular Biology*, **214**, 171–182.

Kowalsi, B.R. (1977) *Chemometrics: Theory and Applications*, ACS Symposium Series, 53, p. 243. American Chemical Society, Washington, DC.

Kvasnicka, V., Sklenak, S. and Pospichal, J. (1993) Neural network classification of inductive and resonance effects of substituents. *Journal of the American Chemical Society*, **115**, 1495–1500.

Lacy, M.E. (1990) Neural network technology and its application in chemical research. *Tetrahedron Computer Methodology*, **3**, 119–128.

Lindeman, L.P. and Adams, J.Q. (1971) Carbon-13 nuclear magnetic resonance spectrometry. Chemical shifts for the paraffins through C_9. *Analytical Chemistry*, **43**, 1245–1252.

Li, Q., Hironi, S. and Mouriguchi, I. (1992) Comparison of the functional-link net and the generalized delta rule net in quantitative structure–activity relationship studies. *Chemical and Pharmaceutical Bulletin*, **40**, 2962–2969.

Livingstone, D.J. (1991a) Pattern recognition methods in rational drug design, in *Methods in Enzymology*, **203**, pp. 613–638. Academic Press, San Diago.

Livingstone, D.J. (1991b) Quantitative structure–activity relationships. In *Similarity Models in Organic Chemistry, Biochemistry and Related Fields* (eds T.M. Krygowski, R. Zalewski and J. Shorter), pp. 557–627. Elsevier, Amsterdam.

Livingstone, D.J. and Salt, D.W. (1992) Regression analysis for QSAR using neural networks. *Bioorganic and Medicinal Chemistry Letters*, **2**, 213–218.

Livingstone, D.J. and Manallack, D.T. (1993) Statistics using neural networks: chance effects. *Journal of Medicinal Chemistry*, **36**, 1295–1297.

Livingstone, D.J., Hesketh, G. and Clayworth, D. (1991) Novel method for the display of multivariate data using neural networks. *Journal of Molecular Graphics*, **9**, 115–118.

Lohninger, H. (1993) Evaluation of neural networks based on radial basis functions and their application to the prediction of boiling points from structural parameters. *Journal of Chemical Information and Computing Sciences*, **33**, 736–744.

Lyman, W.J., Reehl, W.F. and Rosenblatt, D.H. (1990) *Handbook of Chemical Property Estimation Methods*. American Chemical Society, Washington, DC.

Maggiora, G.M., Elrod, D.W. and Trenary, R.G. (1992) Computational neural networks as model-free mapping devices. *Journal of Chemical Information and Computing Sciences*, **32**, 732–741.

Manallack, D.T. and Livingstone, D.J. (1992) Artificial neural networks: applications and chance effects for QSAR data analysis. *Medicinal Chemistry Research*, **2**, 181–190.

Manallack, D.T. and Livingstone, D.J. (1994a) Neural networks — a tool for drug design. In *Advanced Computer-Assisted Techniques in Drug Discovery* (ed. H. Van de Waterbeemd). VCH, Weinheim. In press.

Manallack, D.T. and Livingstone, D.J. (1994b) Limitations of functional-link nets as applied to QSAR data analysis. *Quantative Structure–Activity Relationships*. **13**, 18–21.

Manallack, D.T., Ellis, D.D. and Livingstone, D.J. (1994) Analysis of linear and non-linear QSAR data using neural networks. *Journal of Medicinal Chemistry*. In press.

McClelland, J.L. and Rumelhart, D.E. (1986) Parallel Distributed Processing, Vol. 1. MIT Bradford Press, London.

McCulloch, W.C. and Pitts, W. (1943) A logical calculus of the ideas immanent in nervous activity. *Bulletin of Mathematical Biophysics*, **5**, 115–133.

McGregor, M.J., Flores, T.P. and Sternberg, M.J.E. (1989) Prediction of β-turns in proteins using neural networks. *Protein Engineering*, **2**, 521–526.

Metfessel, B.A., Saurugger, P.N., Connelly, C.P. and Rich, S.R. (1993) Cross-validation of protein structural class prediction using statistical clustering and neural networks. *Protein Science*, **2**, 1171–1182.

Muskal, S.M., Holbrook, S.R. and Kim, S-H. (1990) Prediction of the disulfide-bonding state of cysteine in proteins. *Protein Engineering*, **3**, 667–672.

Nilsson, N.J. (1965) *Learning Machines: Foundations of Trainable Pattern Classifying Systems*. McGraw-Hill, New York.

Presnell, S.R. and Cohen, F.E. (1993) Artificial neural networks for pattern recognition in biochemical sequences. *Annual Reviews of Biophysics and Biomolecular Structure*, **22**, 283–298.

Qian, N. and Sejnowski, T.J. (1988) Predicting the secondary structure of globular proteins using neural network models. *Journal of Molecular Biology*, **202**, 865–884.

Reibnegger, G., Werner-Felmayer, G. and Wachter, H. (1993) A note on the low-dimensional display of multivariate data using neural networks. *Journal of Molecular Graphics*, **11**, 129–133.

Rost, B. and Sander, C. (1993a) Prediction of protein secondary structure at better than 70% accuracy. *Journal of Molecular Biology*, **232**, 584–599.

Rost, B. and Sander, C. (1993b) Improved prediction of protein secondary structure by use of sequence profiles and neural networks. *Proceedings of the National Academy of Sciences USA*, **90**, 7558–7562.

Ruggiero, C., Sacile, R. and Rauch, G. (1993) Peptides secondary structure prediction with neural networks: a criterion for building appropriate learning sets. *IEEE Transactions in Biomedical Engineering*, **40**, 1114–1121.

Salt, D.W., Yildiz, N., Livingstone, D.J. and Tinsley, C.J. (1992) The use of artificial neural networks in QSAR. *Pesticide Science*, **36**, 161–170.

Sanger, T.D. (1989) Optimal unsupervised learning in a single-layer linear feedforward neural network. *Neural Networks*, **2**, 459–473.

Shneider, G. and Wrede, P. (1993) Development of artificial neural filters for pattern recognition in protein sequences. *Journal of Molecular Evolution*, **36**, 586–595.

Siegel, R.A. (1993) Commentary on 'Neural networks in pharmacodynamic modeling. Is current modeling practice of complex kinetic systems at a dead end?', *Journal of Pharmacokinetics and Biopharmaceutics*, **20**, 413–416.

Song, X-H. and Yu, R-Q. (1993) Artifical neural networks applied to the quantitative structure–activity relationship study of dihydropteridine reductase inhibitors. *Chemometrics and Intelligent Laboratory Systems*, **19**, 101–109.

Syu, M-J., Tsai, G-J. and Tsao, G.T. (1993) Artificial neural network modeling of adsorptive separation. *Advances in Biochemical Engineering and Biotechnology*, **49**, 97–122.

Tetko, I.V., Luik, A.I. and Poda, G.I. (1993) Applications of neural networks in structure–activity relationships of a small number of molecules. *Journal of Medicinal Chemistry*, **36**, 811–814.

Topliss, J.G. and Edwards, R.P. (1979) Chance factors in studies of quantitative structure–activity relationships. *Journal of Medicinal Chemistry*, **22**, 1238–1244.

Veng-Pedersen, P. (1993) Response to Siegel's commentary. *Journal of Pharmacokinetics and Biopharmaceutics*, **20**, 417–418.

Veng-Pedersen, P. and Modi, N.B. (1993) Neural networks in pharamacodynamic modeling. Is current modeling practice of complex kinetic systems at a dead end? *Journal of Pharmacokinetics and Biopharmaceutrics*, **20**, 397–412.

Vieth, M. and Kolinski, A. (1991) Prediction of protein secondary structure by an enhanced neural network. *Acta Biochimica Polonica*, **38**, 335–351.

Villemin, D., Cherqaoui, D. and Cense, J.M. (1993) Neural networks studies: quantitative structure–activity relationship of mutagenic aromatic nitro compounds. *Journal of Chemical Physics*, **90**, 1505–1519.

Wade, R.C., Bohr, H. and Wolynes, P.G. (1992) Prediction of water binding sites on proteins by neural networks. *Journal of the American Chemical Society*, **114**, 8284–8285.

Weigel, U.-M. and Herges, R. (1992) Automatic interpretation of infrared spectra: Recognition of aromatic substitution patterns using neural networks. *Journal of Chemical Information and Computing Sciences*, **32**, 723–731.

Wikel, J.H. and Dow, E.R. (1993) The use of neural networks for variable selection in QSAR. *Bioorganic and Medicinal Chemistry Letters*, **3**, 645–651.

Zupan, J. and Gasteiger, J. (1991) Neural networks: a new method for solving chemical problems or just a passing phase? *Analytica Chemica Acta*, **248**, 1–30.

Zupan, J. and Gasteiger, J. (1993) *Neural Networks for Chemistry*. VCH, Cambridge.

9 Molecular similarity and complementarity based on the theory of atoms in molecules

P.L.A. POPELIER

9.1 Introduction

Quantum chemistry has reached the stage where it has become a tool for a wide variety of chemists. Indeed, whereas extremely accurate calculations can only be performed on small molecules of typically astrophysical or atmospherical interest, reliable theoretical information can now be obtained for fairly large organic molecules of pharmaceutical and biological importance. These results are typically generated by essentially solving the Schrödinger equation using widely available *ab initio* programs and an ever increasing computation power. This remarkably compact and powerful equation yields a wealth of information comprised in the wavefunction which is used to predict the expectation values of several operators of interest.

Probably the most intensively studied expectation value is the energy (and its derivatives). For some reason the expectation value of the density operator or the charge density ρ (Szabo and Ostlund, 1989) met less considerable attention, although it has been used as a basis for drug design (Jeffrey and Piniella, 1991). However, this three-dimensional function is the starting point for our notion of molecular similarity and complementarity. It is a function in real space describing objects in real space (i.e. molecules) and more importantly it is experimentally accessible via diffraction techniques. The principles of our charge-density analysis — which is part of a more encompassing quantum mechanical theory called 'atoms in molecules' (AIM) (Bader, 1990) — are not affected by the way the charge density is acquired. Its source may be a high-resolution X-ray measurement, a Hartree–Fock wavefunction (or beyond) or the charge density itself as obtained by the now increasingly more popular density functional methods (Labanowski and Andzelm, 1991). Recently, experimental densities are being studied within the AIM framework (Destro *et al.*, 1991; Destro and Merati, 1993; Klooster *et al.*, 1992) and most recently even in an attempt to explain the difference in pharmacological activity of two diphenylhydantoin derivatives (Ciechanowicz-Rutkowska *et al.*, 1994).

In the following sections, the theory of AIM is reviewed entirely in the light of molecular similarity and complementarity and the introduced concepts are illustrated by the well-known neuroleptic haloperidol (Reuben and Wittcoff, 1989). Molecular similarity will be defined by a direct comparison of fragment properties, both local and average. It will be shown that in limiting cases similarity can be great enough to enable transfer of a fragment from one system to another (Bader, 1986, 1991). Also of great importance in drug design is an understanding of the closely related topic of complementarity. Two types of complementarity will be distinguished, one related to the charge density and the other to the Laplacian of the charge density, $\nabla^2\rho$. In this manner AIM provides a unified and physical basis for similarity measures to be further processed in some mathematical framework.

9.2 The charge density as starting point

9.2.1 *Why the charge density?*

The charge density is a complicated function which contains an enormous amount of information about the molecule. It is well-defined (McWeeny, 1989) and experimentally accessible and therefore we believe that the analysis of chemical features of a molecule should start here, rather than at the level of molecular orbitals. The question is how to extract this information.

Before the 1970s crystallographers used the charge density mainly to solve the structure of a molecule, i.e. assign nuclear positions to the centers of charge density peaks. Figure 9.1 shows the charge density in the plane of one of the aromatic rings of haloperidol. This picture clearly shows that the density very near the nuclei dominates the function and that chemical information in, for example, the bonding zone is apparently lost in noise.

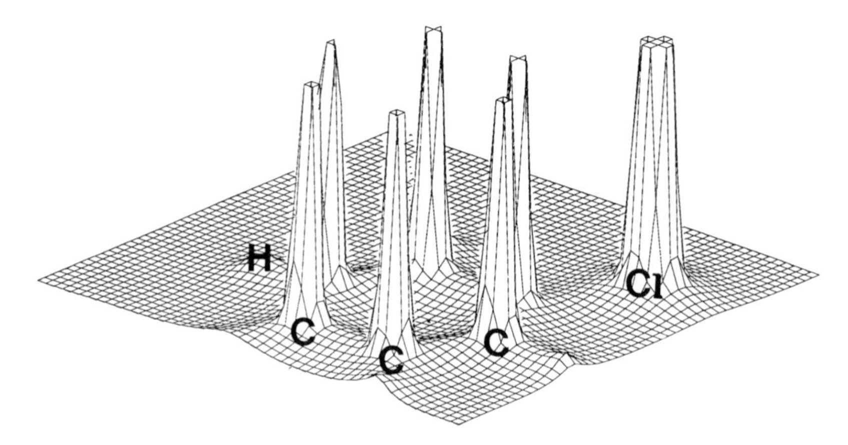

Figure 9.1 The charge density in the plane of the chlorosubstituted benzene ring of haloperidol. The peaks on the nuclei are cut off at an arbitrary value.

Nevertheless, all we need to know is there. In an attempt to unravel chemically interesting features from the charge density the so-called 'deformation density' (Coppens, 1977) was introduced. This function is obtained by subtracting the promolecular density from the molecular density. The promolecule is a superposition of isolated spherical atomic densities with the same nuclear coordinates as the original molecule. Although this approach is widely used in modern high-resolution crystallography it invokes an arbitrary reference density. As will become clear there is no need to introduce an external density: one can reveal the chemistry in the charge density via 'introspection'. This is accomplished by means of the gradient operator, which is an internal difference. The gradient vector field and the Laplacian of the charge density (see below) are two powerful examples of the idea of internal reference.

9.2.2 *A conventional way to compute the charge density*

In this section, a few selected formulae relevant for the definition of atomic properties will be reviewed. If we denote a solution of the Schrödinger stationary-state equation for N electrons and a fixed arrangement of nuclei by $\Psi(\mathbf{x}; \mathbf{X})$ where $\mathbf{x}$ is the set of electronic and spin coordinates and $\mathbf{X}$ the set of nuclear coordinates, then the corresponding (electronic) charge density is given by:

$$\rho(\mathbf{r}; \mathbf{X}) = N \sum_{\text{spins}} \left[\int d\tau_2 \ldots \int d\tau_N \Psi^*(\mathbf{x}; \mathbf{X}) \Psi(\mathbf{x}; \mathbf{X}) \right] \tag{9.1}$$

This equation bridges the gap between the rather abstract concept of a multidimensional wavefunction and a more intuitive three-dimensional function in real space (Ludeña and Kryachko, 1990). Schrödinger himself worried a lot about the probabilistic interpretation of his wavefunction concept as the interested reader will find in Moore (1989). The path from the wavefunction to the charge density involves several steps (Bader and Wiberg, 1987, 1990; McWeeny, 1989) the mathematical details of which are omitted here. To understand the charge density function better, it is helpful to know that $\Psi^*(\mathbf{x}; \mathbf{X})\Psi(\mathbf{x}; \mathbf{X})\, d\mathbf{x}_1\, d\mathbf{x}_2 \ldots d\mathbf{x}_N$ is the probability of finding each one of N electrons in a particular (abstract) volume element $d\mathbf{x}_i = dx_i dy_i dz_i \sigma_i$. Loosely speaking, this quantity is the chance that a given 'snapshot' ocurs where each electron is somewhere in the molecule. The charge density is a result of an integration (and summation) of this detailed information: all electrons contribute in some way to the presence of electronic charge in space.

Equation (9.1) can be abbreviated as follows:

$$\rho(\mathbf{r}) = N \int d\tau' \Psi^* \Psi \tag{9.2}$$

where $d\tau'$ represents summation over all spin coordinates and integration over the Cartesian coordinates of all electrons but one. This mode of integration is central to our development of atomic properties and is carried through in the definition of other density functions. For that purpose one inserts an operator $\hat{A}(\mathbf{r})$ between the wavefunctions in equation (9.2). It is clear then that the charge density is the particular case when $\hat{A}$ equals the unit operator. Note that when equation (9.2) is integrated once more over the remaining Cartesian coordinates, we recover the total number of electrons. In similar vein one can integrate the charge density over a portion of space only. Much of the review below will be devoted to showing that the charge density itself provides us in a natural way with meaningful portions of space to integrate over.

9.2.3 *What wavefunctions are within reach?*

Most *ab initio* programs approximately solve the Schrödinger equation based on the SCF–LCAO–MO (self-consistent field–linear combination of atomic orbitals–molecular orbitals) formalism. This approach to obtain the charge density is not a prerequisite but it is a standard and recommended route. To the author's knowledge only the *ab initio* programs[1] GAUSSIAN (M.J. Frisch *et al.*), CADPAC (R.D. Amos *et al.*) and GAMESS (M.W. Schmidt *et al.*) automatically produce a wavefunction file which acts as an interface between the *ab initio* results and the charge-density analysis software. For the latter the user has a choice between the suite of programs called AIMPAC (Bader *et al.*, 1994) or the single program MORPHY (Popelier, 1994). The latter program has been written from scratch in a highly modular and transparent style and yields a more automated analysis including the generation of postscript plots.

The wavefunction for the drug haloperidol or 4-(4-hydroxy-4-*p*-chlorophenyl-piperidino)-4′-fluorobutyrophenone was obtained using the 6-31G** basis set (Francl *et al.*, 1982) on the 4-21G (Pulay *et al.*, 1979) optimized structure (van Alsenoy *et al.*, 1989). A simple ball and stick representation of this molecule as a free base in its *trans* conformation (with respect to the C_{13}–C_{14} bond) is shown in Figure 9.2. Crystal structures have been reported of the free base and of the corresponding HBr salt (Reed and Schaefer, 1973; Datta, 1979). This neutral molecule has no symmetry, contains 49 nuclei and 198 electrons. The computation of its wavefunction was performed with 509 basis functions (or 913 Gaussian primitives) which required 65 central processor unit hours on a Silicon Graphics Indigo workstation. The SCF procedure converged (within the default convergence criterion of CADPAC) in 13 cycles to an energy of -1571.753799 a.u. Due to the direct-SCF technique (J. Almlöf

[1] Programs are listed on p. 240. Initialled authors refer to these programs.

(a)

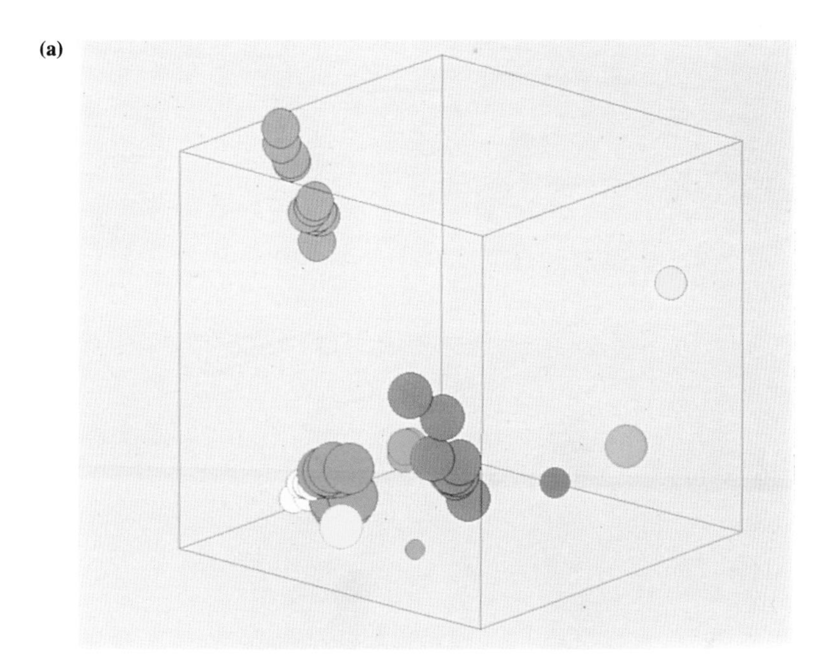

(b)

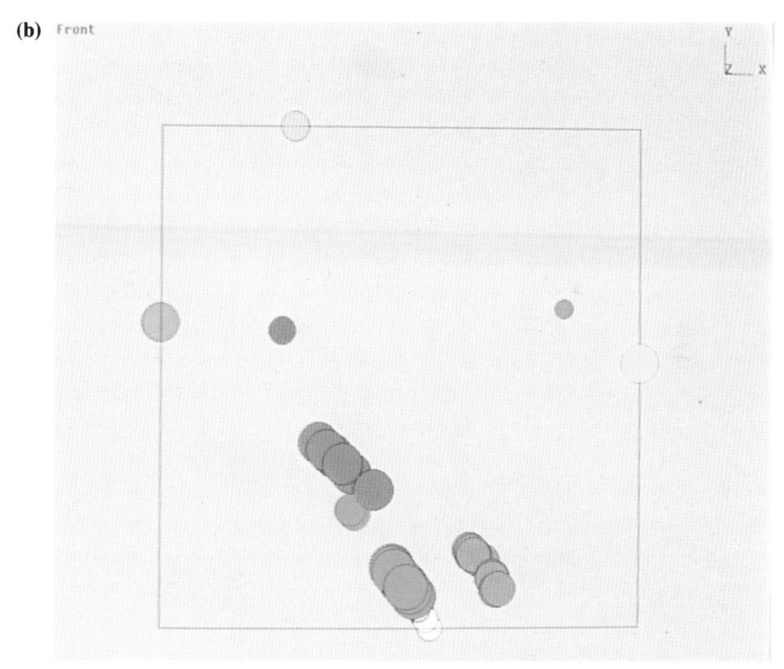

Plate 9.1 (a) A 3D representation of the 59 bond critical points in haloperidol, most of which are given in Table 9.1. The origin of the axis system is the vertex of the cube hidden behind the brown spheres. The cube edge pointing to the left is the *x*-axis, the one to the right is the *y*-axis. The vertical edge constitutes the *z*-axis. These three axes correspond to ϱ_b, $\nabla^2 \varrho_b$ and ε, respectively. (b) The (ϱ_b, $\nabla^2\varrho_b$) projection of the representation in (a). The lower left corner of the square box is the origin for this projection. Colour code: see Plate 9.2.

(a)

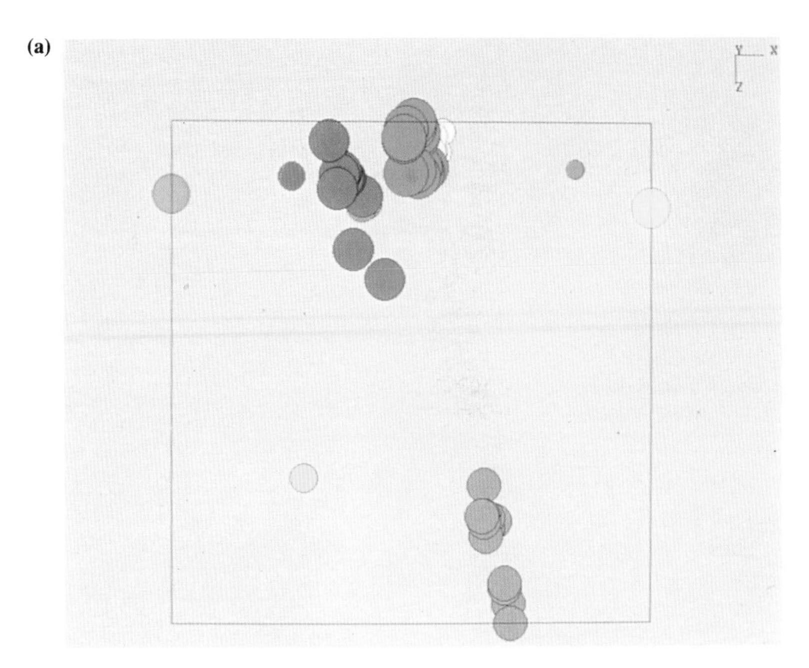

(b)

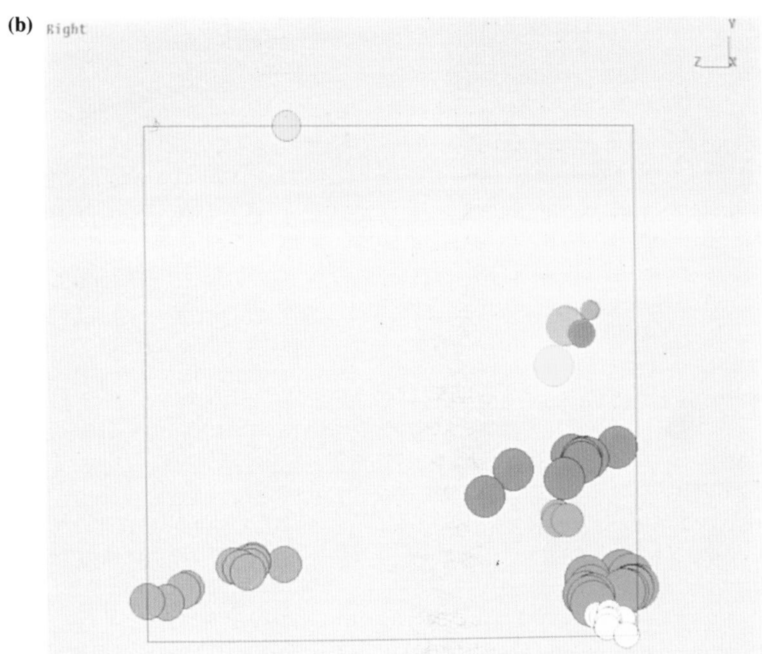

Plate 9.2 (a) The (ϱ_b, ε) projection of the representation in Plate 9.1(a). The upper left corner is the origin. (b) The ($\nabla^2\varrho_b$, ε) projection of the representation in Plate 9.1(a). The lower right corner is the origin. Colour code: C = O (yellow), C–O (red), C–N (blue), C_{arom}–H (white), C_{alif}–H (purple), C_{arom}–C_{arom} (grey), C_{alif}–C_{alif} (brown). O–H (dark green), C–Cl (bright green), C–F (orange).

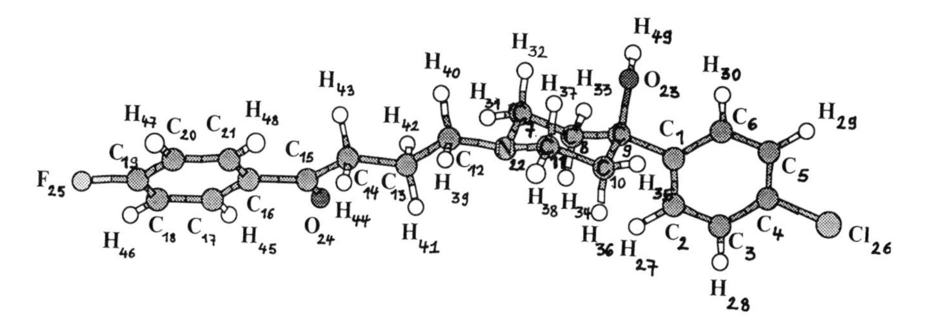

Figure 9.2 A simple representation of haloperidol. This molecule can be regarded as being constructed from four capped fragments: fluorobenzene, chlorobenzene, 4-hydroxypiperidine and butyraldehyde.

et al., 1982) this computation required a peak filespace usage of only 50 Mbytes. These technical data can hopefully convince one that an *ab initio* calculation at this level of theory is quite feasible.

9.3 Selected topics from AIM

9.3.1 *The topology of the charge density*

The topological properties (Bader and Beddall, 1972; Bader *et al.*, 1979a,b) of the gradient vector field $\nabla\rho$ are revealed by tracing gradient paths. These are curves in real space such that at every point of the path the gradient vector is tangent to the curve. This is expressed by the following equation:

$$\frac{\mathrm{d}\mathbf{r}}{\mathrm{d}l} = \frac{\nabla\rho(\mathbf{r})}{\|\nabla\rho(\mathbf{r})\|} \tag{9.3}$$

where l is the path length. Solving this system of ordinary differential equations with purely analytical methods fails because of convergence problems but a combined Chebyshev–Fourier fit to a numerically obtained set of gradient paths has been proven successful (Popelier, 1994). The whole topology of a molecule is hidden in equation (9.3) and can be generated from it using numerical integrators like the Runge–Kutta or Bulirsch–Stoer method (Press *et al.*, 1986).

The gradient paths are everywhere perpendicular to the family of isodensity contour surfaces since $\nabla\rho$ always points in the direction of maximum change in charge density. Using this general property, it is helpful to view a gradient path as the smooth limiting case of a broken line which pierces perpendicularly through a set of isodensity surfaces. Furthermore, gradient paths have an

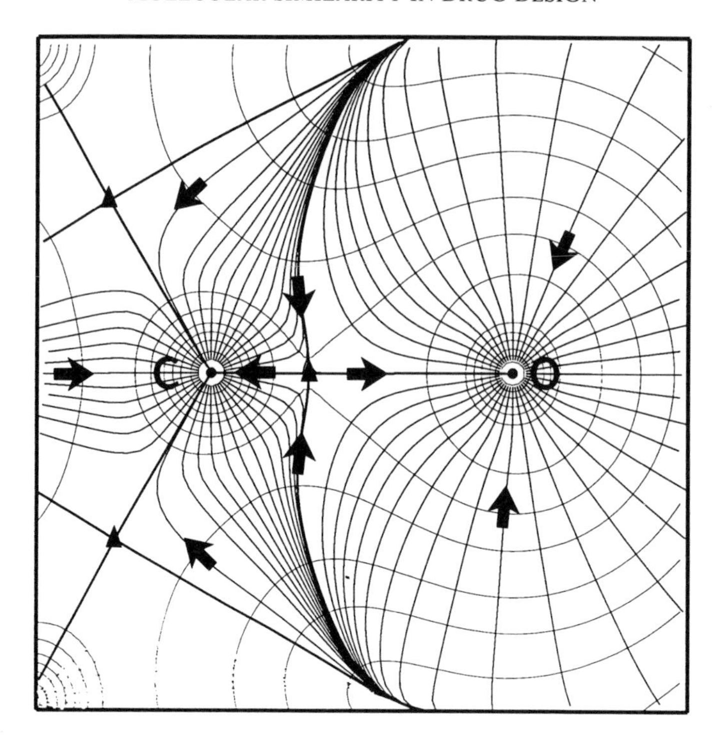

Figure 9.3 A superposition of the gradient vector field on a contour map of the charge density in the neighbourhood of the carbonyl group in haloperidol. The outer contour value corresponds to 0.001 a.u. and the values increase in order 2×10^n, 4×10^n, 8×10^n with n increasing in steps of unity from $n = -3$. The bond critical points or $(3, -1)$ are denoted by triangles and arrows indicate the sense of various gradient paths. Only trajectories attracted by the keto oxygen and carbon are shown (terminated arbitrarily near the nucleus) as well as two special sets of trajectories: interatomic surfaces and bond paths. See text for further explanation.

origin (which can be at infinity) and have a terminus. The gradient vector field of the keto group in haloperidol is shown in Figure 9.3. The observation that each nucleus attracts a subset of gradient paths intuitively leads to the concept of an *atomic basin.* Indeed the gradient vector field partitions itself in a natural way into atoms, which are defined as the union of an attractor (the nucleus) and its associated basin. From Figure 9.3 it can also be seen that there is a special subset of gradient paths which are not attracted to a nucleus but to particular points in between the nuclei. This set is called the *interatomic surface* and is described by the following equation:

$$\nabla\rho \cdot \mathbf{n}(\mathbf{r}) = 0 \qquad \forall\, \mathbf{r} \in \text{surface} \tag{9.4}$$

This equation states that the gradient of the electron density is perpendicular to the normal $\mathbf{n}$ of the surface. Thus the gradient of ρ is parallel to the surface, or finally, the surface *is* a bundle of gradient paths. Since exactly one gradient path passes through every (non-critical) point in space, this bundle can not be crossed by other paths and therefore the interatomic surface is often referred to as a *zero flux surface*.

The particular points in between the nuclei are critical points since here $\nabla\rho = 0$ and because of their geometrical position they are called *bond critical points*. They are denoted by $(3, -1)$ where the first integer (or rank) refers to the number of non-zero eigenvalues of the Hessian (or second derivative matrix) of ρ. The second integer (or signature) is the 'sums of signs' of the eigenvalues after according $+1$ to a positive eigenvalue and -1 to a negative. In general, the coordinate independent eigenvalues of the Hessian are used to classify the four types of (rank three) critical points that can occur in a three-dimensional scalar function like the charge density. So far we have encountered $(3, -1)$ and $(3, -3)$ where the latter is a maximum (rather than a saddle point) and coincides with the nucleus. In the center of the benzene ring for example we find a *ring critical point* or $(3, +1)$. All gradient paths inside the ring and in the plane of the ring originate at this point and terminate at the ring nuclei or the associated bond critical points. The remaining type is a true minimum designated by $(3, +3)$ and is known as *cage critical point*.

The other special subset of gradient paths appearing in Figure 9.3 is the *molecular graph*. For a given configuration of nuclei it is defined as the network of bond paths linking pairs of neighbouring nuclear attractors. A better understanding of a bond path follows from realizing that it is made up by two gradient paths originating at a given $(3, -1)$ critical point and terminating at the neighbouring nuclei or attractors. They define a line (which is sometimes very curved) linking the nuclei along which $\rho(\mathbf{r})$ is a maximum with respect to any neighbouring line. If instead of the charge density we were to study a temperature function, the bond path would be like a 'hot wire' connecting red-hot spots in space. Such a line is found between every pair of nuclei whose atomic basins share a common interatomic surface, and in the general case, it is referred to as an *atomic interaction line* (Runtz *et al.*, 1977; Bader and Essén, 1984). A bond path can now be more rigorously designated as an atomic interaction line for an equilibrium geometry because then this line satisfies the necessary and sufficient condition that atoms be bonded to one another (Bader and Essén, 1984).

9.3.2 Local properties

In this section we introduce and interpret four local quantities constituting a compound similarity index. The solution of the system in equation (9.3) reveals the topology of the molecule, i.e. the presence and positions of critical points $\mathbf{r}_c$. For a characterization of a molecular system by local properties we are guided by the topology because it tells us at which single point we have to evaluate a given function (for example, the charge density). The bond critical points, for example, are special points where the value of ρ yields information about the bond strength (Boyd and Choi, 1985) even of hydrogen bonds (Carroll and Bader, 1988). This quantity denoted by ρ_b can be fitted to an expression to obtain bond orders of 1.0, 1.6, 2.0 and 3.0 for ethane, benzene, ethene and ethyne, respectively (Wiberg *et al.*, 1987).

Another property giving useful information about bonds is $\nabla^2\rho$. This function is in line with the idea of 'introspection' introduced above and again avoids the needs for arbitrary external references. Experimenting with a one-dimensional function $f(x)$ illustrates that where the curvature of $f(x)$ is negative, or $d^2f(x)/dx^2 < 0$, the value of $f(x)$ is greater than the average of its values at the neighbouring points, $x - dx$ and $x + dx$. This statement is also valid in three dimensions, i.e. the value of $\rho(\mathbf{r})$ is greater than the average of its values over an infinitesimal sphere centered on $\mathbf{r}$ when the sum of the three curvatures of ρ is negative or when $\nabla^2\rho < 0$. Accordingly charge is locally depleted when $\nabla^2\rho > 0$.

This function when evaluated at the $(3, -1)$ critical point is denoted by $\nabla^2\rho_b$ and discriminates between two types of bonds: *shared interactions* and *closed-shell interactions*. The former class encompasses the covalent or polar bonds where electronic charge is shared by the two nuclei. The latter class contains hydrogen bonds, van der Waals complexes, ionic bonds and noble gas repulsive states. This distinction can be readily understood by viewing the Laplacian as the trace of the Hessian of the charge density. Since we consider a bond critical point, two curvatures (eigenvalues λ_1 and λ_2) of the Hessian are negative and one is positive (λ_3). The curvature along the interaction line is positive because charge density is locally depleted at $\mathbf{r}_c$ relative to the neighbouring points along the line and ρ is a minimum along this line. On the other hand, the two negative eigenvalues express the fact that ρ is a maximum at $\mathbf{r}_c$ in the interatomic surface where it is locally concentrated with respect to other points on the surface. The formation of a bond is the result of a competition between the perpendicular contractions of ρ towards the bond path and the parallel expansion of ρ away from the interatomic surface. If the latter effect dominates then a separate concentration of charge density occurs in each of the atomic basins and the Laplacian is positive at $\mathbf{r}_c$ since the positive eigenvalue is larger than the sum of the negative ones. Hence the name closed-shell interaction. On the other hand, if the perpendicular contractions of ρ dominate, the Laplacian is negative. In that case, charge is concentrated in the internuclear region and therefore we coined this a shared interaction.

Given that $|\lambda_1| > |\lambda_2|$ the ellipticity of the bond or ε is defined as:

$$\varepsilon = \lambda_1/(\lambda_2 - 1) \tag{9.5}$$

It provides a measure for the extent to which charge is preferentially accumulated in a given plane (Bader *et al.*, 1983). Obviously for cylindrically symmetric bonds ε vanishes and this measure increases with increasing

π-character. In fact ε recovers the anticipated consequences of the conjugation and hyperconjugation models of electron delocalization (Bader *et al.*, 1983).

Finally, there are some properties related to the geometry of the bond path and the position of the bond critical point. Trends in the distance of the (3, −1) critical point (denoted by $r_b(A)$ for an arbitrary atom A) from the respective nuclei can be understood via substitution effects to common functional groups (Slee, 1986a,b; Slee *et al.*, 1988). Also, the length of a traced bond path can be contrasted to the distance (along a straight line) between the nuclei. The bond path angle is probably better suited to characterize the degree of bending of a bond path, which is observed in electron-deficient systems (Bader and Legare, 1992) or small ring hydrocarbons (Runtz *et al.*, 1977; Boyd and Choi, 1985; Wiberg *et al.*, 1987). This bond path angle, α_b is defined as the angle subtended at a nucleus by the pair of bond paths linking it to the two nuclei which define the corresponding geometrical bond angle α_e.

Let us conclude this brief review on local properties with the two following remarks. The positions of the critical points are determined by the total charge density and therefore reflect the properties of the whole system. That is, although we focus on functions locally evaluated at these critical points, we still recover properties of the whole system and our local properties are influenced by changes anywhere in the system. Secondly, topological information combined with quantitative evidence for a set of local properties can demonstrate the existence or absence of a hydrogen bond in a molecule. A convincing example is the intramolecular hydrogen bond in creatine and carbamoyl sarcosine (Popelier and Bader, 1992).

9.3.3 *Integrated properties*

The topology also determines the regions of space to which an atom is confined. By an integrated property we mean a set of numbers (scalar, vector, tensor) associated with an atom after integrating some corresponding function over the volume of the atom. Again, this approach is very direct since the integration procedure happens in real space rather than in an abstract multidimensional Hilbert space.

The atomic subspace is a complete space just like ordinary Cartesian space since it is described by three independent coordinates ranging in principle from $-\infty$ to $+\infty$ such as the path parameter s which is related to the path length l. The two other coordinates are the usual spherical polar angles θ and ϕ which because of their cyclic nature need to range from 0 to π (or 2π) only. The atom is in fact a mapping from (s, θ, ϕ) to (x, y, z) space. In a molecule, every atomic mapping coexists with any other and each mapping describes a finite part of real space, such that Cartesian space is completely filled. One of the attractive properties of the idea of an atomic subspace is that it satisfies the virial theorem, just like the total system does. This will become clear below.

The expectation value of an operator $\hat{A}$ is defined as:

$$\langle \hat{A} \rangle = \int \mathrm{d}\tau \Psi^* \hat{A} \Psi \tag{9.6}$$

If this operator is Hermitian then

$$\int \mathrm{d}\tau \Psi^* (\hat{A}\Psi) = \int \mathrm{d}\tau (\hat{A}\Psi)^* \Psi \tag{9.7}$$

For atomic properties this is no longer true even if $\hat{A}$ is Hermitian. Indeed we have:

$$N \int_{\Omega} \mathrm{d}\tau \int \mathrm{d}\tau' \Psi^* (\hat{A}\Psi) \neq N \int_{\Omega} \mathrm{d}\tau \int \mathrm{d}\tau' (\hat{A}\Psi)^* \Psi \tag{9.8}$$

where $N \int_{\Omega} \mathrm{d}\tau \int \mathrm{d}\tau'$ is the label of an atomic integration. With an eye on equation (9.2) this integration procedure becomes immediately clear since the only new element is $\int_{\Omega} \mathrm{d}\tau$ which corresponds to a volume integral of the property density $\rho_A = N \int \mathrm{d}\tau' \Psi^* (\hat{A}\Psi)$.

The reason for the violation of equation (9.6) for atomic properties is due to one of Green's theorems which can take the following form $\int_{\Omega} \mathrm{d}\tau \int \mathrm{d}\tau' [\Psi^* \nabla^2 \Psi - (\nabla^2 \Psi)^* \Psi] \mathrm{d}\tau = \oint \mathrm{d}S (\Psi^* \nabla \Psi - \Psi \nabla \Psi^*) \cdot \mathbf{n}$. So if the integration is limited to a subsystem Ω bounded by a surface $S(\Omega)$ then a surface term arises. Because of this non-vanishing surface term or asymmetry in expectation values, we must define the atomic property as the mean of the two sides of equation (9.6):

$$A(\Omega) \equiv \langle A \rangle_{\Omega} = N \int_{\Omega} \mathrm{d}\tau \int \mathrm{d}\tau' [\Psi^* (\hat{A}\Psi) + (\hat{A}\Psi)^* \Psi] \tag{9.9}$$

If $\hat{A}$ is set to the unit operator we obtain the electron population of an atom in a molecule or its average number of electrons $N(\Omega) = \int_{\Omega} \mathrm{d}\tau \rho(\mathbf{r})$. The net charge on an atom $q(\Omega)$ is then given by the sum of its nuclear charge $Z_{\Omega} e$ and its average electronic charge $-N(\Omega)e$:

$$q(\Omega) = [Z_{\Omega} - N(\Omega)]e \tag{9.10}$$

Higher moments of the atom's charge distribution can readily be defined by setting the operator $\hat{A}$ to the appropriate expression constructed from the set $\{x_{\Omega}, y_{\Omega}, z_{\Omega}, r_{\Omega}\}$ where $\mathbf{r}_{\Omega}$ is the position vector of an electron with respect to the nucleus of an atom. Its components are $(x_{\Omega}, y_{\Omega}, z_{\Omega})$ and its magnitude r_{Ω}. The first moment of an atom's charge distribution $\mathbf{M}_{\Omega}$ is then:

$$\mathbf{M}(\Omega) = -e \int_{\Omega} \mathrm{d}\tau \mathbf{r}_{\Omega} \rho(\mathbf{r}) \tag{9.11}$$

This quantity tells us in which direction and to what extent the atom's charge density is (dipolarly) polarized by determining the displacement of the atom's centroid of negative charge from the position of its nucleus. In this chapter we will only quote the magnitude of this vector, $M(\Omega)$, because it is independent of the choice of coordinate system. An example of a component of the atomic quadrupolar polarization tensor is $\mathbf{Q}_{zz}(\Omega) = -e\int_\Omega d\tau \frac{1}{2}(3z_\Omega^2 - r_\Omega^2)$. Rather than exhaustively discuss all the integrated properties that have been derived so far (Bader and Popelier, 1993), we will focus on the electronic energy of an atom $E(\Omega)$. The expression for this important quantity follows from the atomic virial theorem which is proven using Green's theorem. Indeed, the essential idea is to compute a volume integral and a surface integral and to equate them by virtue of Green's theorem. The virial theorem relates the kinetic energy of a system to its potential energy and it takes the following form for an atom:

$$-2T(\Omega) = V_b(\Omega) + V_s(\Omega) + L(\Omega) \tag{9.12}$$

where

$$T(\Omega) = N\int_\Omega d\tau \int d\tau'(-\hbar^2/4m)[\Psi^*\nabla^2\Psi + (\nabla^2\Psi^*)\Psi] \tag{9.13}$$

$$V_b(\Omega) = N\int_\Omega d\tau \int d\tau' \Psi^*(-\mathbf{r}\cdot\nabla\hat{V})\Psi \tag{9.14}$$

$$V_s(\Omega) = \oint dS(\Omega)\mathbf{r}\cdot\sigma\cdot\mathbf{n}(\mathbf{r}) \tag{9.15}$$

$$L(\Omega) = 0 \tag{9.16}$$

In this set of equations $\hat{V}$ is the potential energy operator, σ is the quantum mechanical stress tensor (see equation (6.12) in Bader, 1990) and $\mathbf{n}$ represents the normal to the surface. The kinetic energy $T(\Omega)$ is definable for an atom since the two existing types of kinetic energy $K(\Omega)$ and $G(\Omega)$ are equal. This is true because in general $K(\Omega) = G(\Omega) - (\hbar^2/4m)\int_\Omega d\tau\nabla^2\rho(\mathbf{r}) = G(\Omega) + L(\Omega)$ and $L(\Omega) = 0$. The potential energy contributed by the atomic basin $V_b(\Omega)$ is the integrated average of the virial of the Ehrenfest force on an electron in the basin of the atom. Its surface counterpart $V_s(\Omega)$ should be added to $V_b(\Omega)$ to yield the total virial (or potential energy) of the atom, $V(\Omega)$. So finally we obtain

$$-2T(\Omega) = V(\Omega) \tag{9.17}$$

This atomic virial theorem is identical to that of the total system which proves the statement made above that the atomic basins are remarkable fragments with respect to energy balance and partitioning. The root of the validity of equation (9.17) is the fact that the Laplacian vanishes when integrated over the atomic basin (or $L(\Omega) = 0$) which is also true for the total system. Finally the energy of an atom $E(\Omega)$ can be equated to $-T(\Omega)$ since this energy is defined as $T(\Omega) + V(\Omega)$.

One last property to be discussed here is the *atomic volume* $v(\Omega)$. Of course this would be infinite if no outer boundaries were imposed on the entire molecule. It is important to point out that this boundary is not determined by some fixed radius which only depends on the atom type (like the typical van der Waals radius). Rather our contour surface should be determined by the quantum mechanics of the *whole* system. A straightforward choice is to delimit the atomic basin by a charge density contour surface. Although the numerical value of ρ for this surface is arbitrary, the 0.001 a.u. envelope yields molecular sizes in agreement with kinetic theory data for gas-phase atoms and molecules. The slightly smaller volume obtained using the 0.002 a.u. contour has been useful in describing the closer packing found in the solid state (Bader and Preston, 1970). We then define an atomic volume as the space enclosed by a particular charge density envelope and the parts of the interatomic surfaces that border the atomic basin within the molecule. It was shown before (Bader *et al.*, 1987) that an approximation to the van der Waals surface in terms of overlap of atomic hard spheres fails to recover important chemical features as was made evident in departures of the envelope from atomic sphericity.

In spite of recent gradual improvements of the original integration procedure (Biegler-König *et al.*, 1981, 1982) obtaining integrated properties is still an expensive process but we believe that the efficiency limit has not yet been reached, which is why work to perform a practically completely analytical integration is currently in progress. A typical way to probe the accuracy of such an integration is the value of $L(\Omega)$ which should ideally be zero. The atomic fragments exhaust space and since the integrated properties are a direct consequence of this real space partitioning, they must be additive. Figure 9.4 shows a three-dimensional picture of a methyl group that may nourish one's intuition in representing groups of atoms as defined by AIM. Therefore group properties can be readily obtained by merely summing the values of the constituent atoms. This principle is applied in order to define peptide fragments in proteins. Indeed it has been shown that for example the glycyl group $|HNCH_2C(=O)|$ is transferable in oligopeptides in both planar and helix conformations (Bader *et al.*, 1992; Chang and Bader, 1992; Popelier and Bader, 1994).

9.4 Similarity

In this section we will inspect the data for haloperidol and discuss some trends. But first we emphasize the nature of our notion of similarity. Carbó (1980) and collaborators published a paper with the title 'How similar is a molecule to another? An electron density measure of similarity between two

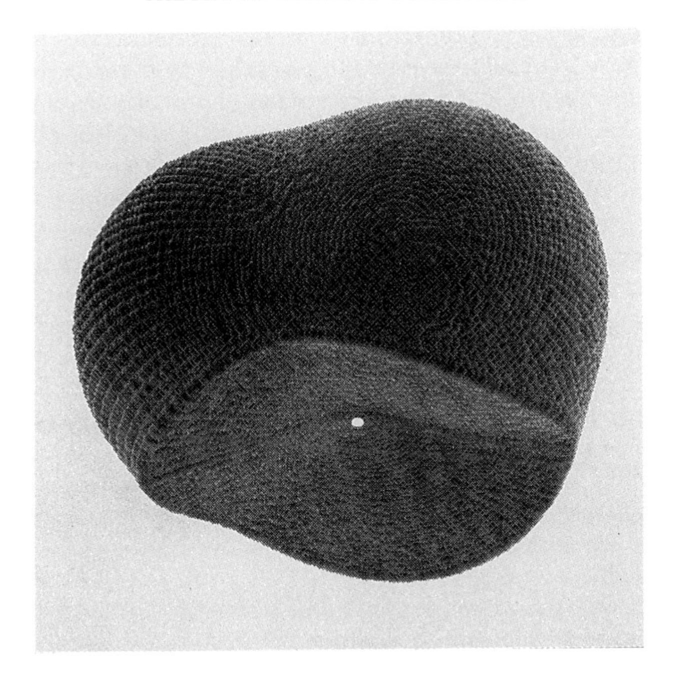

Figure 9.4 A 3D representation of the methyl group in methanol. This group consists of four joint atomic basins bounded by the 0.001 a.u. charge-density contour surface and the interatomic surface separating the carbon and the oxygen atom. The corresponding bond critical point is marked by a dot.

molecular structures'. In the introduction they state '... As far as we know there is no answer from a quantum chemical viewpoint to this question ...'. In order to address this central question they defined a matching measure of the electron distribution between molecules A and B as $\varepsilon_{AB} = \int_V |\rho_A - \rho_B|^2 \, dV$. Note that this measure depends on three rotation angles and three translation vector components. In the end, Carbó's algorithm comes down to searching for a maximum value of the integral $\int_V \rho_A \rho_B \, dV$. When first proposed, this similarity index could not be used other than with crude densities. Later, however, this idea was reapplied with better densities (Bowen-Jenkins *et al.*, 1985) and very recently revisited in the context of density functional theory (Lee and Smithline, 1994) and even in the context of AIM (Cioslowski and Nanayakkara, 1993). These indices together with some similar variants (see references in Bowen-Jenkins *et al.*, 1985) all have in common that the two molecules have to be aligned in some way as to maximize their similarity index. This is a cumbersome procedure leading to several problems: the similarity index is dependent on the method chosen for molecular matching (Dean, 1990), many local maxima occur requiring a sophisticated global maximization algorithm (Lee and Smithline, 1994), problems arise if structures with large groups are added because it is not clear if the comparison should be limited to small regions or performed over the whole molecule (Bowen-Jenkins *et al.*, 1985). Aside from these technical difficulties there is the fundamental objection that these indices are dominated by the overlap between the very peaked core densities (Richard and Rabinowitz, 1987).

Abandoning the aforementioned philosophy, we believe similarity indices should rather be definable and computable for a single molecule and *directly* comparable with those of other molecules. In this process we use the insight AIM offers us, i.e. via the topology it tells where and how the chemistry can be found in the charge density. Indeed, if one does not know about critical points nor atomic basins, one is forced to compare the whole of the charge density of two molecules. Moreover there is no need to perform a calculation for every comparison: similarity is a set of 'monomer' (single molecule) properties and not a 'pair' property as in the Carbó-like indices. Although these methods are complete in the sense that they allow a comparison of *all* points in space, the resulting indices are biased by the comparison of many irrelevant points.

A second principle of our present approach is that similarity is basically a multicomponent quantity. Many references to vectors of chemical descriptors can be found under the heading 'product spaces' of Johnson's review on mathematical spaces underlying molecular similarity analysis (Johnson, 1989). An example of an AIM chemical descriptor vector is $(\rho_b, \nabla^2\rho_b, \varepsilon, r_b(A), r_b(B))$. Proximity measures like the city-block, euclidean metric, correlation coefficient and Tanimoto coefficient (Johnson, 1989) can be used to classify single atoms or groups of them according to similarity.

The final principle is that AIM, when used as a tool, acts like an interface between the wavefunction and chemical concepts. In view of its depth (see chapter 8 of Bader, 1990) it provides a rigorous physical basis (Srebrenik *et al.*, 1978; Bader *et al.*, 1978; Bader, 1988) to fuel a mathematical structure to organize, compare and filter the provided data. As explained in the next section AIM can supply usual information as a starting point for graph theory. Because of this physical basis a pharmacological prediction using AIM should not suffer from the well-known drawback of QSAR, that is being unable to suggest novel (non-congeneric) classes of structures binding to a specified site.

9.5 Complementarity

9.5.1 *Basic tenet*

We do not regard complementarity as the opposite of similarity: two identical molecules may be complementary to each other. We reserve the word 'dissimilarity' for a lack of similarity. Complementarity parallels similarity and attempts to elucidate how molecules *interact* with another one. Put bluntly, complementarity measures are similarity measures for intermolecular interaction. In the broadest sense this interaction includes acid–base behaviour, hydrogen bonding, ligand–receptor docking but also solid-state packing. We will show that AIM can be used to reveal the intrinsic features of a single molecule in connection with its interaction with other molecules. It is therefore possible to compare the complementarity of two molecules.

In order to understand packing a molecular shape provided by the 0.002 a.u. contour surface was used before (Barrett and Meyer, 1967). This type of complementarity was previously coined *van der Waals complementarity* (Bader *et al.*, 1992). The other type is called *Lewis complementarity* and is based on the Laplacian of the charge density rather than on ρ itself. It has been shown before (Carroll and Bader, 1988) that a van der Waals surface is of no help in deciding in Lewis complementarity because these surfaces are mutually interpenetrated in, for example, the formation of weak hydrogen bonds.

9.5.2 *The topology of the Laplacian*

In analogy with the charge density extrema, the (scalar) Laplacian function can be searched for and classified according to rank and signature. The critical points occur when $\nabla(\nabla^2\rho) = \mathbf{0}$ and the eigenvalues of the Hessian of $\nabla^2\rho$ are the principal curvatures of $\nabla^2\rho$ at the critical point. In the outer quantum shell of an atom, one can find a region where $\nabla^2\rho < 0$, which is called the *valence-shell charge concentration* (VSCC) (Bader *et al.*, 1984, 1988). In the following we adopt the convention to refer to the negative of the Laplacian $(-\nabla^2\rho)$ so that a positive value corresponds to a local charge concentration. In a free atom the VSCC is a sphere which becomes (very) distorted if the atom is in chemical combination. The surface of the VSCC is no longer one of uniform concentration but will contain 'hills and valleys'. In a two-dimensional surface we can only encounter $(2, -2)$, $(2, 0)$ and $(2, +2)$ type of critical points. Since we are confined to a surface of maximum charge concentration we can formally add $(1, -1)$ to the aforementioned types of rank 2 critical points. We then obtain $(3, -3)$ or a maximum in $-\nabla^2\rho$, $(3, -1)$ or a saddle point and $(3, +1)$ or a minimum.

The electronic charge density does not offer any evidence of the localized bonded and non-bonded pairs of electrons evoked in the well-known Lewis model of electronic structure (Bader, 1991). The Laplacian, however, does. It has been observed in numerous cases that maxima or $(3, -3)$ points correspond in number, location and magnitude to a localized pair of electrons. In fact the Laplacian provides the physical basis for the valence-shell electron pair repulsion (VSEPR) model of Gillepsie (1972), which is a natural extension of the localized electron pair model of Lewis. This is a gratifying result since VSEPR theory is an empirical theory based on a vast amount of data.

The $(3, +1)$ critical points in $-\nabla^2\rho$ correspond to a 'hole' in the VSCC and since charge is very much depleted in the zone around it, this point characterizes an acidic zone in a molecule. On the other hand the $(3, -3)$ points correspond to basic zones. Since the Lewis model encompasses chemical reactivity as well as generalized acid–base reaction, we can re-interpret the $(3, +1)$ critical point as a center of electrophilic attack and $(3, -3)$ as a center of nucleophilic attack. It is on this basis that the Laplacian becomes the central vehicle in a unified theory for reactivity. Electrostatic potential maps have been used to make similar predictions. However these maps fail to predict nucleophilic attack without artificial assumptions and extra constraints. A thorough theoretical explanation of why the Laplacian performs so well has not yet been given to our knowledge (Carroll and Bader, 1988; Carroll *et al.*, 1988). Indeed, the use of the Laplacian for these purposes relies on two assumptions: firstly, molecules tend to align themselves such as to put the non-bonded maximum and minimum on one line with the associated nuclei; secondly, when more non-bonded charge concentrations are present, interaction with widely different values for the Laplacian interaction will preferentially occur with the one with the highest absolute value of $\nabla^2\rho$ (Slee and Bader, 1992).

The following examples just indicate the power of this method: preferential protonation of the keto oxygen in formamide and substituted derivatives (Slee and Bader, 1992), angle of attack of approximately 110° (with respect to the carbon–oxygen axis) for the nucleophilic approach to C=O in the solid state (Bürgi and Dunitz, 1983), successful predictions for the Michael addition reaction with acrylic acid derivatives (Carroll *et al.*, 1989) and prediction of sites of electrophilic attack in a series of substituted benzenes (Bader and Chang, 1989).

9.5.3 *The atomic graph*

It is convenient to construct a graph to express the topology of the VSCC. This analogue of the molecular graph is called the *atomic graph.* Since in a strict mathematical sense a graph is only a representation of a connectivity scheme we can represent the 'atomic graph' in a formal way without reference to its actual three-dimensional shape. It has been proven (Bader *et al.*, 1992) that any atomic graph can be drawn as a *planar graph.* Because of this canonical way of displaying atomic graphs, it is appealing to create an exhaustive library of them. For some examples we refer the reader to Bader *et al.* (1992) and Bader and Chang (1989).

9.6 Example: Haloperidol

9.6.1 Local properties

The information presented here is not meant to form a complete study of haloperidol but merely to illustrate the concepts introduced above. Data will be contrasted in two ways: internally or within the haloperidol molecule and in comparison with the corresponding data of the four fragment molecules constituting haloperidol. Indeed, it is straightforward to discern the following fragments in haloperidol: benzyl fluoride, butyraldehyde, 4-hydroxypiperidine and benzyl chloride. Wavefunctions were generated for these fragments at

Table 9.1 Bond critical point data (in atomic units) for haloperidol. See text for explanation of the symbols

Type	A–B	ρ_b	$\nabla^2\rho_b$	ε	r_b (A)	r_b (B)
C_{arom}–C_{arom}	C_1–C_2	0.325	−0.996	0.226	1.311	1.315
	C_2–C_3	0.327	−1.014	0.237	1.293	1.320
	C_3–C_4	0.336	−1.085	0.265	1.223	1.373
	C_4–C_5	0.336	−1.090	0.267	1.375	1.218
	C_{16}–C_{17}	0.326	−1.009	0.208	1.303	1.324
	C_{17}–C_{18}	0.331	−1.035	0.229	1.298	1.306
	C_{18}–C_{19}	0.338	−1.126	0.277	1.190	1.412
	C_{19}–C_{20}	0.339	−1.125	0.288	1.404	1.194
C_{arom}–H	C_2–H_{27}	0.301	−1.181	0.017	1.244	0.779
	C_3–H_{28}	0.302	−1.202	0.018	1.255	0.766
	C_{18}–H_{46}	0.300	−1.189	0.021	1.255	0.765
	C_{21}–H_{48}	0.302	−1.200	0.009	1.250	0.773
	C_6–H_{30}	0.304	−1.224	0.018	1.261	0.758
	C_5–H_{29}	0.302	−1.201	0.017	1.255	0.766
C–Cl	C_4–Cl_{26}	0.175	−0.269	0.042	1.433	2.000
C–F	C_{19}–F_{25}	0.239	0.350	0.204	0.821	1.760
C_{alif}–C_{alif}	C_1–C_9	0.263	−0.721	0.073	1.467	1.426
	C_7–C_8	0.256	−0.682	0.031	1.449	1.452
	C_9–C_{10}	0.259	−0.699	0.033	1.506	1.399
	C_{10}–C_{11}	0.257	−0.687	0.029	1.445	1.453
	C_{12}–C_{13}	0.255	−0.677	0.039	1.469	1.437
	C_{13}–C_{14}	0.251	−0.652	0.012	1.404	1.499
	C_{14}–C_{15}	0.267	−0.750	0.043	1.371	1.504
	C_{15}–C_{16}	0.278	−0.803	0.090	1.438	1.391
C–N	C_{11}–N_{22}	0.267	−0.867	0.048	1.058	1.737
	C_{12}–N_{22}	0.269	−0.880	0.042	1.065	1.724
	C_7–N_{22}	0.269	−0.880	0.047	1.067	1.722
C–O	C_9–O_{23}	0.233	−0.295	0.032	0.894	1.865
C=O	C_{15}–O_{24}	0.407	0.397	0.049	0.746	1.554
O–H	O_{23}–H_{49}	0.370	−0.220	0.027	1.468	0.356
C_{alif}–H	C_{10}–H_{35}	0.292	−1.100	0.004	1.247	0.796
	C_{10}–H_{36}	0.292	−1.098	0.004	1.243	0.800
	C_{12}–H_{39}	0.295	−1.123	0.027	1.246	0.801
	C_{12}–H_{40}	0.288	−1.066	0.030	1.252	0.811
	C_{13}–H_{41}	0.294	−1.112	0.009	1.252	0.789
	C_{13}–H_{42}	0.294	−1.111	0.007	1.245	0.795

the same level of theory and for a nuclear geometry identical to that of haloperidol. A first important question is whether our similarity measures recover the various classes of chemical bonds. The data in Table 9.1 have been divided into ten different classes of bonds, a division which is supported by Plates 9.1 and 9.2 representing all 51 bonds in haloperidol by spheres in a 3D space spanned by $(\rho_b, \nabla^2\rho_b, \varepsilon)$. It is clear from this picture that the five classes with more than one member, i.e. C_{arom}–C_{arom}, C_{arom}–H, C_{alif}–C_{alif}, C–N and C_{alif}–H, each form a tight cluster. A closer inspection of the C_{arom}–C_{arom} class shows that this cluster is in fact split in two: the smaller subcluster represents the two pairs of benzene carbon–carbon bonds adjacent to the C–F or C–Cl bond. These four bonds show a somewhat higher ellipticity than the other members of their class which can be explained by the fact that halogens are π-donors. Moreover, this fine tuning is even correct in predicting that fluorine is a stronger π-donor than chlorine, since the former causes the largest increase in ε. We can explain the two outsiders of the C_{alif}–C_{alif} cluster along the same lines. They are identified as the bonds connecting the benzene rings to the rest of the haloperidol skeleton (C_1–C_9 and C_{15}–C_{16}). Again, the main reason for their edge position in the cluster is an increased ε which is essentially due to hyperconjugation (Bader *et al.*, 1983).

A further look at Table 9.1 reveals that the bond critical points of the aromatic C–C bonds adjacent to the halogen atoms are shifted away from the atoms bound to the halogens (C_4 and C_{19}). Combined with the increased atomic population on these atoms this effect is indicative of a charge transfer towards the halogens which in turn can be explained by their σ-acceptor character (Slee *et al.*, 1988). It is clear from projections onto the $(\rho_b, \nabla^2\rho_b)$, (ρ_b, ε) and $(\nabla^2\rho_b, \varepsilon)$, planes of all 51 points that ε is the most powerful discriminator for the details of the clusters as mentioned above.

At a quantitative level we recover a typical C_{arom}–C_{arom} bond order (n) range of 1.60 to 1.75 when using an expression given by Wiberg *et al.* (1987): $n = \exp[6.458(r_b - 0.252)]$. Also the remarkable positive value for $\nabla^2\rho_b$ for C–F has been noted before and corroborates the peculiarity of this bond (see p. 274 of Bader, 1990). It has been observed before that C–F is very similar in different chemical environments and that the electronegativity of the fluorine pushes the C–F bond critical point right up against the core of the carbon atom (Slee, 1986b) as can be deduced from r_b(A) in Table 9.1.

A comparison between Tables 9.1 and 9.2 proves that the small variations in local properties guarantee the transferability of information from the fragment molecules to the full molecule haloperidol. Quite expectedly, the largest differences occur near the fragment molecule's hydrogen cap replacing the rest of haloperidol. These differences will be excluded in the following discussion since they can be unreasonably large and not very meaningful.

The least sensitive parameter is r_b showing a maximum difference at C_{14}–C_{15} of only 0.005 a.u. or about 2% relative to the average magnitude of r_b at this point. A more sensitive probe is $\nabla^2\rho$, which shows a maximum

Table 9.2 Bond critical point data (in atomic units) for the fragments constituting haloperidol

Fragment	A–B	ρ_b	$\nabla^2\rho_b$	ε	r_b (A)	r_b (B)
Benzyl chloride	C_1–C_2	0.325	−1.007	0.242	1.341	1.285
	C_2–C_3	0.327	−1.010	0.235	1.302	1.312
	C_3–C_4	0.336	−1.087	0.265	1.220	1.376
	C_2–H_{27}	0.301	−1.195	0.016	1.252	0.771
	C_3–H_{28}	0.302	−1.203	0.019	1.257	0.764
	C_4–Cl_{26}	0.175	−0.270	0.040	1.437	1.996
4-Hydroxypiperidine	C_9–C_{10}	0.260	−0.708	0.028	1.523	1.378
	C_{10}–C_{11}	0.255	−0.676	0.028	1.444	1.454
	C_{11}–N_{22}	0.264	−0.853	0.040	1.031	1.765
	C_{10}–H_{35}	0.292	−1.104	0.006	1.251	0.792
	C_{10}–H_{36}	0.293	−1.103	0.007	1.247	0.796
	C_9–O_{23}	0.237	−0.414	0.155	0.902	1.857
	O_{23}–H_{49}	0.371	−0.221	0.027	1.468	0.356
Butyraldehyde	C_{12}–C_{13}	0.251	−0.658	0.020	1.480	1.426
	C_{13}–C_{14}	0.250	−0.651	0.019	1.400	1.503
	C_{14}–C_{15}	0.262	−0.732	0.049	1.332	1.543
	C_{15}–O_{24}	0.407	0.425	0.092	0.744	1.556
	C_{13}–H_{41}	0.295	−1.118	0.011	1.250	0.792
	C_{13}–H_{42}	0.295	−1.126	0.011	1.253	0.788
Benzyl fluoride	C_{16}–C_{17}	0.326	−1.008	0.244	1.332	1.295
	C_{17}–C_{18}	0.330	−1.027	0.238	1.301	1.303
	C_{18}–C_{19}	0.338	−1.111	0.291	1.206	1.395
	C_{18}–H_{46}	0.300	−1.186	0.021	1.253	0.767
	C_{19}–F_{25}	0.239	0.377	0.170	0.821	1.760

discrepancy of 0.028 a.u. or 7% at the C–F bond and which is remarkable since this bond is remote from the hydrogen cap. However, this effect can be understood by regarding the fluorobenzene as a conjugated system including the fluorine. To some extent, the differences in ε are parallel to those in $\nabla^2\rho$ but are more pronounced percentage-wise, which means that ε is the most sensitive indicator.

9.6.2 Integrated properties

Transferability is also preserved in the integrated properties as can be seen from Table 9.3. The average discrepancy in net atomic charge of the displayed atoms is only 0.012 *e*, with a maximum of 0.022 *e* on the keto oxygen, which is again due to the effect of the hydrogen cap. The values of the net charge on carbon, hydrogen and fluorine are in perfect agreement with earlier work on substituted benzenes (Bader and Chang, 1989). The magnitude of the dipole moment differs on average by 0.003 a.u., the volume by 0.33 a.u. and the energy by 94 kJ/mol, excluding the keto oxygen. It is worth mentioning that the chlorine is hardly affected by the absence of the rest of haloperidol since the volume and energy are virtually invariant from the fragment to the whole molecule. The energies of the glycyl groups in formylglycylamide and triglycine differ by only 12 kJ/mol (Bader *et al.*, 1992).

Table 9.3 Comparison of integrated atomic properties (in atomic units) for haloperidol and the corresponding fragment molecules

Type	Atom	Haloperidol molecule	$q(\Omega)$	$M(\Omega)$	$v(\Omega)$	$E(\Omega)$
C_{arom}	C_3	benzylchloride	0.121	0.167	76.83	−37.8003
			0.129	0.163	76.85	−37.7950
C_{alif}	C_{10}	4-hydroxypiperidine	0.212	0.064	51.80	−37.7474
			0.230	0.078	53.15	−37.8281
C_{keto}	C_{15}	butyraldehyde	1.157	0.965	40.71	−37.0456
			1.122	1.177	55.35	−36.9462
O_{keto}	O_{24}	butyraldehyde	−1.334	0.680	125.9	−75.4872
			−1.312	0.789	130.4	−75.7027
F	F_{25}	benzyl fluoride	−0.738	0.381	99.23	−99.7168
			−0.741	0.374	99.20	−99.8346
Cl	Cl_{26}	benzyl chloride	−0.298	0.102	172.3	−459.895
			−0.290	0.100	172.2	−459.897
H_{arom}	H_{28}	benzyl chloride	−0.008	0.096	47.11	−0.6406
			−0.002	0.096	46.80	−0.6377
H_{alif}	H_{43}	butyraldehyde	−0.062	0.110	49.62	−0.6593
			0.043	0.112	49.61	−0.6509

9.6.3 Complementarity

Here we focus on the topology of the Laplacian. In Table 9.4 the non-bonded critical points in this scalar function are listed for three atoms which are most likely involved in intermolecular interaction: the nitrogen atom, the keto oxygen and the hydroxyl oxygen. It is known that the antagonist haloperidol binds to the dopamine D_2 receptor (Tsuchhashi, 1992) and previously reported models for the closely related D_1 (agonist) binding site (Vinter and Gardner, 1994) have as a common feature interaction with hydroxy and amine groups. The pyramidal nitrogen has two non-bonded critical points one of which occurs in a tetrahedral arrangement around this atom together with the three carbons attached to nitrogen. In the crystal (Datta *et al.*, 1979) this nitrogen is protonated and the N...Br distance suggests a hydrogen bond. The orientation of the *N* lone pair was considered an important feature for dopaminergic activity and most receptor models described since have highlighted this observation. The other non-bonded

Table 9.4 Non-bonded critical points in $-\nabla^2\rho$ for selected atoms in haloperidol

Parameter	O_{24}	O_{24}	O_{23}	O_{23}	N_{22}	N_{22}
$\nabla^2\rho$	−6.79	−6.76	−6.70	−6.64	−3.48	−1.59
r	0.631	0.631	0.632	0.633	0.730	0.765
α*	105	105	100	100	106†	63, 55, 109

*This angle (in degrees) is defined as A − B − *nb* where *nb* is the non-bonded critical point belonging to the VSCC of atom B and atom A is attached to B.
†The three possible α angles are identical to the quoted accuracy.

critical point lies almost in the plane formed by the nitrogen and its two bonded critical points in the piperidine ring. It is situated roughly in between the atoms C_1 and C_5 *trans* to the other non-bonded critical point. The atomic graph of the pyramidalized nitrogen in a formamide rotamer studied before (Laidig and Bader, 1991) contained only one non-bonded critical point but two occurred in the planar rotamer.

As shown in Figure 9.5 the keto oxygen has two practically equivalent non-bonded maxima symmetrically positioned at either side of the C=O axis and lying in the C_{14}–C_{15}–C_{16} plane. The corresponding directions are typical candidates for hydrogen bonding. Finally, two equivalent non-bonded critical points are found on the hydroxyl oxygen atom, one of which lies in the O_{23}–C_9–C_1 plane. The angle formed by these two critical points and the hydroxyl oxygen is 140°. Interaction between a D_1 receptor site and the hydroxyl of the agonist 3′,4′-dihydroxynomifensine has been reported by Dandridge *et al.* (1984).

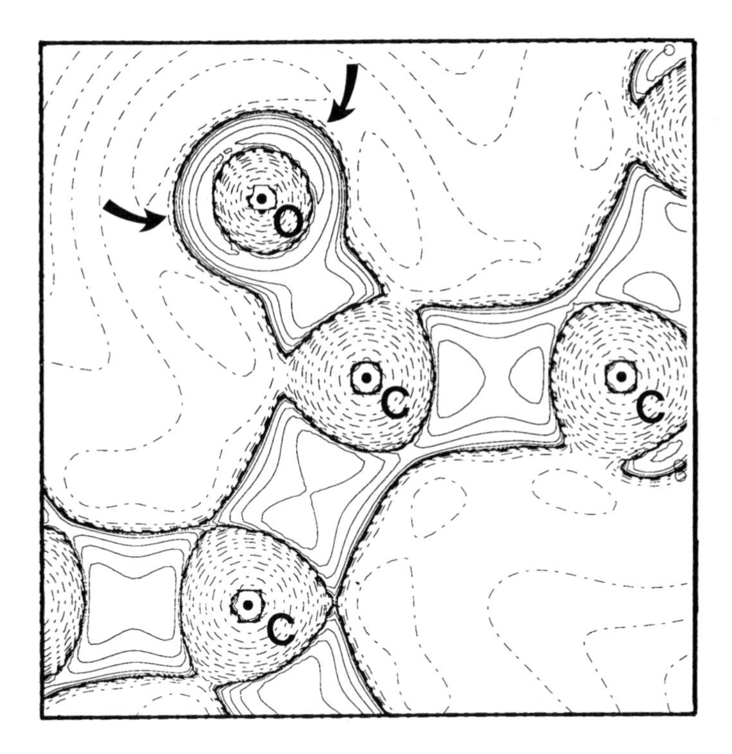

Figure 9.5 Contour map of the negative of the Laplacian of the charge density of the carbonyl group in haloperidol. Dashed contours designate regions of charge depletion ($\nabla^2\rho > 0$) and solid lines regions of charge concentration ($\nabla^2\rho < 0$). Solid (open) circles denote nuclei (not) lying in the plotted plane. The non-bonded charge concentrations on the ketonic oxygen atom are indicated by arrows.

9.7 Conclusion

The theory of 'atoms in molecules' extracts chemical information from the charge density obtained from any source without invoking an arbitrary reference density. With today's computing power, reliable charge densities can be obtained for a surprisingly vast set of drugs. Our proposed similarity measure is not a single number but a vector with components given by AIM. The latter involve both local and integrated properties, which are definable exactly because of the topology of the charge density. Therefore, indices based on a total comparison of charge densities everywhere in space are unnecessary regardless of to what extent their concomitant technical difficulties will be overcome. Furthermore, AIM also provides a physical model for complementarity via the Laplacian of the charge density and reveals the propensity and geometry of interaction between the ligand and the binding site.

9.8 Future developments

In principle, AIM could be applied to densities for semi-empirical wavefunctions of large biomolecules (Minhhuy *et al.*, 1994) and could then be used to study docking of a ligand in a known binding site (Popelier and Bader, 1992). Most likely the process of drug modification or even *de novo* design could be steered by a feedback from chemical descriptors (both similarity and complementarity indices) based on AIM. Modern chemical graph theory (Trinajstić, 1983) could be provided with a more advanced physical basis than that developed in the previous century. In this respect, it is interesting to note that the recent extension of graphs (Koča *et al.*, 1989) to incorporate virtual vertices (in the context of the synthon approach) can be identified with molecular graphs endowed with atomic graphs. It is hoped that the AIM approach will further prove to be fruitful in the vast and challenging area of drug design (Dean, 1987) as a bridge between a charge density and the chemistry it is responsible for.

Acknowledgement

Gratitude is expressed to the European Community for a Human Capital and Mobility Scheme Fellowship.

List of computer programs

AIMPAC A suite of programs written by the Bader group at McMaster University, Canada (1982). Available via bader@mcmail.cis.mcmaster.ca.

CADPAC5 The Cambridge Analytic Derivatives Package Issue 5, Cambridge (1992). A suite of quantum chemistry programs developed by R.D. Amos with contributions from I.L. Alberts, J.S. Andrews, S.M. Colwell, N.C. Handy, D. Jayatilaka, P.J. Knowles, R. Kobayashi, N. Koga, K.E. Laidig, P.E. Maslen, C.W. Murray, J.E. Rice, J. Sanz, E.D. Simandiras, A.J. Stone and M.D. Su.

GAMESS M.W. Schmidt, K.K. Baldridge, J.A. Boatz, S.T. Elbert, M.S. Gordon, J.H. Jensen, S. Koseki, N. Matsunaga, K.A. Nguyen, S. Su, T.L. Windus, M. Dupuis and J.A. Montgomery, Jr.

GAUSSIAN92 (1992) M.J. Frisch, G.W. Trucks, H.B. Schlegel, P.M.W. Gill, M.W. Wong, J.B. Foresman, B.G. Johnson, M.A. Robb, E.S. Replogle, R. Gomperts, J.L. Andres, K. Raghavachari, J.S. Binkley, C. Gonzalez, R.L. Martin, D.J. Fox, D.J. DeFrees, J. Baker, J.P. Stewart and J.A. Pople, Gaussian Inc., Pittsburgh, Pennsylvania.

MORPHY A program written by P.L.A. Popelier, McMaster University, Canada and Cambridge University. Available via plap100@cus.cam.ac.uk.

References

Almlöf, J., Faegri, K., Jr and Korsell, K. Principles for a direct SCF approach to LCAO–MO *ab initio* calculations. *Journal of Computational Chemistry*, **3**, 385–399.

Bader, R.F.W. (1975) Molecular fragments or chemical bonds? *Accounts of Chemical Research*, **8**, 34–40.

Bader, R.F.W. (1985) Atoms in molecules. *Accounts of Chemical Research*, **18**, 9–15.

Bader, R.F.W. (1986) Properties of atoms and bonds in carbocations. *Canadian Journal of Chemistry*, **64**, 1036–1041.

Bader, R.F.W. (1988) From Schrödinger to atoms in molecules. *Pure and Applied Chemistry*, **60**, 145–155.

Bader, R.F.W. (1990) *Atoms in Molecules: A Quantum Theory*. Clarendon, Oxford.

Bader, R.F.W. (1991) A quantum theory of molecular structure and its applications. *Chemical Reviews*, **91**, 893–928.

Bader, R.F.W. and Preston, H.T.J. (1970) Determination of the charge distribution of methane by a method of density constraints. *Theoretica Chimica Acta*. **17**, 384–395.

Bader, R.F.W. and Beddall, P.M. (1972) Viral field relationship for molecular charge distributions and the spatial partitioning of molecular properties. *Journal of Chemical Physics*, **56**, 3321–3329.

Bader, R.F.W. and Essén, H. (1984) The characterization of atomic interactions. *Journal of Chemical Physics*, **80**, 1943–1960.

Bader, R.F.W. and Wiberg, K.B. (1987) A bond energy from quantum mechanics. In *Density Matrices and Density Functionals* (eds R. Erdahl and V.H. Smith), pp. 677–692. D. Reidel, Dordrecht.

Bader, R.F.W. and Chang, C. (1989) Properties of atoms in molecules: electrophilic aromatic substitution. *Journal of Physical Chemistry*, **93**, 2946–2956.

Bader, R.F.W. and Legare, D.A. (1992) Properties of atoms in molecules: structures and reactivities of boranes and carboranes. *Canadian Journal of Chemistry*, **70**, 657–676.

Bader, R.F.W. and Popelier, P.L.A. (1993) Atomic theorems. *International Journal of Quantum Chemistry*, **45**, 189–207.

Bader, R.F.W., Srebrenik, S. and Nguyen-Dang, T.T. (1978) Subspace quantum dynamics and the quantum action principle. *Journal of Chemical Physics*, **68**, 3680–3691.

Bader, RF.W., Anderson, S.G. and Duke, A.J. (1979a) Quantum topology of molecular charge distributions. 1. *Journal of the American Chemical Society*, **101**, 1389–1395.

Bader, R.F.W., Nguyen-Dang, T.T. and Tal, Y. (1979b) Quantum topology of molecular charge distributions. 2. Molecular structure and its change. *Journal of Chemical Physics*, **70**, 4316–4329.

Bader, R.F.W., Slee, T.S., Cremer, D. and Kraka, E.J. (1983) Description of conjugation and hyperconjugation in terms of electron distribution. *Journal of the American Chemical Society*, **105**, 5061–5068.

Bader, R.F.W., McDougall, P.J. and Lau, C.D.H. (1984) Bonded and nonbonded charge concentrations and their relation to molecular geometry and reactivity. *Journal of the American Chemical Society*, **106**, 1594–1605.

Bader, R.F.W., Carroll, M.T., Cheeseman, J.R. and Chang, C. (1987) Properties of atoms in molecules: atomic volumes. *Journal of the American Chemical Society*, **109**, 7968–7979.

Bader, R.F.W., Gillespie, R.J. and McDougall, P.J. (1988) A physical basis for the VSEPR model of molecular geometry. *Journal of the American Chemical Society*, **110**, 7329–7336.

Bader, R.F.W., Popelier, P.L.A. and Chang, C. (1992) Similarity and complementarity in chemistry. *Journal of Molecular Structure (Theochem)*, **255**, 145–171.

Bader, R.F.W., Popelier, P.L.A. and Keith, T.A. (1994) Theoretical definition of a functional group and the molecular orbital paradigm. *Angewandte Chemie International Edition*, **33**, 620–631.

Barrett, C.S. and Meyer, L. (1967) Molecular packing, defects, and transformations in solid oxygen. *Physical Reviews*, **160**, 694–697.

Biegler-König, F.W., Nguyen-Dang, T.T., Tal, Y., Bader, R.F.W. and Duke, A.J. (1981) Calculation of the average properties of atoms in molecules. *Journal of Physics B: Atomic and Molecular Physics*, **14**, 2739–2751.

Biegler-König, F.W., Bader, R.F.W. and Tang, T.H. (1982) Calculation of the average properties of atoms in molecules. II. *Journal of Computational Chemistry*, **13**, 317–328.

Bowen-Jenkins, P.E., Cooper, D.L. and Richards, W.G. (1985) *Ab initio* computation of molecular similarity. *Journal of Physical Chemistry*, **89**, 2195–2197.

Boyd, R.J. and Choi, S.C. (1985) A bond length–bond order relationship for intermolecular interactions based on the topological properties of molecular charge distributions. *Chemical Physics Letters*, **120**, 80–85.

Bürgi, H.B. and Dunitz, J.D. (1983) From crystal statics to chemical dynamics. *Accounts of Chemical Research*, **16**, 153–161.

Carbó, R., Leyda, L. and Arnau, M. (1980) How similar is a molecule to another? An electron density measure of similarity between two molecular structures. *International Journal of Quantum Chemistry*, **17**, 1185–1189.

Carroll, M.T. and Bader, R.F.W. (1988) An analysis of the hydrogen bond in BASE–HF complexes using the theory of atoms in molecules. *Molecular Physics*, **65**, 695–722.

Carroll, M.T., Chang, C. and Bader, R.F.W. (1988) Prediction of the structures of hydrogen-bonded complexes using the Laplacian of the charge density. *Molecular Physics*, **63**, 387–405.

Carroll, M.T., Cheeseman, J.R., Osman, R. and Weinstein, H. (1989) Nucleophilic addition to activated double bonds: predictions of reactivity from the Laplacian of the charge density. *Journal of Physical Chemistry*, **93**, 5120–5123.

Chang, C. and Bader, R.F.W. (1992) Theoretical construction of a polypeptide. *Journal of Physical Chemistry*, **96**, 1654–1662.

Ciechanowicz-Rutkowska, M., Kieć-Kononowicz, K., Howard, S.T., Lieberman, H. and Hursthouse, M.B. (1994) Molecular and electronic structures of two (anticonvulsant) diphenylhydantoin derivatives. *Acta Crystallographica*, **B50**, 86–96.

Cioslowski, J. and Nanayakkara, A. (1993) Similarity of atoms in molecules. *Journal of the American Chemical Society*, **115**, 11213–11213.

Coppens, P. (1977) Experimental electron densities and chemical bonding. *Angewandte Chemie. International Edition*, **16**, 32–40.

Dandridge, P.A., Kaiser, C., Brenner, M., Gaitanopoulos, D., Davis, L.D., Webb, R.L., Foley, J.J. and Sarau, H.M. (1984) Synthesis, resolution, absolute stereochemistry and enantioselectivity of 3′,4′-dihydroxynomifensine. *Journal of Medicinal Chemistry*, **27**, 28–35.

Datta, N., Mondal, P. and Pauling, P. (1979) The structure of haloperidol hydrobromide [4-[4-(4-chlorophenyl)-4-hydroxypiperidine]-4′-fluorobutyrophenone HBr]. *Acta Crystallographica*, **B35**, 1486–1488.

Daudel, R., Leroy, G., Peeters, D. and Sana, M. (1983) *Quantum Chemistry* (John Wiley, Chichester).

Dean, P.M. (1987) *Molecular Foundations of Drug–Receptor Interaction.* Cambridge University Press, New York.

Dean, P.M. (1990) Molecular recognition: the measurement and search for molecular similarity in ligand–receptor interaction. In *Concepts and Applications of Molecular Similarity* (eds M.A. Johnson and G.M. Maggiora), pp. 211–238. John Wiley, New York.

Destro, R. and Merati, F. (1993) Preliminary results of a 20K X-ray study of citrinin. *Zeitschrift für Naturforschung. A. A Journal of Physical Sciences*, **48**, 99–104.

Destro, R., Bianchi, R., Gatti, C. and Merati, F. (1991) Total electronic charge density of L-alanine from X-ray diffraction at 23K. *Chemical Physics Letters*, **186**, 47–52.

Francl, M.M., Pietro, W.J., Hehre, W.J., Binkley, J.S., Gordon, M.S., DeFrees, D.J. and Pople, J.A. (1982) Self-consistent molecular orbital methods. 23. A polarization-type basis set for second-row elements. *Journal of Chemical Physics*, **77**, 3654–3664.

Gillespie, R.J. (1972) *Molecular Geometry.* Van Nostrand Reinhold, London.

Jeffrey, G.A. and Piniella, J.F. (eds) (1991) *The Application of Charge Density Research to Chemistry and Drug Design.* Plenum, New York.

Johnson, M.A. (1989) A review and examination of the mathematical spaces underlying molecular similarity analysis. *Journal of Mathematical Chemistry*, **3**, 117–145.

Klooster, W.T., Swaminathan, S., Nanni, R. and Craven, B.M. (1992) Electrostatic properties of 1-methyluracil from diffraction data. *Acta Crystallographica*, **B48**, 217–227.

Koča, J., Kratochvi, M., Kvasnicka, V., Matyska, L. and Pospichal, J. (1989) *Synthon Model of Organic Chemistry and Synthesis Design, Lecture Notes in Chemistry*, **51**. Springer-Verlag, Berlin.

Labanowski, J.K. and Andzelm, J.W. (eds) (1991) *Density Functional Methods in Chemistry.* Springer, New York.

Laidig, K.E. and Bader, R.F.W. (1991) Distorted amides: correlation of their enhanced solvolysis with local charge depletions at the carbonyl carbon. *Journal of the American Chemical Society*, **113**, 6312–6313.

Lee, C. and Smithline, S. (1994) An approach to molecular similarity using density functional theory. *Journal of Physical Chemistry*, **98**, 1135–1138.

Ludeña, E.V. and Kryachko, E.S. (1990) *Energy Density Functional Theory of Many-Electron Systems.* Kluwer Academic, Dordrecht.

McWeeny, R. (1989) *Methods of Molecular Quantum Mechanics*, 2nd edn. Academic Press, San Diego.

Minhhuy, H., Schmider, H., Edgecombe, K.E. and Smith, V.H. (1994) Topological analysis of valence electron charge distributions from semiempirical and *ab initio* methods. In press.

Moore, W. (1993) *Schrödinger. Life and Thought*, pp. 219 and 435. Cambridge University Press, Cambridge.

Popelier, P.L.A. (1994) An analytical expression for interatomic surfaces in the theory of atoms in molecules. *Theoretica Chimica Acta*, **87**, 465–476.

Popelier, P.L.A. and Bader, R.F.W. (1992) The existence of an intramolecular C–H–O hydrogen bond in creatine and carbamoyl sarcosine. *Chemical Physics Letters*, **189**, 542–548.

Popelier, P.L.A. and Bader, R.F.W. (1994) Effect of twisting a polypeptide on its geometry and electron distribution. *Journal of Physical Chemistry*, **98**, 4473–4481.

Press, W.H., Flannery, B.P., Tenkolsky, S.A. and Vetterling, W.T. (1986) *Numerical Recipes.* Cambridge University Press, Cambridge.

Pulay, P., Fogarasi, G., Pang, F. and Boggs, J.E. (1979) Systematic *ab initio* gradient calculation of molecular geometries, force constants, and dipole moment derivatives. *Journal of the American Chemical Society*, **101**, 2550–2560.

Reed, L.L. and Schaefer, J.P. (1973) The crystal and molecular structure of haloperidol, a potent psychotropic drug. *Acta Crystallographica*, **B29**, 1886–1890.

Reuben, B.G. and Wittcoff, H.A. (1989) *Pharmaceutical Chemicals in Perspective.* John Wiley, New York.

Richard, A.M. and Rabinowitz, J. (1987) Modified molecular charge similarity indices for choosing molecular analogues. *International Journal of Quantum Chemistry*, **31**, 309–323.

Runtz, G.R., Bader, R.F.W. and Messer, R. (1977) Definition of bond paths and bond directions in terms of the molecular charge distribution. *Canadian Journal of Chemistry*, **55**, 3040–3045.

Slee, T.S. (1986a) Inductive effects on the electron distribution of the vinyl group: a correlation between substituent electronegativity and bond point shift. *Journal of the American Chemical Society*, **108**, 606–612.

Slee, T.S. (1986b) On the correspondence between simple orbital concepts and molecular electron distributions. *Journal of the American Chemical Society*, **108**, 7541–7548.

Slee, T.S. and Bader, R.F.W. (1992) Properties of atoms in molecules: protonation at carbonyl oxygen. *Journal of Molecular Structure* (*Theochem*), **255**, 173–188.

Slee, T.S., Larouche, A. and Bader, R.F.W. (1988) Properties of atoms in molecules: dipole moments and substituent effects in ethyl and carbonyl compounds. *Journal of Physical Chemistry*, **92**, 6219–6227.

Srebrenik, S., Bader, R.F.W. and Nguyen-Dang, T.T. (1978) Subspace quantum mechanics and the variational principle. *Journal of Chemical Physics*, **68**, 3667–3679.

Szabo, A. and Ostlund, N.S. (1989) *Modern Quantum Chemistry. Introduction to Advanced Electronic Structure Theory*, 1st rev. edn. McGraw-Hill, New York.

Trinajstić, N. (1983) *Chemical Graph Theory*, Vol. 1. CRC Press, Boca Raton, Florida.

Tsuchhashi, H., Sasaki, T., Kojima, S. and Nagatomo, T. (1992) Binding of [^{3}H]haloperidol to dopamine D_2 receptors in the rat striatum. *Journal of Pharmacy and Pharmacology*, **44**, 911–914.

van Alsenoy, C., Lenstra, A.T.H. and Geise, H.J. (1989) The gradient-optimized geometry of haloperidol at the 4-21G level. *Journal of Computational Chemistry*, **10**, 302–308.

Vinter, J.G. and Gardner, M. (eds) (1994) *Molecular Modelling and Drug Design*, pp. 340–343. Macmillan, London.

Wiberg, K.B., Bader, R.F.W. and Lau, C.D.H. (1987) Theoretical analysis of hydrocarbon properties. 1. Bonds, structures, charge concentrations, and charge relaxations. *Journal of the American Chemical Society*, **109**, 1001–1013.

10 Methods of molecular shape similarity and topological shape design

P.G. MEZEY

10.1 The role of shape analysis in drug design

The shape of molecules has an important role in biochemical processes. Molecular shape effects in the vicinity of a given functional group can both enhance and hinder the reactivity of the functional group. The shape and size of local molecular regions are of primary importance in enzyme–drug interactions. In many biochemical systems involving large molecules, such as proteins, shape is an important selectivity principle: even if some functional groups are inherently reactive if accessible to reactants, the spatial limitations of otherwise inert parts of the molecules may selectively enhance or prevent specific reactions. Systematic and detailed analysis of molecular shape has an important role in the elucidation of quantitative structure–activity relationships (QSAR), in studies on drug–receptor interactions, and in contemporary efforts aimed at rational drug design. Both experimental and theoretical elucidation of the role of shape have contributed to important advances in molecular modelling and in the study of molecular similarity (see references quoted in Johnson and Maggiora (1990) and Mezey (1993)).

A better understanding of the role of shape in biochemical interactions can much enhance the success of rational drug design. However, even without a detailed understanding of how a given shape feature can result in a specific drug action, molecular shape analysis can provide important leads and suggest novel targets for future synthesis. In most instances, the overall activity of a drug molecule depends on a very large number of factors. At present, all of these factors can hardly be accounted for in detail, even by the most advanced, state-of-the-art biochemical models. Nevertheless, the mere similarity of the shapes of two molecules can already serve as a basis for predictions: if the shapes of two molecules are similar, we can expect similarities in their biochemical activities. The recognition and evaluation of similarity, representing only the first steps on the long road to understanding, can already be valuable tools in their own right.

10.1.1 Molecular similarity and the prediction of molecular properties

One of the elementary observations of chemistry, that 'similar molecules have similar properties', borders on tautology. Phrased somewhat more precisely, if one family of properties of two molecules are similar, then we may expect similarities in some other properties as well. In particular, if the shapes of two molecules are similar in the vicinity of a molecular moiety participating in a given type of reaction, then we may expect similarities in their reactivities. Molecules showing common shape features often act in a similar manner in biochemical processes. One of the essential tasks is finding precise and reproducible methods for the evaluation and comparison of molecular shapes.

In particular, the development of computer-based techniques for non-visual, algorithmic evaluation of molecular shape similarity in large sequences of molecules is of importance in drug-design applications. These methods are suitable for similarity ranking of molecular sequences, and important predictions can be made on the expected biochemical activity and drug potency.

The fuzzy, electron-density cloud of a molecule is a continuum, and even simplified shape representations require a considerable amount of data. In order to characterize in some detail the shape of a molecule, a single number is insufficient. Even a rudimentary description of molecular shape requires several numbers. Whereas this fact is well recognized, in many current applications of molecular modeling to drug design, the emphasis is placed on one-dimensional correlations between a specified type of biochemical activity and molecular shape. According to the customary procedure, in correlating molecular-shape properties with a specified biochemical activity in a series of potential drug molecules, a molecular ordering is established based on experimentally observed activities. The usual goal is to establish a parallel or nearly parallel ordering of molecular activities with some well-defined shape property. If the attempt is successful and such a correlation is found for a set of experimentally tested molecules, then the identified shape features are used as a predictive tool. If the given shape feature can be enhanced in some related molecule, then for the new molecule a high biochemical activity is predicted. This approach is essentially a one-dimensional 'shape calibration' that does not take advantage of the multidimensional features and the wealth of detail of molecular shapes.

Most drug-design efforts to date have been focused on enhancing the main effects of potential drug molecules, while suppressing side-effects. Frequently, all potential side-effects are considered within a single category of 'undesired features' to be eliminated as best as possible, and little inference is drawn from the vast amount of information represented by the varied degrees of a large number of simultaneous side-effects. This information is valuable, and algorithmic shape-analysis techniques can be used to exploit it. Non-visual shape-analysis methods are applicable for multiple similarity ranking based on several local and overlapping molecular-shape features, to be correlated with levels of a large number of simultaneous side-effects.

The possibility of identifying and individually tracking separate shape features by non-visual, algorithmic methods can be exploited for a systematic analysis of both main effects and of several potential side-effects in large sequences of potential drug molecules. In a multiple side-effect analysis, several different shape orderings of the molecular sequence can be generated, based on information supplied by a few molecules of experimentally known side-effects. Such multiple shape orderings can be defined in terms of shape motifs and combinations of shape motifs which appear to enhance a particular side-effect. In computational terms, the biochemical problem is converted into a multidimensional correlation problem between the various orderings of the molecular sequence according to the main effects and side-effects, and the orderings according to the shape features of the selected shape motifs.

In another sense, this formulation corresponds to a discrete shape optimization problem in a space spanned by formal coordinates assigned to various side-effects and local shape features. One can visualize this approach as follows. A 'multiparameter shape–multiple biochemical activity analysis' requires experimental (or, possibly theoretical) information on the main biochemical effect and also on various side-effects for a series of molecules. Assume that the total number of experimentally measured biochemical properties, including both the main effects and side-effects, is k. These k types of biochemical activities define a formal k-dimensional parameter space. The molecular shapes occurring in the given molecular family are assigned to various points of this parameter space, depending on the actual activities of molecules exhibiting these shape features. The analysis of the distribution, clustering, and various patterns of the arrangements in the parameter space provides a detailed multiparameter correlation between shape features and multiple biochemical effects. Alternatively, a parameter space can be defined in terms of a family of shape motifs, and activities according to each side-effect considered can be mapped in this parameter space. A combination of these two approaches uses abstract coordinates including both those assigned to side-effects and shape motifs, and a full, multidimensional correlation pattern can be generated. The number of coordinates is usually much more than three, and visualization of the patterns which emerge is seldom attempted. The results, however, can be interpreted by algorithmic techniques. These techniques of multiparameter shape analysis provide more information than individual single-effect correlations. The fact that all the selected shape and activity properties are treated in an integrated manner enhances the predictive value of the approach.

A major component of contemporary biochemical research and drug-design efforts is aimed at the rationalization of molecular behavior based on molecular-shape properties. In this context it is important to distinguish molecular shape from molecular conformation. Stereochemical, conformational models and diagrams usually refer to the nuclear arrangements in three dimensions, and the orientations of formal bonds represented by lines. A molecular conformation is fully specified if the nuclear positions are given, and the bonding pattern is a powerful tool used to convey stereochemical information. Yet, molecular conformation, as specified by the nuclear configuration, represents the molecular shape only indirectly. Molecules are fuzzy bodies of electronic clouds, where the nuclear arrangement and the formal bonding pattern represents only a molecular skeleton. The electronic cloud that surrounds this formal skeleton represents the actual shape; molecular interactions, molecular recognition, chemical reactions, and most aspects of molecular behavior are directly dependent on this shape. In this chapter some of the topological methods of molecular shape similarity and shape complementarity analysis will be reviewed, and some novel approaches will be suggested.

10.1.2 Three-dimensional representations of molecular shape

How do we describe the shapes of molecules? It is natural to attempt to use concepts and methods we employ when characterizing objects we encounter in our everyday life. For most ordinary objects of our macroscopic world, the shape features can be described by concepts used for bodies with well-defined boundaries. A potato appears as an object that occupies a well-defined region of the three-dimensional (3D) space; its surface is well recognizable, its volume can be measured rather accurately, the dents and bumps on it provide us with a clear perception of its shape. Unfortunately, none of these concepts is directly applicable to molecules. A molecular electron distribution does not occupy a well-defined region of the space; the fuzzy electronic cloud fades away only gradually with the distance from the nuclei. Consequently, a molecule does not have a surface in the macroscopic sense. On a more fundamental level, the quantum mechanical uncertainty relation plays an important role in molecular behavior. The Heisenberg uncertainty relation is seldom revealed in macroscopic experiments, and one certainly does not need to think of it when evaluating the shape of a potato. By contrast, the very fact that electrons are delocalized within molecules indicates the paramount importance of quantum mechanical uncertainty within molecules.

Fortunately, some macroscopic analogies still work well for molecules. The analogy with formal clouds is a useful one, and in designing molecular shape analysis techniques, we should not forget the fuzzy aspect of molecular bodies.

Additional concern is the non-rigidity of molecules. Nuclear arrangements are never fixed, molecules undergo continuous vibrational motions as well as conformational changes. In fact, nuclei are also subject to quantum mechanical uncertainty, although the much greater mass of nuclei, when compared to electrons, allows us to use classical analogies with more confidence. Nevertheless, the dynamic aspects of the nuclear arrangements must be considered when discussing molecular shape. One of the usual questions asked is the following: 'What is the most stable conformation of the molecule?' Whereas for small molecules both experiment and theory can give reasonably relevant answers to this question; for large molecules, with a whole range of possible nuclear arrangements of nearly identical energy, the question appears somewhat misplaced. The relevance of this question might be judged by the following macroscopic analogy: what is the most stable conformation of a chemist? There are many comfortable positions I can take, and my comfort level can be very similar while holding my limbs in rather different positions. Furthermore, my identity is not affected at all if I stretch my arm. For people as well as for molecules, a whole range of the low energy conformations are important, since depending on the surroundings that usually vary rather randomly, a variety of arrangements may be slightly preferred over some others. The usual preoccupation with the most stable conformation of a macromolecule does not seem to be justified, except, perhaps, in highly controlled and constrained situations, such as within crystal lattices.

The dynamic aspects of molecular shape analysis are of particular importance in drug design, where most biochemical action is initiated by mutual shape recognition and shape adjustment of two or several molecules.

The conventional ball-and-stick models are among the simplest tools used for molecular shape representation. Whereas these models provide valuable stereochemical information, they portray poorly the actual space requirements of molecules.

The space-filling models, and one of their variants, the fused-sphere van der Waals models, describe the formal molecular bodies more faithfully; however, they, too, suffer from rather dramatic oversimplifications. The artificial features of fused spheres at the seams of interpenetrating spheres misrepresent the internuclear regions. Secondary interactions between the fuzzy electron distributions of molecular parts not formally linked by bonds are poorly described. Furthermore, hydrogen bonds, double and triple bonds, aromatic character, hyperconjugation, and many other details of the actual molecular shapes are not properly represented by such fused-sphere models.

Electron densities provide a valid representation of molecular shape. Experimental electron densities, obtained, for example, from X-ray diffraction studies, do not have at present sufficient resolution within the low-electron density ranges of molecules to provide a detailed enough shape description. Quantum chemical computations, on the other hand, can provide sufficiently detailed and reasonably accurate shape representations. Using an LCAO *ab initio* wavefunction computed for a molecule of some fixed conformation K, the electronic density $\rho(\mathbf{r})$ can be computed as follows. If n is the number of atomic orbitals $\varphi_i(\mathbf{r})$ $(i = 1, 2, \ldots, n)$, $\mathbf{r}$ is the 3D position vector variable, and $\mathbf{P}$ is the $\mathbf{n} \times \mathbf{n}$ density matrix, then the electronic density $\rho(\mathbf{r})$ of the molecule is obtained as

$$\rho(\mathbf{r}) = \sum_{i=1}^{n} \sum_{j=1}^{n} P_{ij}\varphi_i(\mathbf{r})\varphi_j(\mathbf{r}) \tag{10.1}$$

This function $\rho(\mathbf{r})$ represents the fuzzy body of the electronic charge cloud, that in turn represents the shape of the molecule. At present, the best shape representation of molecules is provided by quantum chemical electron density calculations.

Unfortunately, conventional, direct *ab initio* calculations of electron densities are feasible only for molecules of moderate size, with a current limit of about a hundred atoms on the fastest supercomputers. Note, however, that some of the main computational difficulties arising for large molecules have been recently circumvented (see sections 10.2.1 and 10.2.2).

One approach, somewhat less representative than electron-density computation but a valid shape description, nonetheless, is based on a rather simple molecular property: the composite nuclear potential. Also called the 'bare nuclear potential' by Parr and Berk (1981), this potential, as a function of position in the 3D space, has been shown to mimic many features of the electronic density of molecules. The potential can be computed by a simple application of Coulomb's law. The bare nuclear potential $V_n(\mathbf{r})$ generated by the collection of the nuclei of the molecule is defined as

$$V_n(\mathbf{r}) = \sum_i Z_i/|\mathbf{r} - \mathbf{R}_i| \qquad (10.2)$$

The value of the nuclear potential $V_n(\mathbf{r})$ generated at point $\mathbf{r}$ is independent of the electronic state of the molecule; it depends only on the nuclear charges Z_i and the formal nuclear positions $\mathbf{R}_i$.

The nuclear potential is an important molecular property in its own right, and a highly reliable tool for assessing molecular shape similarity. The computational advantages of nuclear potentials are particularly important for large systems, and these potentials have been advocated for shape similarity analysis of biological macromolecules (Mezey, 1993).

In this chapter, the emphasis is placed on the electronic density, and no specific techniques aimed at alternative shape representations will be discussed. Note, however, that most of the methods discussed are equally applicable for the composite nuclear potential.

10.1.3 The need for numerical similarity measures and complementarity measures

Visual similarity assessment is often subjective and not necessarily reproducible. These disadvantages are avoided if non-visual, computer-based, algorithmic similarity analysis techniques are used. The development of these techniques requires precise definitions of shape representations, precise criteria for the evaluation of shape features, and algorithmic techniques for shape comparisons. It is of special importance to formulate these definitions and criteria so that the algorithmic techniques built upon them are able to evaluate the chemically relevant aspects of molecular shape and similarity.

The potential of discovering such shape features is much enhanced if systematic shape analysis can be performed for large molecular sequences. Hence, the results of algorithmic, non-visual shape analysis can be used to make better predictions of biochemical activities of new molecules. If the shape analysis is detailed enough, then the mere presence of a given shape feature can already be used to predict a given chemical property, even without knowing how a given shape feature leads to a particular biochemical action. For suggesting new leads, it is sufficient to know what the important shape features are, and by combining these features in computer-designed molecular models, we are able to propose new molecular structures.

This approach to the computer-based design of new molecules, drugs and industrial materials represents a change of emphasis when compared to the conventional QSAR approach. The new approach, quantitative shape–activity relations, or QShAR (Mezey, 1992), uses shape, instead of the structural features of the bond skeleton of molecules, to arrive at correlations between molecular properties and biochemical activities.

Algorithmic molecular modelling methods can be used to construct 3D molecular models from fundamental physical laws. For example, using well-established computational methods, we can obtain detailed molecular electronic-charge distributions. An important aspect of molecular modelling is its versatility: any subsequent topological shape analysis of computed electronic densities is equally applicable to any existing molecule or to molecules which have not been synthesized yet. This is precisely where the predictive power of advanced molecular modelling lies. Using these methods, we can select the most likely candidates from a large sequence of thousands of possible molecules, based on detailed shape analysis and integrated main and side-effect correlations. The actual, expensive synthetic work and chemical tests of new molecules can be focused on the most promising candidates. This approach leads to important savings in human efforts, animal experiments, and financial resources.

10.2 Electron-density computations for large molecules

For small and medium size molecules, containing less than about a hundred atoms, the conventional, *ab initio* quantum chemical computational techniques are applicable for the generation of reasonably representative electron densities, as long as the molecular conformations do not contain unusual bonding arrangements. Near the upper limit of this molecular range, however, the computations can become rather expensive, requiring many hours or even days of central processor unit time on some supercomputer. This computational difficulty has been hindering the application of electron-density calculations for drug-design purposes, and such calculations have been restricted to smaller molecules.

A new development in computational methodology, described in the next section, is likely to change this.

10.2.1 The MEDLA approach: molecular electron-density lego assembler as 'computational microscope'

A set of topological rules and a new computational method, the molecular electron-density lego assembler (MEDLA) technique, were developed for the construction of accurate (*ab initio* quality), yet inexpensive electron densities of large molecules (Walker and Mezey, 1993a,b). The MEDLA method builds the electron densities of large molecules by constructing an interpenetration pattern of precalculated, *ab initio* electron densities of fuzzy molecular fragments and functional groups, obtained from direct *ab initio* calculations on small molecules and stored in a fragment density databank. The density fragments have no boundaries; they are realistic, fuzzy electron distributions. If the fragments originate from the given molecule, then the reconstructed MEDLA electron density is exact (within the given *ab initio* framework); non-exact, but highly accurate densities, can be obtained in all other cases.

The computation time required for the construction of MEDLA densities grows (essentially) linearly with molecular size, hence the method is easily applicable to large molecules. A databank of less than 200 fragments is sufficient to build electron densities for many biomolecules. The MEDLA method has been tested on several small molecules and already applied to proteins and other large biomolecules. This new technique has the potential of significantly increasing the scope of accurate molecular modelling.

According to a simple electron-density fragment additivity principle, described earlier (Walker and Mezey, 1993a), reasonably accurate electron densities of large molecules can be built from molecular fragment densities obtained from high quality, *ab initio* calculations for small molecules. The fragment densities are stored in a databank, and can be retrieved and used by the MEDLA program, MEDLA93 (Walker and Mezey, 1993b), to 'build' electronic densities for various molecules. The 'lego'-type fragment combination process, that has motivated the name of the technique, can be applied to truly large molecules, including proteins and DNA.

The MEDLA method, and its computer implementation using the program MEDLA93, are based on an electron-density fragment additivity principle that can be formulated in several ways. One, rather natural scheme for the implementation of this principle has been described by Walker and Mezey (1993a), and is reviewed below.

Consider a small molecule of some fixed conformation K, and an LCAO *ab initio* wavefunction of the molecule, given in terms of a suitable atomic orbital basis set. In terms of the atomic orbitals $\varphi_i(\mathbf{r})$ of the basis set and the computed $n \times n$ *ab initio* density matrix, **P**, the electronic density $\rho(\mathbf{r})$ of the molecule can be calculated using equation (10.1).

A formal fragment of the electronic density is defined by an arbitrary collection of the nuclei from the molecule, and by a corresponding subset of the elements of the above density matrix. For the calculation of the kth fragment $\rho^k(\mathbf{r})$ of the electron density $\rho(\mathbf{r})$, the following criterion is used to generate the corresponding $n \times n$ fragment density matrix P^k_{ij}:

$P^k_{ij} = P_{ij}$ if both $\varphi_i(\mathbf{r})$ and $\varphi_j(\mathbf{r})$ are atomic orbitals centered on nuclei of the fragment,
$= 0.5\,P_{ij}$ if precisely one of $\varphi_i(\mathbf{r})$ and $\varphi_j(\mathbf{r})$ is centered on a nucleus of the fragment,
$= 0$ otherwise (10.3)

Within the above scheme, both the complete density matrix **P** of the molecule, and the fragment density matrix $\mathbf{P}^k$ of the kth fragment have the same dimensions, $n \times n$.

Using the fragment density matrix $\mathbf{P}^k$, the electron density of the kth density fragment is defined (Walker and Mezey, 1993a) as

$$\rho^k(\mathbf{r})^n = \sum_{i=1}^{n} \sum_{j=1} P^k_{ij}\varphi_i(\mathbf{r})\varphi_j(\mathbf{r}) \tag{10.4}$$

Consider m mutually exclusive families of nuclei, dividing the molecule into m fragments. Since the sum of the fragment density matrices is equal to the density matrix of the molecule, the sum of the fragment densities is equal to the density of the molecule:

$$P_{ij} = \sum_{k=1}^{m} P^k_{ij} \tag{10.5}$$

and

$$\rho(\mathbf{r}) = \sum_{k=1}^{m} \rho^k(\mathbf{r}) \tag{10.6}$$

The electron-density fragment additivity rules (10.5) and (10.6) are exact on the given *ab initio* LCAO level: the reconstruction of the electronic density of the given small molecule from the corresponding fragment electron densities is exact.

The electron-density fragment additivity principle (Walker and Mezey, 1993a) can be used for constructing densities for additional molecules. Precalculated electron density fragments can be combined to form approximate electron density for a different molecule, by selecting and arranging fragment densities so that the nuclear positions closely match those in the target molecule. The main advantage of the above additivity scheme lies in the possibility of building approximate electron densities for large molecules.

Using MEDLA93, the technique has been shown (Walker and Mezey, 1993a) to produce good quality approximate electron densities that are quantitatively very similar to densities obtained by conventional, direct 6-31G** *ab initio* calculations. According to extensive test calculations, the molecular isodensity contours (MIDCO) obtained by the MEDLA and direct *ab initio* methods are found visually indistinguishable for a family of small molecules. The MEDLA method has also been applied for the construction of detailed electron densities of proteins (Walker and Mezey, 1994), providing the first technique to produce realistic, detailed images of macromolecules of this size. The MEDLA method extends the scope of earlier local and global shape analysis methods of molecules (Mezey, 1993) to macromolecules of biological importance.

The above electron-density fragmentation scheme (Walker and Mezey, 1993a), described by equations (10.3) to (10.6), is a special case of a more flexible scheme, based on a more general molecular electron-density fragment additivity principle. In the scheme of these equations, the interfragment electron density is distributed by taking a weight of 0.5 for the inter-fragment density matrix elements, and the resulting quantity is included in the fragment density matrix. However, the interfragment electron density can be distributed by other weighting schemes, while preserving the general additivity properties. We may justify a scheme based on comparisons of fragment charges calculated in the parent molecules, or some other scheme based on simple electronegativity comparisons.

A more general electron-density fragment additivity principle and the associated, more flexible scheme described below covers most alternatives:

$P^k_{ij} = P_{ij}$ if both $\varphi_i(\mathbf{r})$ and $\varphi_j(\mathbf{r})$ are atomic orbitals centered on nuclei of fragment k,

$= f(k, i, j)P_{ij}$ if precisely one of the atomic orbitals $\varphi_i(\mathbf{r})$ and $\varphi_j(\mathbf{r})$ is centered on a nucleus of fragment k, where for the weighting factor the relations $f(k, i, j) > 0$, and $f(k, i, j) + f(k', i, j) = 1$ hold, and where the fragment k' contains the nuclear center of the other atomic orbital,

$= 0$ otherwise. (10.7)

There are restrictions on the choice of function $f(k, i, j)$ which must be observed in order to obtain a valid, additive scheme. For example, one may select a scalar property $A(i)$ that can be assigned to atomic orbitals and does not change sign. Formal electronegativity is such a scalar property. For any such scalar property $A(i)$, the following simple choice for function $f(k, i, j)$ fulfills the conditions:

$$f(k, i, j) = A(i)/|[A(i) + A(j)] \quad (10.8)$$

Here we have assumed that orbital $\varphi_i(\mathbf{r})$ is centered on a nucleus that belongs to the kth fragment.

Most of the experience with the MEDLA approach is restricted to the simple additive scheme given by equations (10.3) to (10.6), that is, indeed, a special case of the general scheme of equations (10.7) and (10.8). This implementation follows the spirit of the usual Mulliken population analysis, and appears to give excellent results.

10.2.2 Databank for electron-density modeling of shapes of large molecules

The MEDLA method is used to generate electron densities from fragment densities stored in a density datbank. For standard or nearly standard nuclear arrangements of a simple set of molecules, such as saturated hydrocarbons, a small number of density fragments are sufficient; we do not expect large distortions of local electron densities of various parts of these molecules, with the exception of highly crowded conformations. By contrast, if strongly polarized molecular moieties are present, or if the nuclear configuration is constrained by spatial restrictions of small rings, or crowding, or hydrogen-bonded interactions, then specialized fragments are required. Such specialized fragments can be obtained from small molecules, artificially constrained to a conformation resembling that in the target molecule. In such cases, a larger electron-density database is required, including specialized density fragments to account for the special steric arrangements. There is virtually no limitation on the number of specialized fragments one may include in the database. A single CD ROM of approximately 660 Mbytes of memory is sufficient for the storage of 165 different density fragments of sufficient accuracy and variety for the calculation of 6-31G** *ab initio* quality electron densities for most proteins.

10.3 The topology of molecular shape: basic concepts

Reliable molecular modelling in drug design requires methods suitable for treating both the fuzzy and the dynamic features of molecular shape. One family of methods is provided by topology.

Topological shape analysis methods, applied to molecules, usually involve two main steps. In the first step, a family of geometrical models is considered, such as individual molecular contour surfaces and the molecular bodies enclosed by them. No attempt is made to select any one object to represent shape. Instead, all these objects are classified using some physical or geometrical criterion, such as the distribution of a range of values of electrostatic potential, or a range of local curvature values. Note that the physical criteria are also represented in terms of geometry, such as the distribution of patches of high electrostatic potential along an isodensity surface of a molecule. In the second step, the common topological properties of families of contour surfaces or the associated bodies are identified, and these properties are used to generate a topological description. In fact, the common, topological properties are defined, for example, by various intervals of possible curvature values or a range of electron-density contour values. This two-step process is usually referred to as 'Geometrical classification and topological characterization' (Mezey, 1993).

Many topological approaches used for molecular similarity analysis are based on a related principle. According to this principle, referred to as the GSTE principle, one regards 'geometrical similarity as topological equivalence' (Mezey, 1993).

Geometrical conditions are used to define a family of geometrical objects, for example, sets of points along a molecular contour surface where the surface satisfies some local geometrical condition, leading to a geometrical classification of the surface points into domains. A typical geometrical condition for a specific domain is that the local curvatures of surface points should fall within a prescribed range. Some of these geometrical objects may be similar or dissimilar. In the next step, a topological characterization of the various interrelations among these domains is carried out. The important observation is that the topological characterization does not change for most small geometrical variations, that is, as long as the geometrical patterns are similar enough, the topological characterization remains invariant. In fact, geometrical similarity is reflected in topological equivalence. In this approach, topology is used to extract the essential shape information that is topologically stable for small enough geometry variations and undergoes a topological change only if greater geometry changes occur.

The common features of most molecular shape analysis methods based on the GSTE principle of treating 'geometrical similarity as topological equivalence' can be summarized as follows:

1. All the possible (in principle, infinitely many) geometrical patterns and arrangements of local shape domains of molecular contours or bodies are classified by some combination of geometrical and topological criteria.
2. The resulting classes are characterized by topological means.

In the following sections, some examples will be discussed in some detail.

10.3.1 Molecular bodies and molecular surfaces

Molecular electron distributions are fuzzy, quantum mechanical entities and they do not possess a body or a surface in the same sense as the macroscopic potato discussed earlier. Nevertheless, formal molecular surfaces and formal molecular bodies can be defined, and these formal objects can be used for shape characterization. According to one approach, an electronic density threshold value a can be selected, and one may consider the contour surface $G(a)$ about the nuclei of the molecule where the density happens to be equal to this value a. Such a surface, clearly defined by the properties of the molecule, can be taken as a formal molecular surface. Using the same argument, the body F(a) enclosed by this surface G(a) can be regarded as a formal molecular body. Of course, for any non-zero value a, the actual molecule does extend beyond the contour G(a); however, if the threshold a is low enough, then 'most of the molecule' falls within G(a).

Clearly, no single contour surface G(a) is sufficient to describe the molecule. However, we may take a whole family of such contour surfaces G(a), for a whole range of possible threshold values a, and use this family to characterize the shape of the molecule. This approach follows the spirit of the topological shape analysis: the common, topological features for various subranges of the threshold values provide a valid description.

In the above sense, the simple, intuitive concepts of a formal molecular body and molecular surface are very useful for the interpretation of molecular size and shape properties, various molecular interactions, conformational changes and chemical reactions.

Whereas electronic density is the most important physical property for shape representation, there are alternative physical properties suitable for representing approximate molecular bodies and surfaces. As mentioned above, the molecular electrostatic potential and the bare nuclear potential are alternatives, and formal bodies and surfaces can be defined in terms of these properties. For specific purposes, additional choices are available: the contours of individual molecular orbitals, such as frontier orbitals, or simple molecular van der Waals surfaces generated by fused atomic spheres, solvent accessible surfaces, and various other surfaces surrounding some, or all, of the nuclei of a molecule.

The distinction between conformation and shape has already been pointed out. The molecular conformation, as specified by the nuclear arrangement, can be described precisely in terms of the $3N - 6$ internal coordinates of the molecule, where we assume that the number of nuclei is N ($N \geqslant 3$). Each of these internal nuclear arrangements can be represented by a point K of a formal nuclear configuration space. This abstract space can be provided with a metric, turning it into a metric space M, where a formal distance function d expresses the dissimilarity between any two nuclear arrangements K_1 and K_2. The molecular energy is a function E(K) of the nuclear arrangement, and can be thought of as a potential energy hypersurface (Mezey, 1987a). For each formal electronic state there is a different potential energy hypersurface; usually, the hypersurface of the electronic ground state is the most important. This energy function describes the relative stabilities of various nuclear arrangements.

Each chemical species is associated with a formal catchment region of the given energy hypersurface (Mezey, 1987a); if a deformation of the nuclear arrangement K is confined to the catchment region, then the identity of the chemical species is preserved. The catchment region model relaxes the classical constraints of rigid nuclear arrangement and avoids the formal conflict with the Heisenberg uncertainty relation (Mezey, 1987a).

Of course, molecular shape is also a function of the nuclear arrangement K, and for dynamic shape considerations one must take into account a whole range of formal nuclear arrangements K accessible to the molecule at the given temperature, subject to spatial and other constraints. In a manner similar to the construction of potential energy hypersurfaces, the electronic density of a specified electronic state can be regarded as a function $\rho(K,\mathbf{r})$ defined over the nuclear configuration space M. The shape of an actual, nonrigid molecule can be represented by an entire family of $\rho(K,\mathbf{r})$ charge-density distributions which occur within the configuration space domain accessible to the molecule. This accessible domain is limited by the catchment region of the molecule (Mezey, 1987a, 1993). Clearly, the representation of molecular surfaces and molecular bodies are dependent on the nuclear configuration K, that can be shown explicitly by the notations $\mathrm{G}(K,a)$, and $\mathrm{F}(K,a)$.

10.3.2 Molecular isodensity contours (MIDCO)

A molecular isodensity contour surface, MIDCO $\mathrm{G}(K,a)$ of nuclear configuration K and density threshold a is defined as

$$\mathrm{G}(K,a) = \{\mathbf{r}:\rho(K,\mathbf{r}) = a\} \tag{10.9}$$

that is, as the collection of all points $\mathbf{r}$ of the 3D space where the electronic density $\rho(K,\mathbf{r})$ is equal to the threshold value a. The $\mathrm{G}(K,a)$ collection of points $\mathbf{r}$ fulfilling equation (10.9) is, indeed, a continuous surface, since the molecular electronic density $\rho(K,\mathbf{r})$ is a continuous function of the position vector $\mathbf{r}$.

Most of the electronic charge cloud is localized within a few ångströms of the cluster of nuclei of the molecule; the electronic density becomes negligible at about 8–10 Å from the nearest nucleus. In the strict sense, the electronic charge-density function $\rho(K,\mathbf{r})$ becomes zero only at infinite distance from the nuclei; however, in an approximate sense, it is sufficient to consider only a finite range of electron densities above some small positive threshold. In this sense, only those regions of the 3D space are regarded as belonging to the molecule where the electronic density $\rho(K,\mathbf{r})$ is larger than this threshold a_{min}. For an appropriate threshold value $a > a_{\mathrm{min}}$, an approximate molecular body is defined as the collection $\mathrm{F}(K,a)$ of all those points $\mathbf{r}$ of the 3D space where the electronic density is greater than a,

$$\mathrm{F}(K,a) = \{\mathbf{r}:\rho(K,\mathbf{r}) > a\} \tag{10.10}$$

Such collections of points where the value of a function stays above some specified level a are called level sets of the function. The set $\mathrm{F}(K,a)$ is a level set of the electronic density for the threshold level a. Although electrons are negatively charged, the density of electrons is positive or zero, and a negative charge means a positive value for the density function $\rho(K,\mathbf{r})$.

Consider a fixed nuclear configuration K and the associated electron distribution, represented by an infinite family of MIDCOs $G(K, a)$ for a range $[a_{min}, a_{max}]$ of density threshold values a. A single MIDCO surface $G(K, a)$ cannot describe all the essential details of the electron distribution; for different electron-density thresholds a, the shapes of the MIDCOs are generally different. Consider a continuous decrease of the threshold from a_{max} to a_{min}. Depending on the threshold, separate parts of the MIDCO may join, holes which appear near the center of aromatic rings get filled up and disappear, and other topologically significant changes may occur. However, for most small changes of the density threshold a, the topology of the MIDCO $G(K, a)$ does not change. For the entire range $[a_{min}, a_{max}]$ of density thresholds, a, there are only a finite number of topologically different MIDCO surfaces. The essential shape features of the fuzzy electron density of the molecule can be described by the topological equivalence classes of MIDCOs. Clearly, the GSTE principle applies.

We may decide to give a more detailed shape description, using the pattern of local curvature domains along the MIDCO surfaces. For example, the locally convex-, concave- and saddle-type domains of each $G(K, a)$ can be used. The pattern of these domains undergoes important changes as the density threshold a changes within the interval $[a_{min}, a_{max}]$; however, for the entire range $[a_{min}, a_{max}]$ there are only a finite number of topologically different patterns. The topological equivalence of these patterns within subranges of the interval $[a_{min}, a_{max}]$ provides an alternative, more detailed shape characterization. The geometrical similarity of these patterns is characterized by their topological equivalence.

Consider the variations of the nuclear arrangements K within the catchment region of the molecule, that is, within the range of limited deformations which are not considered to alter the chemical identity of the molecule. These variations may alter the size, the location, and even the very existence of shape features, for example, the local curvature domains on the MIDCO surfaces. Nevertheless, for most small changes of the nuclear arrangement within the catchment region, the topological pattern of these shape domains remains invariant: there exists some small range of geometrical changes of the nuclear arrangement where the topological pattern of shape domains is invariant on the corresponding MIDCO surfaces. The infinitely many geometrical arrangements falling within this small range all have the same topological pattern, and are regarded as belonging to a single topological equivalence class. The characterization of these classes by topological means corresponds to a dynamic shape characterization of the molecule.

The same topological approach can be applied when comparing two different molecules. If the shape domain characterization of their MIDCOs gives equivalent topological results, then the two molecules are similar in a geometrical sense. The geometrical similarity of the two molecules corresponds to a topological equivalence (Mezey, 1993).

Alternative shape representations can be analysed by very similar techniques. The bare nuclear potential contour surfaces, NUPCOs, have many properties analogous to those of MIDCOs; their dynamic shape analysis can follow the treatment outlined above. The molecular electrostatic potential contours, MEPCOs, are also valid tools for shape representation. However, there are topologically significant differences between MIDCOs and MEPCOs. The MEPCOs of a molecule are not necessarily closed surfaces, and MEPCOs are defined for both positive and negative threshold values.

10.3.3 Density domains analysis (DDA) and functional groups

An electronic density domain $DD(K, a)$ of some density threshold a is a formal body that includes a molecular isodensity contour (MIDCO) surface $G(K, a)$ as its boundary. The interior of the density domain $DD(K, a)$ is the level set $F(K, a)$ of the same density threshold a. Formally, the density domains $DD(K, a)$ are defined as

$$DD(K, a) = \{\mathbf{r} : \rho(K, \mathbf{r}) \geqslant a\} \tag{10.11}$$

For a given electronic state of the molecule, the shape and size of density domains $DD(K, a)$ depend on the nuclear arrangement K and on the threshold value a.

The fact that density domains include their boundaries provides a close analogy with the bodies of ordinary, macroscopic objects. The terminology should not be taken to imply that they are necessarily domains in a mathematical sense; for example, a formal density domain $DD(K, a)$ of a high density threshold value a may have several disconnected parts. In some instances, if this is evident from the context, the individual pieces are also referred to as density domains.

The density domains provide a more natural representation of chemical bonding in molecules than the conventional stereochemical 'skeletal' bonding patterns represented by lines of formal single, double and triple bonds. In density domain analysis (DDA) of chemical bonding (Mezey, 1992, 1993), the interfacing and mutual interpenetration of local, fuzzy charge density clouds are used to describe bonding. Considering the relevant range $[a_{min}, a_{max}]$ of density thresholds, and the associated family of density domains $[DD(K, a_{min}), DD(K, a_{max})]$, the topological changes as a function of the threshold a can be monitored, providing a detailed description of the actual pattern of bonding within the molecule. Starting at high-density thresholds, only disjoined, local nuclear neighborhoods appear. At intermediate thresholds the separate pieces of density domains gradually join, and change into a series of topologically different bodies. At low threshold values, we find a single, usually simply connected body.

For a topological analysis, it is sufficient to take one representative from each of the equivalence classes of all topologically different density domains within the range $[a_{min}, a_{max}]$ of density thresholds. This finite family of density domains gives a detailed description of the essential features of both the bonding pattern and the molecular shape. The topological DD approach gives a more detailed and descriptive alternative to the conventional 'skeletal model' of chemical bonding.

The density domain approach has been suggested for a quantum chemical representation of formal functional groups (Mezey, 1992, 1993). Consider a single connected density domain $DD(K, a)$ and the nuclei enclosed by it. The very fact that this subset of the nuclei of the molecule is separated from the rest of the nuclei by the boundary $G(K, a)$ of the density domain $DD(K, a)$ indicates that these nuclei, together with the local electronic density cloud surrounding them, represent a sub-entity of the molecule, with individual identity. It is natural to regard this density domain $DD(K, a)$ as a representative of a formal functional group. Although this concept of 'density domain functional group' deviates somewhat from the usual, somewhat intuitive concept of functional group used by chemists, it has been found that the density-based definition provides interesting insight (Mezey, 1993, 1994).

The pattern of density domains follows some general trends in most molecules. At high density thresholds a near the value a_{max}, one usually finds a subrange where only individual nuclear neighborhoods appear as disconnected density domains. In this subrange there is precisely one nucleus within each density domain which appears. This subrange of density thresholds is the atomic range. The atomic range is usually subdivided into two subranges. These subranges are the 'strictly atomic range' and the 'prebonding range'. In the strictly atomic range all the atomic density domains are convex sets, and in the prebonding range at least one density domain is not convex, due to a local 'growth' that reaches out to eventually join a neighboring density domain.

Within a lower subrange of the density threshold, some nuclear neighborhoods join and form density domains containing two or more nuclei; however, not all nuclei of the molecule are contained within a single-density domain. In this subrange, called the 'functional group range', various functional groups appear as individual density domains. This subrange shows the pattern of interconnection of density domains that represents the bonding pattern; hence, this subrange is also called the 'bonding range for density domains'.

At lower density thresholds, all of the nuclei of the molecule are found within a common density domain. In this subrange, called the molecular density range, the molecular pattern of bonding is established. The molecular density range usually contains further subranges. Near the higher limit of the molecular range, the density domain usually has at least one local 'neck' region, where the density domain $DD(K, a)$ is not locally convex. In this subrange the molecular body $DD(K, a)$ is 'skinny'; this range is called the 'skinny molecular range'. At lower densities, no such neck regions occur, but the density domain $DD(K, a)$ usually has at least one local non-convex region.

This subrange is the 'corpulent molecular range'. At low enough thresholds, the density domain is convex for all molecules. The corresponding subrange is the 'quasi-spherical molecular range'. The atomic and the functional group ranges together constitute the localized range, whereas the molecular density range is regarded as the global density range.

10.3.4 The shape group methods

The reader may find a detailed introduction into the shape group methods of molecular shape characterization by Mezey (1993); here we shall present only a brief overview of these methods.

The shape groups are algebraic groups, describing various aspects of shape (Mezey, 1986, 1987b, 1988). These groups are not related to point symmetry groups, although the presence of symmetry may influence the shape groups. In technical terms, the shape groups are the homology groups of truncated objects, where the truncation is determined by local shape properties.

In most applications of shape groups, the local shape properties are specified in terms of shape domains; for example, in terms of the local convex-, concave-, or saddle-type regions of MIDCOs, mentioned earlier.

The use of local shape domains of contour surfaces appears necessary for a detailed enough shape description of most molecules. Most contour surfaces of small molecules, including their MIDCOs, NUPCOs and MEPCOs, are topologically rather simple objects. Within the chemically interesting threshold ranges, many of these surfaces are topologically equivalent to a sphere, or to a doughnut or to objects composed of a few 'fused' doughnuts. Direct topological characterization of such simple objects provides only a crude shape characterization and leads to nearly trivial results in a similarity analysis.

However, by identifying and topologically classifying the patterns of local shape domains on such contour surfaces, a detailed topological shape description is obtained. In general, some geometrical or physical conditions (denoted by μ) are used to define local shape domains on a contour surface $G(K,a)$. These local shape domains are denoted by the symbol D_μ. For example, a MIDCO surface $G(K, a)$ can be compared to a plane. If the plane is moved along the MIDCO as a tangent plane, then the local curvature properties of the MIDCO is compared to the plane. Each point $\mathbf{r}$ of the MIDCO is characterized by the local relation between the tangent plane and the density domain enclosed by the surface. The tangent plane may fall on the outside, on the inside, or it may cut into the given density domain within any small neighborhood of the surface point $\mathbf{r}$, depending on whether at point $\mathbf{r}$ the MIDCO is locally convex-, locally concave-, or locally saddle-type, respectively. The tangent plane is used to characterize the local curvature of the contour surface. By carrying out this characterization for all points $\mathbf{r}$ of the MIDCO, we obtain a subdivision of the molecular contour surface into locally convex-, locally concave-, and locally saddle-type shape domains, denoted by the symbols D_2, D_0, and D_1, respectively.

More detailed shape description is obtained if the tangent plane is replaced by some other objects, for example, if the MIDCO is compared to a series of tangent spheres of various radii r, or to a series of oriented tangent ellipsoids T, if a characterization involving some reference directions is required. In the case of oriented tangent ellipsoids, they can be translated but not rotated as they are brought into tangential contact with the MIDCO surface $G(K, a)$. The tangent object T may fall locally on the outside, on the inside, or it may cut into the given density domain $DD(K, a)$ within any small neighborhood of the surface point $\mathbf{r}$ where the tangential contact occurs. These differences lead to a new family of local shape domains D_2, D_0, and D_1, respectively, relative to the new tangent object T. If tangent spheres are used, orientation cannot play any role, and one may use the curvature of the sphere, $b = 1/r$, for specification. In the latter case, the local shape domains D_2, D_0, and D_1, represent the local relative convexity domains of the MIDCO, relative to the reference curvature b. Note that $b = 0$ corresponds to the case of the tangent plane.

For any specific reference curvature b, the local shape domains D_2, D_0, and D_1 generate a partitioning of the MIDCO surface $G(K, a)$. All D_μ domains of a specified type μ (for example, all the locally convex domains D_2 relative to b), can be excised from the MIDCO surface $G(K, a)$, leading to a new, topologically more interesting object, a truncated contour surface $G(K, a, \mu)$ that inherits some essential shape information from the original MIDCO surface $G(K, a)$. This shape information is now accessible by simple topological means. A topological analysis of the truncated surface $G(K, a, \mu)$ corresponds to a (somewhat limited) shape analysis of the original MIDCO surface $G(K, a)$. If this procedure is repeated for a whole range of reference curvature values b, then a detailed shape analysis of the MIDCO is obtained. It is noteworthy that for the whole range of possible reference curvature values b there are only a finite number of topologically different truncated MIDCO $G(K, a, \mu)$. These truncated surfaces can be characterized by their topological invariants, and these invariants provide a numerical shape characterization.

The primary tools used for such characterization are the homology groups of truncated surfaces (Mezey, 1993). Homology groups of algebraic topology are topological invariants, expressing important features of the topological structure of bodies and surfaces. The ranks of these groups are the Betti numbers, themselves important topological invariants.

By definition (Mezey, 1993), the shape groups of the original MIDCO $G(K, a)$ are the homology groups $H^p_\mu(a, b)$ of the truncated surfaces $G(K, a, \mu)$. The $b^p_\mu(a, b)$ Betti numbers of the $H^p_\mu(a, b)$ shape groups are used to generate numerical shape codes for molecular electron density distributions. For each reference curvature b and shape domain and truncation pattern μ of a given MIDCO $G(K, a)$ of density threshold a, there are three shape groups, $H^0_\mu(a, b)$,

$H^1_\mu(a, b)$, and $H^2_\mu(a, b)$. The formal dimensions p of these three shape groups are 0, 1, and 2, collectively expressing the essential shape information of the MIDCO. Accordingly, for each (a, b) pair of parameters and for each shape domain truncation type μ, there are three Betti numbers, $b^0_\mu(a, b)$, $b^1_\mu(a, b)$, and $b^2_\mu(a, b)$. For a detailed introduction and mathematical derivations of shape groups see Mezey (1993) and Chapter 11. Several examples for the actual calculation of the shape groups and their Betti numbers are also given by Mezey (1993).

Shape groups can also be defined directly for the entire, 3D electronic charge distribution, where subdivisions of the 3D space are obtained first for a range of parameters $p_1, p_2, \ldots, p_t$, describing the local gradient and second derivative properties of the charge distribution at each point $\mathbf{r}$. A general range of these parameters is denoted by μ. By excising 3D domains characterized by some range μ of these parameters, new objects are obtained. The homology groups of the truncated three-dimensional charge distribution define the shape groups $H^0_\mu(p_1, p_2, \ldots, p_t)$, $H^1_\mu(p_1, p_2, \ldots, p_t)$, $H^2_\mu(p_1, p_2, \ldots, p_t)$, and $H^3_\mu(p_1, p_2, \ldots, p_t)$, of the entire electronic charge density, with their Betti numbers $b^0_\mu(p_1, p_2, \ldots, p_t)$, $b^1_\mu(p_1, p_2, \ldots, p_t)$, $b^2_\mu(p_1, p_2, \ldots, p_t)$, and $b^3_\mu(p_1, p_2, \ldots, p_t)$, respectively, being informative topological invariants. For visualization of shape analysis, however, the shape group method, as applied to individual contour surfaces, appears more suitable.

The shape group method of topological shape description combines the advantages of geometry and topology, and follows the spirit of the GSTE principle. The local shape domains and the truncated MIDCOs $G(K, a, \mu)$ are defined in terms of geometrical classification of points of the surfaces, using local curvature properties. In turn, the truncated surfaces $G(K, a, \mu)$ are characterized topologically by the shape groups and their Betti numbers.

10.4 Numerical shape codes and measures of molecular similarity and complementarity

One of the advantages of the topological shape analysis techniques is the numerical representation of shape information. The results of a shape group analysis can be represented by a finite family of Betti numbers $b^p_\mu(a, b)$ for all the shape groups which occur for a given molecule, and these numbers form a numerical shape code. These shape codes can be compared by a computer, providing a well-defined numerical measure of molecular shape similarity. By a suitable transformation, the shape codes can be used for generating a numerical measure of shape complementarity. These similarity and complementarity measures can be generated by algorithmic methods using a computer. This eliminates the subjective element of visual shape comparisons. Algorithmic methods appear particularly advantageous if large sequences of molecules are to be compared. Such numerical shape codes or alternative numerical shape information can be generated for practically all molecules in the databanks of most drug companies, and local and global similarity analysis based on such shape codes may prove to be a valuable practical tool for drug design.

10.4.1 The (a, b)*-parameter maps as shape codes*

In this section we shall focus on the most common type of shape groups: those generated for the electronic density of molecules, using curvature-based shape domains on the MIDCO surfaces, for a range of curvature parameters b. According to the usual convention, a positive b value indicates that a tangent sphere, as reference object, is placed on the exterior side of the molecular surface, whereas a negative b value indicates that the sphere is placed on the interior side of the MIDCO. For a fixed nuclear arrangement K, the shape groups of a molecule depend on two parameters: on the electronic-density threshold a and on the reference curvature b. The ranges of these two parameters, a and b, define a formal, two-dimensional map, called the (a, b)-map, and the shape group distribution along this map provides a detailed shape characterization of the electronic density of the molecule.

Two types of (a, b)-maps have been studied (Mezey, 1993). In the first type, the molecular electron density is regarded as a single object, even if in the high-density regions (for high threshold values a) this object consists of several disjoint pieces. Such density thresholds belong to the localized range of density domains. In the second type of (a, b)-maps, each disjoint piece of the electronic density at the given threshold a is regarded as a separate object, and the shape analysis is performed separately for each such piece. Whereas this latter approach has some advantages (Mezey, 1993), here we shall be concerned only with the first type of (a, b)-maps.

Separate (a, b)-maps are generated for each of the three types of Betti numbers, $b_\mu^0(a, b)$, $b_\mu^1(a, b)$, and $b_\mu^2(a, b)$, according to the dimension of the shape groups. The Betti numbers obtained for a given pair of values of parameters a and b are assigned to the given location of the (a, b) parameter map. The most important Betti numbers, conveying the chemically most relevant shape information, are those of type $b_\mu^1(a, b)$ for shape groups of dimension 1.

The (a, b)-map of Betti numbers is usually generated in a discretized form, by considering a grid of a and b values within the interval $[a_{\min}, a_{\max}]$ of density thresholds and the interval $[b_{\min}, b_{\max}]$ of reference curvature values. In most applications the choice

$$b_{\min} = -b_{\max} \tag{10.12}$$

is used.

If the range of these parameters covers several orders of magnitude, then it is advantageous to use logarithmic scales, where for negative curvature parameters b the $\log|b|$ values are taken (Mezey, 1993). In one variant of this technique, a 41×21 grid is used, where the ranges are [0.001, 0.1 a.u.] (a.u. = atomic unit) for the density threshold values a, and $[-1.0, 1.0]$ for the curvature b of the test spheres against which the local curvatures of the MIDCOs are compared (Mezey, 1993).

In either the direct or the logarithmic representations, the values of the Betti numbers at the grid points (a, b), or at the points $(\log(a), \log|b|)$ of the logarithmic map, form a matrix, $\mathbf{M}^{(a,b)}$. This matrix is a numerical shape code for the fuzzy electron density of the molecule.

10.4.2 Shape similarity measures from shape codes

If the matrix form $\mathbf{M}^{(a,b)}$ of the (a, b)-map of Betti numbers $b^p_\mu(a, b)$ is available for a family of molecules, then these matrices, as shape codes, can be used for calculating numerical similarity measures between molecules. The simplest of these approaches involves a direct numerical comparison. Considering two molecules, A and B, both in some fixed nuclear configuration, and taking their shape codes in their matrix forms $\mathbf{M}^{(a,b),\mathrm{A}}$ and $\mathbf{M}^{(a,b),\mathrm{B}}$, respectively, a numerical shape similarity measure can be defined as

$$\mathrm{s}(\mathrm{A}, \mathrm{B}) = m[\mathbf{M}^{(a,b),\mathrm{A}}, \mathbf{M}^{(a,b),\mathrm{B}}]/t \tag{10.13}$$

Here $m[\mathbf{M}^{(a,b),\mathrm{A}}, \mathbf{M}^{(a,b),\mathrm{B}}]$ is the number of matches between corresponding elements in the two matrices, and t is the total number of elements in either matrix. Clearly,

$$t = n_a n_b \tag{10.14}$$

where n_a and n_b are the number of grid divisions for parameters a and b, respectively.

If the 41×21 grid (Mezey, 1993) is used for the ranges [0.001, 0.1 a.u.] for the density threshold values a, and $[-1.0, 1.0]$ for the reference curvature b, then the elements of the 41×21 matrix $\mathbf{M}^{(a,b)}$ can be stored as an integer vector $\mathbf{C}$ of 861 components. In this case the shape similarity measure $\mathrm{s}(\mathrm{A}, \mathrm{B})$ can be expressed as

$$\mathrm{s}(\mathrm{A}, \mathrm{B}) = \sum_{i=1}^{861} \delta_{j(i),k(i)}/861 \tag{10.15}$$

where $\delta_{j,k}$ is the Kroenecker delta, with indices

$$j(i) = C_i(\mathrm{A}) \tag{10.16}$$

and

$$k(i) = C_i(\mathrm{B}) \tag{10.17}$$

Alternative, somewhat more complex similarity measures, also based on various (a, b)-map representations have been described by Mezey (1993).

In many instances, shape comparisons of local regions of molecules are more important than the evaluation of the global similarities of molecules, using, for example, the global similarity measure of equation (10.13). This is the case if the molecule is large and the chemical feature we are studying has been established to have local character.

The electron-density fragment additivity principle, described by Walker and Mezey (1993a), provides a simple approach for analysing and evaluating local shape similarity of molecules. Following the same steps as those applied in the MEDLA approach for the construction of fragment electron densities, the local electron density of any molecular fragment can be generated. Note that these fragments have formal 'dangling' electron-density components at locations where the fragments formally join the rest of the molecule. For any LCAO-based quantum chemical electron density, the fragment electron densities F are well defined, hence it is meaningful to consider the family of MIDCOs, the shape groups, as well as the shape codes $\mathbf{M}^{(a,b),F}$ for such fragments F. The similarity measure

$$s(F, F') = m[\mathbf{M}^{(a,b),F}, \mathbf{M}^{(a,b),F'}]/t \qquad (10.18)$$

of any two fragments, F and F', is defined the same way as the similarity measure of equation (10.13) for two molecules A and B.

It should be pointed out that shape similarity measures of NUPCOs or other molecular surfaces can be obtained by the same steps, using their shape groups and the corresponding shape codes.

Local shape similarities of both MIDCO and MEPCO surfaces of the reacting molecules can give important hints in the study of the initial stages of biochemical reactions. The shapes of MEPCOs are often used to interpret the essential biochemical interactions of a given ligand with a receptor site of an enzyme. The study of the local shapes and local shape similarities of molecules showing similar biochemical activities is an important task in rational drug design.

10.4.3 *Shape complementarity measures from shape codes*

Molecular recognition usually depends on the complementarity of local regions of molecules, where complementarity may refer to electron distributions, polarizability properties, or electrostatic potentials. Topological techniques are suitable for the quantification of the degree of molecular complementarity and can be used as tools for the prediction of molecular recognition. In the previous section, a technique was described for the construction of non-visual shape-similarity measures of molecules as well as molecular fragments, using the numerical shape code method. A transformation of these local shape codes generates a representation suitable for a direct evaluation of local shape complementarity. The techniques applied for the construction of local shape similarity measures are also applicable for the construction of local shape complementarity measures.

When applied to electron densities, represented by MIDCOs, shape complementarity implies complementarity of two families of properties. One of these properties is represented by the density threshold, the other by the reference curvature parameter. Shape complementarity of two MIDCOs involves complementary curvatures, as well as complementary values of the charge density contour parameters a. On the one hand, a locally convex domain relative to a reference curvature b shows some degree of shape complementarity with a locally concave domain relative to a reference curvature of $-b$. On the other hand, shape complementarity between the lower electron-density contours of one molecular fragment and the higher electron-density contours of another molecular fragment is required, as implied by the partial interpenetration of interacting molecular fragments.

In general, the interacting molecules penetrate each other only to a limited extent; for a stronger interaction a more pronounced mutual interpenetration is expected. For a given interaction, a formal contact density threshold a_0 can be defined. Consider the MIDCO sequences of both of the interacting molecules. Take the MIDCOs as they occur for the isolated molecules, and place them according to the nuclear positions of the interacting molecules. Take a series of common density thresholds for the two families of MIDCOs. At high-density values, the MIDCOs of the two molecules have no contacts, whereas at low-density thresholds they intersect each other. Consider gradually decreasing common threshold values, the threshold where the first contact occurs is the formal contact density a_0.

If the interacting conformations of molecules M_1 and M_2 are K_1 and K_2, respectively, and if for the given interaction a contact density value a_0 is found, then the local shape complementarity between the MIDCO $G(K_1, a_0)$ and $G(K_2, a_0)$ is of importance. Of course, shape complementarity should not be limited to MIDCOs of a single-density threshold value a_0, and generally one considers the local shape complementarities of all MIDCO pairs $G(K_1, a_0 - a')$ and $G(K_2, a_0 + a')$ in some narrow density interval

$$[a_0 - \Delta a, a_0 + \Delta a] \tag{10.19}$$

The complementarity of the local shapes of MIDCO pairs of threshold values deviating in the opposite sense from the contact density a_0 is investigated.

Shape complementarity implies matches between locally concave and locally convex domains, as well as matches between properly placed saddle-type domains. Instead of the simple $D_\mu(K, a)$ notation, the more elaborate notation $D_{\mu(b),i}(K, a)$ is used when studying the complementarity of local shape domains. This notation includes the relative convexity specification $\mu(b)$, taking values of 2 for convex, 1 for saddle, and 0 for concave domains with respect to reference curvature b, and a serial index i for the shape domains of the given type. The three types of complementary local matches between curvature domain pairs can be written as

$$D_{0(b),i}(K_1, a_0 - a'),\ D_{2(-b),i}(K_2, a_0 + a') \tag{10.20}$$

$$D_{1(b),i}(K_1, a_0 - a'),\ D_{1(-b),i}(K_2, a_0 + a') \tag{10.21}$$

and

$$D_{2(b),i}(K_1, a_0 - a'),\ D_{1(-b),i}(K_2, a_0 + a') \tag{10.22}$$

Both aspects of complementarity, expressed by the pairings $(b, -b)$ and $(a_0 - a', a_0 + a')$, can be incorporated within a single condition using the (a, b) parameter map approach. Furthermore, the complementarity of curvature domain types, for example, D_2 and D_0, can be ensured by taking complementary shape groups defined by complementary truncations of the MIDCO surfaces. In general, this leads to an (a, b)-map for the $b^p_\mu(a, b)$ Betti numbers of the $H^p_\mu(a, b)$ shape groups of molecular fragment F_1 and to an (a, b)-map for the complementary $b^p_{2-\mu}(a, b)$ Betti numbers of the $H^p_{2-\mu}(a, b)$ shape groups of molecular fragment F_2.

As an example, consider the most informative one-dimensional shape groups for both molecules: the shape group $H^1_\mu(a, b)$ with reference to the $\mu = 2$ truncation for molecular fragment F_1, and the shape group $H^1_{2-\mu}(a, b)$ for the complementary $\mu' = 2 - \mu = 0$ truncation for molecular fragment F_2. We obtain the (a, b)-map of the $H^1_2(a, b)$ shape groups of molecular fragment F_1 and to the (a, b)-map of the $H^1_0(a, b)$ shape groups of molecular fragment F_2.

In these maps, the curvature types for the truncation in the two fragments are complementary. However, the above two (a, b) maps cannot yet be compared directly, since if a direct comparison is made by simply overlaying these maps, identical, and not complementary, a and b values occur for the two molecular fragments. However, a simple transformation of one of these maps ensures complementarity of the density threshold and reference curvature values. A central inversion of the (a, b) parameter map of molecular fragment F_2 with respect to the point $(a_0, 0)$ of the map ensures a proper match between complementary parameter values. A direct comparison of the original (a, b)-map of fragment F_1 and the centrally inverted (a, b)-map of fragment F_2 is suitable to evaluate shape complementarity. For example, the locally convex domains of the MIDCO $G(K_1, a_0 - a')$ of fragment F_1 relative to the reference curvature b are tested for shape complementarity against the locally concave domains of the MIDCO $G(K_2, a_0 + a')$ of fragment F_2 relative to a reference curvature $-b$.

The centrally inverted map method (CIMM) of molecular shape complementarity analysis (Mezey, 1993) relies on the method used for similarity measures. The problem of shape complementarity is replaced by a problem of similarity between the original (a, b) parameter map of shape groups $\mathrm{H}_{\mu}^{p}(a,b)$ of fragment F_1 and the centrally inverted (a,b) parameter map of the complementary $\mathrm{H}_{2-\mu}^{p}(a,b)$ shape groups of fragment F_2.

10.4.4 *Pattern graphs of* (a, b)-*parameter maps*

The direct comparison of two shape code matrices $\mathbf{M}^{(a,b),\mathrm{A}}$ and $\mathbf{M}^{(a,b),\mathrm{B}}$ gives a natural measure of similarity. However, minor shifts of shape features along either the electron density or reference curvature coordinate axis of the (a,b)-map may cause a significant lowering of the calculated value for similarity measure. Whereas these shifts represent valid shape changes, alternative methods and similarity measures, which lessen the emphasis on such shape variations, are also needed.

One such method is based on a graph that describes the topological pattern of the Betti number distribution along the (a,b)-map. The (a,b)-map is partitioned into domains $\mathrm{D}(b_{\mu}^{p},i)$ where for each point (a,b) within $\mathrm{D}(b_{\mu}^{p},i)$ the Betti number $b_{\mu}^{p}(a,b)$ is the same, and where i is the serial number of the domain. The domain $\mathrm{D}(b_{\mu}^{p},i)$ is the ith maximum connected component of the subset of the (a,b)-map where the Betti number has the given value b_{μ}^{p}.

The graph $\mathrm{g}(b_{\mu}^{p},\mathrm{A})$ of molecule or fragment A is defined as follows. The vertices of $\mathrm{g}(b_{\mu}^{p},\mathrm{A})$ are identified with the $\mathrm{D}(b_{\mu}^{p},i)$ domains of the (a,b)-map, and an edge is defined between two vertices if the corresponding two $\mathrm{D}(b_{\mu}^{p},i)$ and $\mathrm{D}(b_{\mu}^{\prime p},j)$ domains are neighbors within the (a,b)-map.

This graph is invariant for small shifts of the domain patterns within the (a,b)-map. A similarity measure is obtained by comparing the graphs $\mathrm{g}(b_{\mu}^{p},\mathrm{A})$ of molecule or fragment A and $\mathrm{g}(b_{\mu}^{p},\mathrm{B})$ of molecule or fragment B. For graph comparisons, one of the standard graph similarity measures (Johnson and Maggiora, 1990) may be applied.

10.4.5 *Resolution-based similarity measures*

Molecular shape-similarity analysis can be based on a different approach, using a numerical measure of spatial resolution of details of shape for characterization. A simple example illustrates the principles of these methods.

Consider three objects, A, B, and C, placed at a great distance from an observer. At this distance they appear indistinguishable, for example, all three objects may appear as mere points. If these objects are placed at some closer distance to the observer, then it is possible that one of the objects, object A, becomes distinguishable from objects B and C; however, B and C may still appear indistinguishable. At a closer distance, however, B and C are also distinguishable. The observer concludes that B and C are more similar to each other than A is to B or A is to C. The dissimilarity of A from B and C is already evident at a medium distance, whereas a closer look is needed to distinguish B from C.

Viewing objects at various distances can be replaced by taking descriptions of various resolutions. We may think of viewing a series of photographs of the same set of objects taken at gradually increasing resolutions. Two, highly dissimilar objects are distinguishable at a low resolution; the more similar the objects, the higher resolution is required to distinguish them. If the shapes of the two objects are identical, then they are indistinguishable even at infinite resolution.

Based on this idea, various similarity measures have been defined (Mezey, 1993), relying on the level of resolution required to distinguish objects. These measures form the family of resolution based similarity measures (RBSM). Numerical characterization of resolutions is a simple task; in some applications a rectangular grid of varying grid size has been used for molecular shape characterization and for the computation of numerical shape similarity measures. In these methods a grid represents the level of resolution, that can be changed in a continuous manner.

10.5 Summary

In this chapter, some of the topological techniques for the analysis of the similarities of various molecular shapes have been outlined. In drug design, one of the goals of shape similarity analysis is the establishment of quantitative shape–activity relations, QShAR, which are expected to become predictive tools. Computer-based multiple side-effect analysis, combined with multiple-shape similarity measures of bioactive molecules, is an important component of QShAR. With the new approaches developed for accurate modelling of the shape of electron distributions of large biomolecules, a new range of biochemical and drug-design problems has become accessible to detailed, algorithmic, computer-based shape similarity analysis.

Acknowledgement

The original research leading to the topological shape analysis techniques described in this chapter was supported by both strategic and operating research grants from the Natural Sciences and Engineering Research Council of Canada.

References

Johnson, M.A. and Maggiora, G.M. (eds) (1990) *Concepts and Applications of Molecular Similarity*. John Wiley, New York.

Mezey, P.G. (1986) Group theory of electrostatic potentials: a tool for quantum chemical drug design. *International Journal of Quantum Chemistry. Quantitative Biology Symposia*, **12**, 113–122.

Mezey, P.G. (1987a) The shape of molecular charge distributions: group theory without symmetry. *Journal of Computational Chemistry*, **8**, 462–469.

Mezey, P.G. (1987b) *Potential Energy Hypersurfaces*. Elsevier, Amsterdam.

Mezey, P.G. (1988) Global and local relative convexity and oriented relative convexity; application to molecular shapes and external fields. *Journal of Mathematical Chemistry*, **2**, 325–346.

Mezey, P.G. (1992) Shape similarity measures for molecular bodies; a 3D topological approach to QShAR. *Journal of Chemical Information and Computer Sciences*, **32**, 650–656.

Mezey, P.G. (1993) *Shape in Chemistry. An Introduction to Molecular Shape and Topology*. VCH, New York.

Mezey, P.G. (1994) Quantum chemical shape: new density domain relations for topology of molecular bodies, functional groups and chemical bonding. *Canadian Journal of Chemistry*, **72**, 928–935.

Parr, R.G. and Berk, A. (1981) The bare nuclear potential as a harbinger of the electron density in a molecule. In *Chemical Applications of Atomic and Molecular Electrostatic Potentials* (eds P. Politzer and D.G. Truhlar), pp. 51–62. Plenum Press, New York.

Walker, P.D. and Mezey, P.G. (1993a) Molecular electron density; a Lego approach to molecule building. *Journal of the American Chemical Society*, **115**, 12423–12430.

Walker, P.D. and Mezey, P.G. (1993b) *Program* MEDLA93. Mathematical Chemistry Research Unit, University of Saskatchewan, Saskatoon, Canada.

Walker, P.D. and Mezey, P.G. (1994) Realistic detailed images of proteins and tertiary structure elements: *ab initio* quality electron density calculations for bovine insulin. *Canadian Journal of Chemistry*, **72**, (In press).

11 The application of molecular topology to drug design — topological descriptions of molecular shape

D. WHITLEY and M. FORD

11.1 Introduction

Molecular similarity is becoming an increasingly important topic in drug design. Methods for determining similarity are required, for example, in order to compare the shapes, symmetries and electronic properties of molecules, or search databases for novel structures with features complementary to a biologically active ligand or a characterized receptor. Many of the procedures for assessing similarity are atom-based and use two- or three-dimensional geometry to superimpose or compare different molecular structures or conformations (Dean and Perkins, 1993; Richards, 1993). Because these methods involve calculations using very large sets of real numbers, they can prove inefficient when searching large databases for structures such as proteins which contain many atoms. If molecular flexibility is also considered, the computational problems can become intractable unless sensible approximations are employed (Dean and Perkins, 1993).

Our approach to the description of molecular similarity is based on topological rather than geometrical ideas. Topology is a branch of mathematics popularly known as 'rubber sheet geometry', dealing in properties of objects which are invariant under continuous transformations. In contrast to geometry, where a transformation between two congruent objects must preserve distances and angles, topology allows a wider range of deformations, including bending, twisting, and stretching. This leads to a description of shape which, while cruder than a geometrical one, is more robust (small changes in atomic positions will not alter the shape), and moreover, is determined by integer invariants.

We will associate with each molecular structure a surface, and describe the shape of the structure in terms of the topology of the surface. Throughout the present work we use the van der Waals surface, though we indicate in the concluding section how these ideas may be applied to other molecular surfaces. The topological type of such a surface is determined by a single integer, the genus of the surface (section 11.2.1). This is clearly insufficient to determine molecular shape by itself, since many molecules, and probably the majority of small molecules, are simply spheres, and have the same topology. However, the van der Waals surface provides us, not only with a surface, but also with a graph drawn on that surface: the graph formed by the intersections of the atomic van der Waals spheres. As the shape of a molecule changes, so will the graph, and it is this surface graph which provides our description of molecular shape. The structures whose shapes we compare here are different conformations of a particular molecule, but the same ideas apply to the comparison of shape between different molecules.

The major advantage of this approach is that problems of molecular shape, initially expressed in terms of geometrical coordinates, are translated into questions in discrete mathematics: the topological invariants we use are integers, and the surface graphs are defined by integer adjacency matrices. Comparison of molecular shape becomes the matching of graphs, and while this is still a difficult combinatorial problem, familiar to chemists from substructure searching, it is much easier than comparing shapes by manipulating sets of geometrical coordinates. Similar methods based on graphs derived from geometry have been studied by Artymiuk *et al.* (1990).

The application of topological methods to the description of molecular shape, under the heading of the 'shape group method', is the subject of a number of earlier papers by Mezey and coworkers (see Chapter 10 for a summary). The approach there is to study the topology of a sequence of subsurfaces S_n, obtained from a molecular surface S by deleting a sequence of subsets D_n. For smooth molecular surfaces, such as surfaces of constant electrostatic potential, the deleted sets D_n may be regions of positive, zero, and negative curvature (Mezey, 1987). When S is a van der Waals surface, choices for D_n include the faces of S ordered by surface area (Arteca and Mezey, 1988a), and sets of faces of S grouped by the number of spherical arcs in the face boundary (Arteca and Mezey, 1988b). Subsequent work by Bradshaw *et al.* (1993) showed that the topological invariants of a sequence of surfaces formed by successively deleting atoms from the van der Waals surface in the unique SMILES order (Weininger *et al.*, 1989), successfully distinguished a group of conformations of a six-membered ring, *N*-methyl piperidine (**I**). The present chapter considers the feasibility of developing some of these ideas into practical tools.

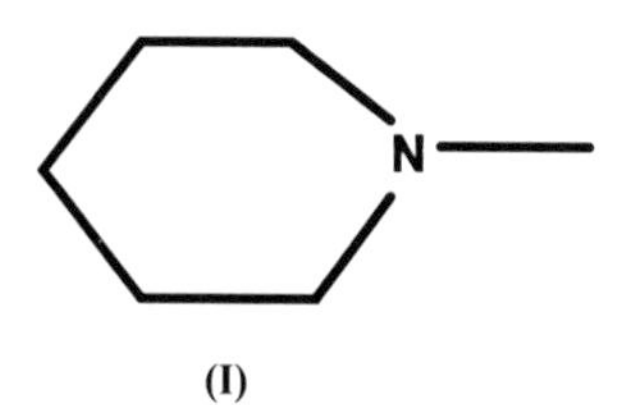

(**I**)

We will begin, in section 11.2, with an informal introduction to the relevant mathematical ideas, and a description of the algorithm by which we obtain the molecular surface graph and topological invariants from a set of atomic coordinates. In section 11.3 we consider a method of generating, from the molecular surface graph, a sequence of integers to represent molecular shape. These shape sequences provide a compact representation of the information contained in the molecular surface graph, but they turn out to depend on the ordering of the atoms in the molecule, and the size of the atomic radii, in a rather unstable way, so in the remainder of the chapter we concentrate on the surface graphs themselves. In section 11.4 we show that the conformations of the *N*-methyl piperidine ring mentioned above are successfully classified by the graphs on the surface formed by just the backbone, non-hydrogen atoms. From the surface graph we derive a shape graph (essentially the geometric dual of the surface graph), and we see that the shape graphs overcome the main difficulties encountered with the shape sequences: the conformational classes identified by the shape graphs are stable with respect to changes in the atomic radii, and their dependence on the atomic labelling is naturally related to the symmetries of the molecule's covalent structure graph. We conclude section 11.4 with an example of a substructure search using the shape graphs for *N*-methyl piperidine as templates.

11.2 Graphs, surfaces and topology

In this section we describe the relationship between surface graphs and surface topology which forms the mathematical basis of our work. We also outline an algorithm for extracting the relevant invariants in the case of a van der Waals surface. We start with a brief and informal discussion of the ideas involved in topology. For a more detailed introduction, however, we recommend the work of Firby and Gardiner (1982), and all the results we employ are contained in works by Massey (1967) and Giblin (1981).

For a molecule M with atoms $a_i = 1, \ldots, n$, we represent each atom by a three-dimensional ball B_i, with some radius $r_i > 0$, and consider M to be the union of these balls

$$M = \bigcup_{i=1}^{n} B_i$$

Our main interest is in the van der Waals surface S which is the topological boundary of M, consisting of all the points of M which do not lie in the interior of any of the balls B_i. This is the 'visible' part of M, a piecewise smooth surface formed by a union of spherical polygons, each of which is a subset of an atomic van der Waals sphere A_i, where A_i is the boundary of B_i. Note that S is an orientable surface; there are no Mobius bands or the like here. Besides the whole surface S, we shall also need to consider subsets of S obtained by removing those parts of the surface belonging to certain atomic van der Waals spheres. The resulting surfaces have boundaries formed by curves surrounding the deleted parts. Therefore the 'surfaces' we shall be concerned with, and whose topological properties we describe below, belong to the class described mathematically as 'compact, orientable surfaces with boundary' (where the boundary may or may not be empty).

11.2.1 Topological equivalence

The basic notion is that two surfaces are topologically equivalent, or homeomorphic, if each can be continuously deformed into the other. The emphasis here is on the continuous nature of the transformation. Informally, the surfaces are thought of as being made from rubber, and they are homeomorphic if they can be transformed into each other by bending, twisting or stretching the rubber; any continuous change is allowed, but the surfaces must not be torn or glued. In particular, the smoothness (i.e. differentiability) of the surface plays no role here; corners or sharp edges are irrelevant, and the surfaces of a cube, a tetrahedron, and an ellipsoid, for example, are all topologically equivalent to a sphere. In the context of our application to van der Waals surfaces, this means that we need not worry about the lack of smoothness along the intersections between adjacent van der Waals spheres.

Changing the topological type of a surface clearly requires quite drastic surgery, and there are effectively only two things we can do. The first possibility is to cut one or more holes in the surface. Starting with a sphere, for instance, if we remove one hole the remaining surface may be (continuously) flattened out into a disc; so a sphere with a hole and a disc are topologically equivalent. Removing two holes from a sphere, or one hole from a disc, leaves an annulus, and if we take one boundary circle of the annulus in each hand and pull them apart, the annulus stretches into the surface of a cylinder. Thus a sphere with two holes is a topological cylinder. Readers may like to convince themselves that a sphere with three holes is a topological pair of pants (Figure 11.1). Since each hole leaves a boundary curve on the surface, the number of holes is often referred to as the number of boundaries of the surface.

The second way to change the topology of a surface is by adding handles. Suppose we remove two holes from, for example, a sphere, and glue to the boundaries of the holes in the sphere the two boundary circles of a cylinder, to form a sphere with a handle. The (rubber) material in the sphere can then be squeezed round into the handle, and we see that a sphere with a handle is topologically the same as a torus, i.e. the surface of a solid ring (Figure 11.2). Adding a second handle produces a double torus, or pretzel. The number of handles is called the genus of the surface.

11.2.2 Classifying surfaces

A basic result in surface topology, the Classification Theorem for Surfaces (Massey, 1967), states that removing holes and attaching handles are the only

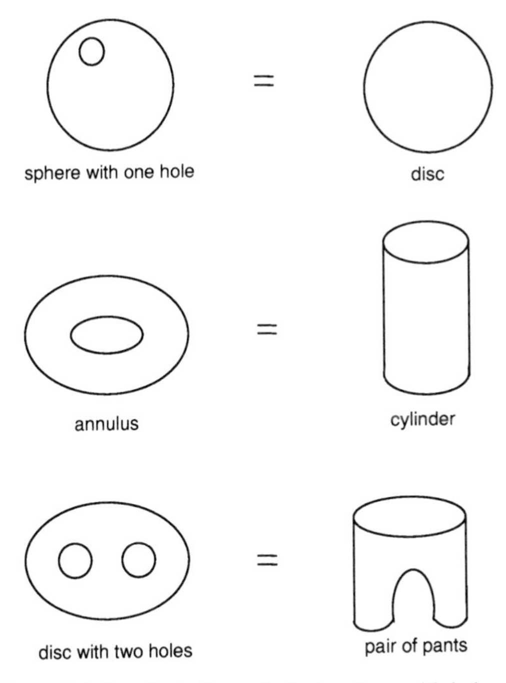

Figure 11.1 Topologically equivalent surfaces with holes.

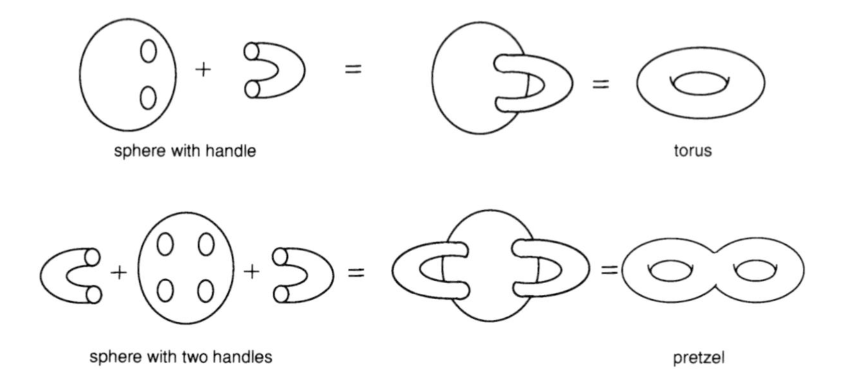

Figure 11.2 Topologically equivalent surfaces with handles.

ways to change the topological type of a surface: two surfaces are homeomorphic if, and only if, they have the same genus and the same number of boundaries. The topology of a surface is therefore completely determined by just two integers: the genus γ and the number of holes or boundary curves β.

To calculate the genus, the surface is triangulated and the numbers of vertices v, edges e, and faces f in the triangulation are counted. The alternating sum

$$\chi = v - e + f \tag{11.1}$$

is independent of the choice of the triangulation, and defines a topological invariant called the Euler Characteristic. (The case $v - e + f = 2$ for the Euclidean solids is a familiar one.) The invariants χ, β and γ are related by

$$\chi = 2 - 2\gamma - \beta \tag{11.2}$$

so any two of these suffice to determine the topology of the surface. The number of boundaries β is often known from some general consideration, and in this case we may calculate χ from a triangulation, then obtain γ from equation (11.2).

In the situation where we have a general graph, not necessarily a triangulation, drawn on the surface, Euler's formula (11.1) has the following generalization

$$\chi + \sum_{i=1}^{n} b_i = v - e + f \tag{11.3}$$

Here v, e, and f are the numbers of vertices, edges, and faces in the graph, and b_i, $i = 1, \ldots, n$, is the first Betti number of the ith face. The Betti numbers are the dimensions of certain algebraic structures associated with the surface, called 'homology groups'. The study of homology, and the proofs of the assertions above, belong to the subject of algebraic topology, which does not directly concern us here. Those interested in pursuing the mathematical details should consult Giblin (1981). We simply note that in the case of van der Waals surfaces, where each face of the surface graph is contributed by an atomic van der Waals sphere, each $b_i = \beta_i - 1$, where β_i is the number of boundary curves on the ith face. For example, the chemical structure graph

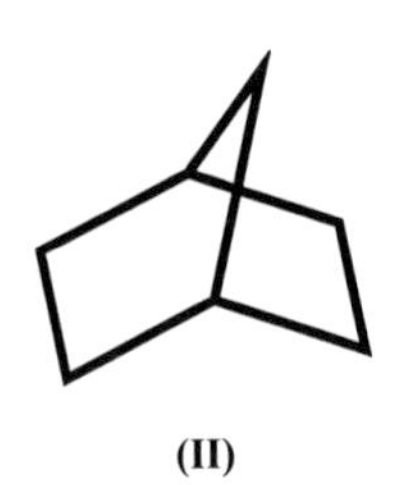

(II)

of norbornane (**II**) is a spherical graph with seven vertices (atoms), eight edges (bonds) and three faces (rings).

11.2.3 An algorithm for surface graphs

The most common example of a molecular surface S, and the one we use here, is the van der Waals surface. For this case, we have developed an algorithm for determining the topological invariants of S, and the molecular surface graph G = G(S) defined by the intersection of the atomic van der Waals spheres. The full details of this algorithm will appear elsewhere. In outline, the faces of G are the spherical polygons which form the visible parts of the atomic surfaces, the edges are the spherical arcs forming the intersections between pairs of overlapping atoms, and the vertices are points on S common to three or more atoms. Typically, only three faces (i.e. three atoms), and three edges meet at a vertex. Since one of the aims of describing molecular shape in topological terms is to provide a robust indicator which is invariant under small deformations of the molecule, it would be inappropriate to consider in detail the non-generic cases where four or more atoms meet in a point on S. We therefore restrict attention to the generic case where the graph G is trivalent at each vertex. (Loops, that is edges without vertices, are allowed, and indeed often occur at the intersections between hydrogen and heavier atoms.) In practice, whenever a non-generic situation is encountered during the algorithm, a perturbation is applied to the molecule and the calculation for that structure is restarted.

The algorithm begins by determining which pairs of atoms (A_i, A_j), and then which triples (A_i, A_j, A_k) have non-empty intersections (Wynn *et al.*, 1993). Points of triple intersection which do not lie inside any other atom are vertices of G. To find the edges of G, we consider in turn each circle of intersection between overlapping atoms. If there are no vertices on the circle, the circle forms a loop of G provided it is not hidden inside other atoms, and to confirm this it suffices to check the visibility of a single point on the circle. (A loop is either wholly visible or wholly hidden; a partially visible circle contains vertices of G and so cannot be a loop.) If the circle contains vertices, these are ordered according to their angles, and the arc on the circle between an ordered pair of vertices will be an edge of G so long as it is not hidden by any other atoms. Again, it suffices to check the visibility of a single point on the arc, and the mid-point of the arc is a convenient choice.

To find the faces of G the edges on each atom are first grouped into cycles: closed loops formed by sequences of edges where the second vertex of the first edge coincides with the first vertex of the second edge, etc. until the loop closes with the second vertex of the last edge equal to the first vertex of the first edge. The cycles must now be partitioned into groups bounding separate faces. In some cases this is straightforward. For instance, if there is only one cycle on an atom, then this bounds a single face which is a disc. The general case however requires a more elaborate procedure such as that described by Connolly (1983).

For a van der Waals surface S, or any subsurface obtained by deleting a collection of atomic spheres A_i, this algorithm determines the vertices, edges, and faces of the molecular surface graph G, so that the Euler Characteristic and genus of the surface may be calculated from equations (11.2) and (11.3). It is also a simple matter to extract the information determining G in the form of the adjacency matrix **A(G)**, defined by $\mathbf{A}(i, j) = 1$ if G has an edge joining vertex i to vertex j, and $\mathbf{A}(i, j) = 0$ otherwise.

An important point here is that although a certain amount of calculation is required to determine the topological and graph theoretic information, this calculation is required only once for each structure. The algorithm reduces the geometrical data defining the conformation to a string of integers, and all comparisons between conformations can now be performed using this integer description. This contrasts with other schemes for comparing molecular similarity which require repeated recalculation as one structure is moved relative to another.

11.3 Shape sequences

Given an ordering for the atoms in the molecule, the molecular surface may be constructed by starting with the visible faces of the first atom, and adding the contributions of the other atoms one by one, according to the ordering. This is essentially the procedure employed by Bradshaw *et al.* (1993), except that here the molecule is constructed by successively adding atoms rather than dismantled by successively removing them. At each stage in the construction we have a partial molecular surface bounded by a number of spherical arcs. This partial surface may consist of several connected components, but each component is a surface whose topology is determined by any two of the three integers χ, β, and γ introduced in the previous section. The shape sequence of the molecular surface S is the sequence of pairs (χ, β) for the partial molecular surfaces. Where the partial surface is disconnected, we include a pair (χ, β) for each component. Since the complete surface has no boundary, the final entry in the sequence has $\beta = 0$, and from equation (11.2) the genus of S is given by $\gamma = (2 - \chi)/2$. In the case where the complete surface is a sphere, i.e. $\gamma = 0$, we have $\chi = 2 - \beta$ for each entry in the shape sequence, so one of the pair (χ, β) is superfluous, and we retain only the Euler Characteristic χ.

Note that this procedure detects cavities hidden within the molecule (cf. Alard and Wodak, 1991). If the final entry in the shape sequence has more than one pair (χ, β), then the surface S has more than one component. Provided the original set of atoms form a connected structure, one component of S is the visible surface of the molecule, and the other components are the surfaces of cavities inside the structure.

The order for building the molecule is given in terms of a partition of the atoms into chemically distinguishable groups, so that valid comparisons between partial surfaces for different conformations can be made. For example, it is impossible to distinguish between the hydrogen atoms in a methyl group—all the hydrogen atoms bonded to any single atom must be considered as a single element in the partition. An entry is added to the shape sequence only after all the atoms in a partition element have been included.

Once the algorithm of section 11.2 has been applied to find the visible faces of the atomic van der Waals spheres A_i, the calculation of the shape sequence for a conformation is largely a matter of book-keeping. Starting with a partial surface whose components are the visible faces of the first atom in the partition, the faces on the second atom are joined to the surface one at a time, and so on for the remainder of the atoms. Now if a face F is joined to a surface Σ to give a new surface Σ', the components of Σ' consist of the components of Σ which are disjoint from F, together with a new component $C' = F \cup C_F$, where C_F is the union of all the components of Σ which intersect F. The set of edges of C′ is the union of the edges of F and the edges of the components in C_F, minus those edges which are common to F and any component C in C_F. The number of boundaries β of C′ is the number of cycles among these edges, and the Euler Characteristic of C′ is given by

$$\chi(C') = \chi(F) + \sum_{C \in C_F} \chi(C) - e(F) + v(F)$$

where $e(F)$ denotes the number of edges common to F and any C in C_F, and $v(F)$ denotes the number of vertices common to F and any C in C_F. In this way, the shape sequence is readily obtained from the information defining the faces of the molecular surface graph.

11.3.1 Shape sequences for N*-methyl piperidine*

As a gentle example, consider the surface S in the upper right of Figure 11.3. This shows the van der Waals spheres $A_i, i = 1, \dots, 7$, for the carbon and nitrogen atoms of *N*-methyl piperidine in an equatorial chair conformation. Each A_i contributes a disc to S. If the atoms are ordered according to the unique SMILES notation, as advocated by Bradshaw *et al.* (1993), we see from the figure that each of the partial surfaces

$$\bigcup_{i=1}^{n} A_i, \qquad n = 1, \dots, 6$$

is a disc with $(\chi, \beta) = (1, 1)$, while S itself is a sphere with $\chi(S) = 2$, $\beta(S) = 0$. This is a zero genus case, so we drop β and write the shape sequence as 1, 1, 1, 1, 1, 1, 2.

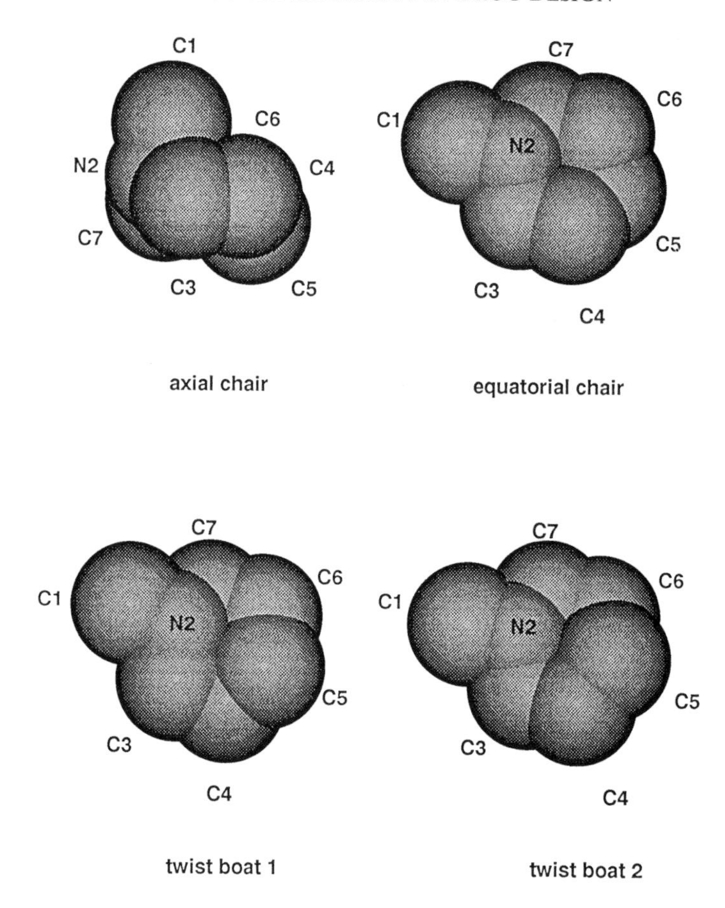

Figure 11.3 Axial chair, equatorial chair and two twist boat conformations of *N*-methyl piperidine, with hydrogen atoms suppressed and the backbone atoms labelled. The equatorial chair and twist boats are viewed from the same direction, the axial chair has been rotated through 90° above the N2–C5 axis.

The shape sequences for the hydrogen-suppressed structures do not in fact distinguish between different conformations of *N*-methyl piperidine; all the conformations we examined had the same shape sequence. However, when the hydrogen atoms are included the shape sequences do vary from one conformation to another. To see how the shape sequence becomes more complex when hydrogen atoms are included, consider Figure 11.4, which shows the same equatorial chair conformation, but this time for all 20 atoms in the molecule. In this case, our strategy for partitioning the atoms is to start with the non-hydrogen atoms, in SMILES order as before, followed by the hydrogens ordered in the same way as the atoms to which they are bonded, with all the hydrogen atoms bonded to a single heavy atom forming

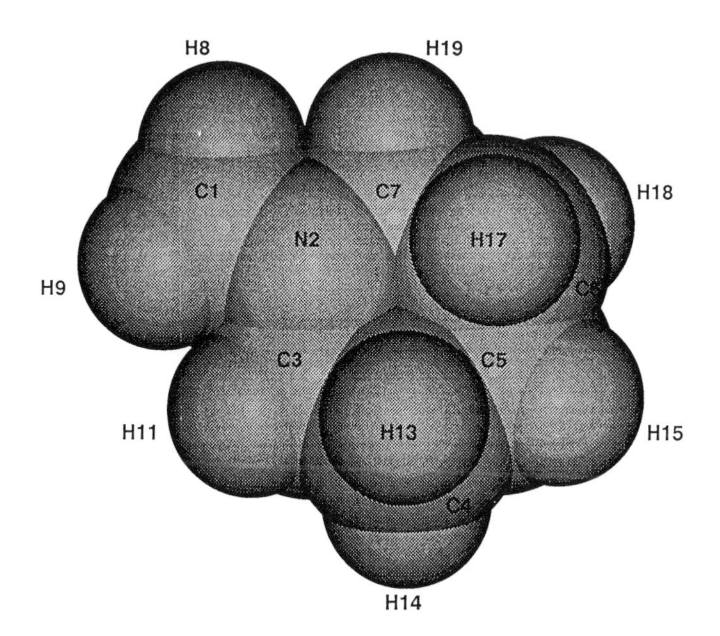

Figure 11.4 Equatorial chair conformation of *N*-methyl piperidine, with hydrogen atoms included and all visible atoms labelled.

a single element in the partition. Thus, with the atoms labelled as in Figure 11.4, the partition has 13 elements:

$$(1)(2)(3)(4)(5)(6)(7)(8,9,10)(11,12)(13,14)(15,16)(17,18)(19,20)$$

and the shape sequence consists of 13 Euler Characteristics:

$$0, 0, -2, -4, -6, -8, -9, -8, -6, -4, -2, 0, 2$$

The overall structure of the shape sequence in this example, first decreasing and then increasing, is typical for spherical molecules using van der Waals radii. As the surface is built, starting with the faces of the non-hydrogen atoms, holes are left behind where the hydrogen atoms will be added later. As the number of holes in the partial surface increases, the Euler Characteristics fall according to equation (11.2), subsequently rising as the hydrogen atoms are added and the holes are filled in. Hence the first part of the shape sequence is a decreasing sequence of integers; the second part an increasing one. For non-spherical surfaces ($\gamma > 0$), the interaction between the changing genus and the changing number of holes as the surface is constructed, may lead to a less simple pattern.

This procedure was applied to a group of 167 conformations of *N*-methyl piperidine, generated from a SYBYL C search (see References), containing axial and equatorial chairs, twist boats, and a single high-energy 'pure' boat conformation. The shape sequences were calculated using the values 1.2(H), 1.7(C), and 1.55(N) for the van der Waals radii, and a total of 18 different sequences were found. In no case was the same shape sequence generated by both a chair and a boat conformation, and moreover, the sequences distinguished axial from equatorial chairs, and twist boats from the pure boat. Of the 18 sequences, eight were generated by equatorial chair conformations, five by axial chairs, four by twist boats, and the pure boat produced a unique sequence all of its own.

11.3.2 Dependence on atomic radii and ordering

By design, a small variation in the values of the radii used for the atomic spheres will not change this result, but larger changes in the radii do alter the graphs G(S) and consequently produce different sets of sequences. The number of shape sequences occurring for the *N*-methyl piperidine sample over a wide range of van der Waals radii are listed in Table 11.1. The first pair of columns in the table show the result of varying the radii by a constant amount; the second pair show the outcome of varying the radii by a percentage of the van der Waals values quoted.

For small radii, the atoms no longer overlap across the centre of the piperidine ring, and the van der Waals surface is a torus rather than a sphere. As there seems no chemical reason for considering this molecule to be a torus, the smallest values at which the molecular surface is spherical, about 10% below the van der Waals radii, were taken as a lower bound for the range of radii considered. For the upper bound, note that as the radii increase proportionately in size, the hydrogen atoms occupy less and less of the van der Waals surface, until the hydrogen atoms are eventually swallowed by the carbon atoms. The point at which the hydrogen atoms disappear, around

Table 11.1 Effect of varying atomic van der Waals radii. The left column shows the effect of varying van der Waals radii by a constant amount; the right column shows the outcome of varying the radii by a percentage of the quoted van der Waals values

Radii ± constant	Sequences	Radii ± percentage	Sequences
van der Waals – 0.1	2	van der Waals – 10%	2
van der Waals – 0.05	7	van der Waals – 5%	4
van der Waals	18	van der Waals + 10%	37
van der Waals + 0.2	45	van der Waals + 20%	56
van der Waals + 0.4	52	van der Waals + 30%	70
van der Waals + 0.6	31	van der Waals + 40%	56
van der Waals + 0.8	15	van der Waals + 50%	44
van der Waals + 1.0	14	van der Waals + 60%	39
van der Waals + 1.2	20	van der Waals + 70%	23
van der Waals + 1.4	13	van der Waals + 80%	32
van der Waals + 1.6	9	van der Waals + 90%	16
van der Waals + 1.8	9	van der Waals + 100%	7
van der Waals + 2.0	8	van der Waals + 110%	2

120% above the van der Waals radii, was chosen as the upper limit for the radii.

For both methods of varying the radii, the overall pattern of the results in Table 11.1 is similar. At radii immediately above the values where the surface is toroidal, there are just two shape sequences. One of these is generated by the axial chairs, the other by the remaining conformations. As the radii increase, the number of shape sequences first grows, becoming quite large for radii in the centre of the range studied, before falling as the radii approach the upper limit. When the radii increase proportionately, immediately below the upper limit we again find only two shape sequences, and, as in the case of small radii, the axial chairs produce the first sequence, the other conformations the second. At these extreme values of the radii the shape sequences are picking out only the grossest distinction between the *N*-methyl piperidine conformations: the axial or equatorial position of the methyl group.

For many intermediate choices of radii the shape sequences discriminate between boat and chair conformations, i.e. boats and chairs have distinct sequences. In the *N*-methyl piperidine sample this is true for radii between the van der Waals values and 2.5Å above these figures when the radii are increased by a constant, and between 100 and 190% of the van der Waals values when the radii are increased proportionately. In addition, axial and equatorial chairs generally produce different sequences, and the sequence for the pure boat almost always differs from those of the twist boats.

Since many molecular similarity algorithms are expensive in terms of the time they take, it is important to note that the generation of shape sequences is relatively fast. Our algorithm, encoded in C and running on a Sun SparcStation, takes about 0.2 seconds to produce the shape sequence for one conformation of *N*-methyl piperidine using the van der Waals radii, and about 1.0 seconds using the solvent radii (van der Waals radii + 1.4 Å). The time increases with the radii because at larger radii there are more atomic intersections, and therefore more potential vertices of the surface graph to be calculated. Since the majority of the time used by the algorithm is occupied with floating point arithmetic, these timings are clearly central processor unit dependent, and the figures will be reduced on a faster processor. In any case, the speed of the algorithm is no obstacle to its practical application.

A single conformational shape is typically represented by a number of sequences, possibly a rather large number depending on the radii used, but the number of representatives of a particular shape may be reduced by cluster analysis, and in some cases a close correspondence between clusters and shapes is successfully obtained (Bradshaw *et al.*, 1993). With the radii at 1.6 Å above the van der Waals values, for example, the shape sequences cluster into three groups; one for the chairs, one for the twist boats, and one containing the single pure boat. However, we have found that this is not the case for all values of the radii investigated above: sequences for different shapes are often clustered before sequences representing the same shape.

Applying the shape sequences to larger molecules revealed a possible drawback: their dependence on the choice of ordering of the atoms used in building the molecule is not well behaved. For *N*-methyl piperidine there is a natural choice for this ordering, which coincides with the unique SMILES notation, so this question does not arise. However, we also examined 200 conformations of a 40 atom pyrethroid insecticide, benzyl (1*R*, 3*R*)chrysanthemate, generated from a molecular dynamics simulation. Using the hydrogen-suppressed structures of 20 backbone atoms, and van der Waals radii, the number of shape sequences varied from 20 to 48. The lower figure arose for a 'linear' atomic ordering, starting at one end of the molecule and working along it, while the higher number came from the SMILES order, which for this compound begins in the centre of the molecule. Moreover, the groups of conformations attached to each sequence are not stable as the order varies; in particular, the groups for the 48 sequences are not a subdivision of the groups for the 20 sequences.

The shape sequences therefore appear to be potentially useful descriptors of 3D molecular shape, but in view of the above difficulties we will not pursue them any further here. Instead, we turn our attention to the underlying molecular surface graphs, which, as we see in the next section, overcome many of these problems.

11.4 Shape graphs

It is clear that conformational shape is a function of the positions of the backbone, non-hydrogen atoms, and that the positions of the hydrogen atoms leave the underlying shape unaltered. The hydrogen-suppressed structures of Figure 11.3, for example, plainly show the differences between the axial chair, equatorial chair and twist boat conformations of *N*-methyl piperidine. In the cases where cluster analysis reduced the sets of shape sequences of the previous section to groups corresponding to conformational shapes, it is the finer level of detail provided by the hydrogen atoms which the cluster analysis filters out. As noted above, the shape sequences for the hydrogen-suppressed molecules do not distinguish between the conformations of *N*-methyl piperidine. However, a closer inspection of the molecular surface graphs for the backbone atoms reveals that there is a variation between graphs for different conformations; a variation which is lost in passing to the shape sequences.

In fact, with the hydrogen atoms removed, the 167 conformations in the *N*-methyl piperidine sample generate only five molecular surface graphs, and these discriminate precisely between the different conformational shapes. (This contrasts with the results of Bradshaw *et al.* (1993), who identified seven conformational shapes for the same data set using single-linkage clustering based on shape sequences.) One graph is produced by the axial chairs, one by the equatorial chairs, one by the pure boat, and two by the twist boats.

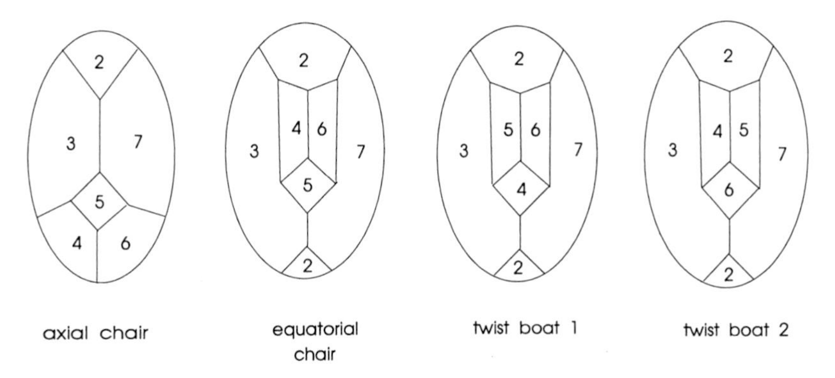

Figure 11.5 Surface graphs for the conformations in Figure 11.3, stereographically projected onto the plane. Faces 2 to 7 of the graphs are numbered in correspondence with the atom labels in Figure 11.3; face 1 (the projection of atom C1) is the unbounded face surrounding each graph.

Four of these graphs, for the chairs and twist boats, are depicted in Figure 11.5. In drawing this figure, we have utilised the correspondence between spherical and planar graphs (Wilson, 1979). Imagining the topologically spherical molecule transformed into a true sphere, stereographically project the molecule from a point p on the carbon atom C1 in the methyl group onto the tangent plane through the antipodal point of p. This transforms the molecular surface graph G to a planar graph G′, with the face of the C1 atom mapped to the unbounded region surrounding G′, and the faces of the other atoms mapped to the bounded faces of G′. To reconstruct the graph on the surface, either reverse the projection, or glue a disc representing the C1 atom to the boundary of the graphs in Figure 11.5. The numbering of the faces of the graphs in Figure 11.5 corresponds to the labelling of the atoms in Figure 11.3. Notice that the nitrogen atom N2 has two faces in the graphs for equatorial chairs and twist boats; in all other cases each atom contributes one face to each graph.

Before proceeding, we make a technical shift in our view of a molecular surface graph G, replacing G by its geometrical dual G* as follows (Wilson, 1979). In each face of G, place a vertex of G*, and join two vertices of G* by an edge if the corresponding faces of G have an edge in common. Note that G* is connected if and only if the molecule contains no cavities. To simplify matters a little, if G has more than one face from a single atom, we replace the corresponding vertices $v_1, \ldots, v_k$ of G* with a single vertex v, and add an edge from v to any vertex which was joined in G* to any of $v_1, \ldots, v_k$. The resulting graph, which we call a molecular shape graph, has one vertex for each visible atom in the molecule, with an edge between vertices whose corresponding atoms have adjoining faces on the molecular surface. Although the collapsing of several vertices into one involves some loss of information, we prefer the shape graphs to the surface graphs because in a molecule where bonded atoms have adjoining faces on the molecular surface, which is typical for small molecules, the shape graph contains the usual structure graph of the molecule as a subgraph. In this case, the shape graphs are a natural generalisation of the structure graphs, incorporating both the connectivity of the molecule and information on its 3D shape. In a large molecule where not all the atoms are visible on the molecular surface, the shape graph will of course have fewer vertices than the structure graph.

Figure 11.6 shows the shape graphs derived from the *N*-methyl piperidine surface graphs of Figure 11.5, with the structure subgraphs emphasized. Here the information discarded by identifying the two vertices for the N2 atom in the dual of the surface graphs for the equatorial chair and twist boat conformations is not crucial, and the shape graphs discriminate between conformations in exactly the same way as the surface graphs.

11.4.1 *Molecular symmetry*

The shape graphs illustrate nicely the reason why there are two graphs for the twist boats, and only one each for the chair conformations. This is because the structure graph of *N*-methyl piperidine is symmetric: the permutation $\pi = (37)(46)$ of the atom labels in Figure 11.5 leaves the structure graph unchanged. This symmetry is shared by the shape graphs (and the surface graphs) for the two types of chair but not by those for the twist boats, where π interchanges the two graphs. The symmetry π corresponds to a reflection about a central vertical axis through the graphs in Figure 11.6. In other words, the twist boat shape graphs are mirror images, representing enantiomeric conformations. We cannot tell which enantiomer a particular conformer represents, for we cannot know if the atoms in the structure were given a labelling l or πl, but the existence of mirror-image shape graphs does reveal the presence of enantiomeric structures in the sample. (In mathematical terms,

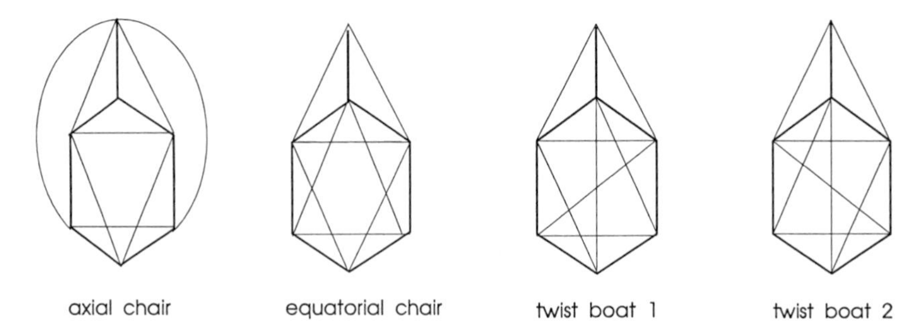

Figure 11.6 Shape graphs for the conformations in Figure 11.3. The vertices of the graphs are the seven end-points of the bold lines. Edges join vertices whose corresponding atoms share a common boundary on the van der Waals surface. Bold edges indicate atoms which are also joined by covalent bonds.

the atom labelling is determined only up to automorphisms of the structure graph, so shape graphs which differ by such an automorphism must represent the same conformational class). We note that optical isomers can also be recognized from mirror image shape graphs, although the absolute configuration of each structure may only be determined using standard methods of stereochemistry.

The role played by the labelling of the atoms, touched on briefly above, is of considerable importance. The surface graphs for the equatorial chairs and twist boats in Figure 11.5 are clearly indistinguishable if the labels on the faces are removed. Also, although it may not be immediately apparent, all the shape graphs of Figure 11.6 are isomorphic. (The permutations (12), (45), and (56), transform the equatorial chair graph into the axial chair, twist boat 1, and twist boat 2 graphs, respectively). Thus the objects we are studying here are not simply graphs, i.e. collections of vertices and edges, but labelled graphs; the surface graphs have labelled faces, the shape graphs labelled vertices. This is a pertinent point, for it means that we cannot use isomorphism invariants of the shape graphs to differentiate between conformations. The adjacency matrix eigenvalues, degree sequences, and other isomorphism invariants (Wilson, 1979) are identical for all the *N*-methyl piperidine shape graphs. In a case where the graphs for different conformations turned out to be non-isomorphic, we could search for an invariant which distinguished between conformations on purely graph-theoretic grounds, irrespective of the atomic labels. For the *N*-methyl piperidine example, however, we must retain the whole labelled shape graph as a conformational invariant. These graphs are defined by their adjacency matrices, with the row and column labels linked to the atom labelling. For example, the shape graph for the axial chair in Figure 11.6 has the 7×7 adjacency matrix

$$\mathbf{A} = \begin{pmatrix} 0111011 \\ 1010001 \\ 1101101 \\ 1010110 \\ 0011011 \\ 1001101 \\ 1110110 \end{pmatrix}$$

where the entry in the ith row and jth column, $\mathbf{A}(i, j) = 1$ if there is an edge between vertices i and j, and $\mathbf{A}(i, j) = 0$ otherwise. Equivalently, recalling that the shape graphs are essentially the duals of the molecular surface graphs, $\mathbf{A}(i, j) = 1$ if atoms i and j have adjoining faces on the van der Waals surface, while $\mathbf{A}(i, j) = 0$ if not. Now the adjacency matrices are symmetric ($\mathbf{A}(i, j) = \mathbf{A}(j, i)$) and have zero diagonal elements, so only the entries above the main diagonal are needed. We chose to store these by concatenating the rows of the upper triangle into a single 21 character string of zeros and ones. (The parsimonious could encode these as 21 bits.) For the matrix above we obtain the string 111011100011101110111. In practice, these strings are the invariants we use to describe the *N*-methyl piperidine conformations.

11.4.2 Stability of shape graphs

Apart from the unavoidable ambiguity in classifying the twist boat enantiomers, the shape graphs above provide a one-to-one correspondence with conformational shapes. This is a clear improvement over the position with shape sequences, where many sequences represent a single conformational shape. The shape graphs also have a much simpler dependence on the size of the atomic radii than the sequences. The graphs described so far are for the standard van der Waals radii, and although the graphs themselves do change as the radii vary, the groups of conformers associated with distinct graphs do not. The number of graphs is almost constant, the only variation being that for radii above 130% of the van der Waals values, there are six graphs not five, with the axial chairs split between two graphs.

The molecular shape graphs were also calculated with all the atoms in the molecule included, for radii varying from 10% below to 50% above the van der Waals values. To avoid the problem of the ambiguous labelling of hydrogen atoms, we followed the procedure adopted in the case of the shape sequences, and considered all the hydrogen atoms bonded to one atom as a single entity, represented by a single vertex in the shape graph. Hence the shape graph for the complete molecule has 13 vertices, with edges joining vertices when any of the associated atoms have intersecting faces on the molecular surface.

The number of shape graphs is much larger for the full structure, with over 100 for the larger radii. However, cluster analysis applied to the shape graphs produced clusters which in most cases are identical to the conformational groups identified by the hydrogen-suppressed shape graphs. In particular, shape graphs for different conformational shapes clustered after graphs for the same conformational shape. (The cluster analysis was carried out by representing the graphs as 0/1 character strings, here of length 78, as described above, using the number of different entries in two strings as the distance between them, and the average linking method for forming the clusters.) Thus the shape graphs accurately portray the role of hydrogen in molecular similarity as a secondary influence, filtered out by cluster analysis.

We note here the relationship between shape graphs and the idea of '*n*-type faces' employed by Arteca and Mezey (1988b). For a molecular surface graph, Arteca and Mezey define the 'type' of a face to be the smallest number of edges in any boundary of the face. If the face has only one boundary cycle, then this is just the number of edges in the boundary, which is also the number of edges emanating from the vertex representing the face in the dual of the surface graph. In graph theoretical terms, this is the degree of the vertex. Thus when each atom contributes one face to the surface graph (so that the dual of the surface graph and the shape graph coincide) and all the faces of the surface graph are topological discs, Arteca and Mezey's sequence of n-type faces is the degree sequence of the shape graph.

11.4.3 A substructure search

As an example of the potential use of shape graphs, we describe the outcome of a search of the Cambridge Structural Database (Allen *et al.*, 1983) for molecules containing *N*-methyl piperidine as a substructure, and the classification of these substructures by their shape graphs. This was carried out in three stages. First, a conventional substructure search for compounds containing *N*-methyl piperidine produced a hit list of 1544 structures for which X-ray crystallographic coordinates are available on the database. Using software provided by Matthew Stahl, of the Artificial Intelligence in Chemistry Group at the University of Arizona, a second search of these 1544 structures was made to identify explicitly the atoms of the *N*-methyl piperidine fragment. This stage is, of course, a simple repetition of the first, but it was necessary since the software used for the initial search only identifies those structures containing a given fragment, and does not make available to the user the actual atom-to-atom mapping—information which is crucial for our purposes. Several of the structures identified in the first stage contain more than one occurrence of *N*-methyl piperidine, and the second stage located a total of 1724 *N*-methyl piperidine substructures. Finally, the shape graphs for the substructures were determined using the algorithm above, and compared with those of Figure 11.6.

The results of this experiment are outlined in Table 11.2. The 1724 substructures generated 63 different shape graphs. The five shape graphs

Table 11.2 Results of substructure search

	Graphs	Substructures
Template matches		
equatorial chairs	1	706
axial chairs	1	18
twist boats (mirror images)	2	28
pure boats	1	11
Flat structures	13	582
Partially flat structures	26	299
Strained structures		
axial chairs	2	28
twisted chairs	2	17
twisted boats	6	21
axial boats	9	14
Totals	63	1724

matching the templates of Figure 11.6 accounted for 763 substructures, with the largest contribution being some 706 equatorial chairs. For each of the other graphs, a visual examination of one of the substructures producing that graph was made. This revealed that a large number of substructures are either wholly or partially flat. Substructures where all the atoms are approximately co-planar are referred to in Table 11.2 as 'flat structures' and will include structures containing pyridine rings. (Pyridine will not be distinguished from piperidine because the substructure search was based solely on atom connectivities and valency was ignored.) There are 582 flat structures, represented by 13 graphs. The reason the flat substructures give rise to so many graphs is that they are close to a perfectly symmetrical structure where all the atoms in the piperidine ring meet in two single points, one above and one below the ring. In this case the van der Waals surface for the ring atoms resembles the surface of an orange, consisting of a number of segments. This is a highly unstable situation and small perturbations of such a structure produce many different patterns of intersection between the ring atoms, leading to many different shape graphs. We also observed many 'partially flat' substructures where a subset of four or more bonded atoms is approximately co-planar: there are 299 of these, represented by 26 shape graphs. The remaining graphs include substructures which adopt different shapes due to the strain imposed on the substructure from a surrounding fused ring system. In particular we observed chair conformations with the C1 substituent between the axial and equatorial positions, axial boat conformations, and various twisted chairs and boats.

11.5 Conclusions and future directions

The preliminary studies described in this chapter suggest that topology can provide a number of useful procedures for comparing the similarity of molecules. For chemists wishing to identify novel structures with similar properties to a lead compound, the description of shape in terms of integer invariants is a major advantage over other methods, offering substantial savings in the time taken to search large databases for matching structures. The compact manner in which the topological information is encoded should offer similar advantages when studying substructures within macromolecules. A number of difficulties must be resolved, however, before the full potential of the topological approach can be realized.

The shape sequences of section 11.3, for example, offer a concise summary of the topological information in the molecular surface graphs, but they require careful choice of the atomic radii and the ordering of the atoms used to build the molecule. Within a sample of conformations, a single conformational shape may be represented by a number of shape sequences and cluster analysis is then required to obtain a one-to-one correspondence between sequences and shapes.

The shape graphs of section 11.4, which were the most successful shape descriptions in the present study, are a straightforward extension of the familiar molecular structure graphs and incorporate sufficient 3D information to distinguish different conformations. In the case of *N*-methyl piperidine, the shape graphs for the hydrogen-suppressed molecules provide a one-to-one correspondence with conformational shapes. A finer level of detail is obtained if the hydrogen atoms are included, with a larger number of distinct graphs from which the major conformational shapes may be recovered by cluster analysis. Another useful feature of shape graphs is their ability to detect the existence of mirror images within a sample, although absolute configuration cannot be assigned from the graphs.

An important potential application of molecular surface graphs is to ligand–receptor binding. A ligand which matches a given receptor may be constructed artificially by packing the receptor site with appropriately-sized spheres. Intuitively, the closest fit between the ligand and receptor surfaces is obtained by placing visible faces of the ligand spheres opposite vertices of the receptor surface and vice versa. In other words, the surface graphs of ligand and receptor are expected to be approximate geometrical duals. A strategy for identifying potential ligands matching a known receptor is thus to search for molecules with surface graphs dual to that of the receptor.

The topological approach adopted here is not restricted to the purely steric description of molecular shape represented by the van der Waals surface. Electronic or electrostatic effects may also be studied from this perspective. For any smooth function on a smooth surface, such as the molecular electrostatic potential on the Connolly surface (Connolly, 1983), there is an interplay between the topology of the surface and the critical points of the function. This relationship forms the subject of a mathematical study called Morse theory. In particular, an integer index is associated with each critical point of a function. A qualitative match between the electrostatic properties of two molecular structures would require identical sets of indices for their electrostatic potentials.

Finally, methods must be developed for comparing the topological graphs of molecules which contain different numbers of atoms before the procedures described in this chapter are of general use.

Acknowledgements

The authors wish to acknowledge the contributions of a number of colleagues associated with the studies reported in this chapter. Dr John Bradshaw, Glaxo Research and Development, initiated the work and has contributed to a number of seminal discussions of the topic. Drs Watcyn Wynn and

David Salt of the School of Mathematical Studies, Portsmouth University have also made valuable contributions. Data for the *N*-methyl piperidine conformations was provided by Dr Darko Butina of Glaxo Research and Development. Neil Hoare of the School of Biological Sciences, Portsmouth University, performed the molecular dynamics simulations for benzyl chrysanthemate and assisted in producing the pictures of the *N*-methyl piperidine conformations in Figures 11.3 and 11.4 using SYBYL. The substructure search in section 11.4.3 was carried out using software kindly supplied by Matthew Stahl of the Artificial Intelligence in Chemistry Group at the University of Arizona. We also acknowledge the use of the SERC funded Chemical Database Service at Daresbury. Financial support was provided by Glaxo Research and Development.

References

Alard, P. and Wodak, S.J. (1991) Detection of cavities in a set of interpenetrating spheres. *Journal of Computational Chemistry*, **12**, 918–922.

Allen, F.H., Kennard, O. and Taylor, R. (1983) Systematic analysis of structural data as a research technique in organic chemistry. *Accounts of Chemical Research*, **16**, 146–153.

Arteca, G.A. and Mezey, P.G. (1988a) A topological characterization for simple molecular surfaces. *Journal of Molecular Structure* (*Theochem*), **166**, 11–16.

Arteca, G.A. and Mezey, P.G. (1988b) Shape characterization of some molecular model surfaces. *Journal of Computational Chemistry*, **9**, 554–563.

Artymiuk, P.J., Rice, D.W., Mitchell, E.M. and Willett, P. (1990) Structural resemblance between the families of bacterial signal-transduction proteins and of G proteins revealed by graph theoretical techniques. *Protein Engineering*, **4**, 39–43.

Bradshaw, J., Wynn, E.W., Salt, D.W. and Ford, M.G. (1993) Topological approaches to three-dimensional shape. In *Trends in QSAR and Molecular Modelling 92* (ed. C.G. Wermuth), Escom, Leiden, pp.220–224.

Connolly, M.L. (1983). Analytical molecular surface calculation. *Journal of Applied Crystallography*, **16**, 548–558.

Dean, P.M. and Perkins, T.D.J. (1993). Searching for molecular similarity between flexible molecules. In *Trends in QSAR and Molecular Modelling 92* (ed. C.G. Wermuth), Escom, Leiden, pp.207–215.

Firby, P.A. and Gardiner, C.F. (1982) *Surface Topology*. Ellis Horwood, Chichester.

Giblin, P.J. (1981). *Graphs, Surfaces and Homology*. Chapman & Hall, London.

Massey, W.S. (1967). *Algebraic Topology: An Introduction*. Harcourt, Brace and World, New York.

Mezey, P.G. (1987) The shape of molecular charge distributions: group theory without symmetry. *Journal of Computational Chemistry*, **8**, 462–469.

Richards, W.G. (1993) Molecular similarity. In *Trends in QSAR and Molecular Modelling* 92 (ed. C.G. Wermuth), pp.203–206. Escom, Leiden.

SYBYL Molecular Modelling System (Version 6.04a). Tripos Associates, St Louis, Missouri.

Weininger, D., Weininger, A. and Weininger, J.L. (1989) SMILES II. Algorithm for generation of unique SMILES notation. *Journal of Chemical Information and Computer Sciences*, **29**, 97–101.

Wilson, R.J. (1979) *Introduction to Graph Theory*, 2nd edn. Longman, Harlow.

Wynn, E.W., Salt, D.W., Ford, M.G. and Bradshaw, J. (1993) Calculation of the topological invariants for the excised van der Waals surface of a molecule. In *Trends in QSAR and Molecular Modelling 92* (ed. C.G. Wermuth), pp.384–385. Escom, Leiden.

12 Comparative molecular field analysis (CoMFA)

K.H. KIM

12.1 Introduction

Comparative molecular field analysis (CoMFA) is a method for three-dimensional (3D) quantitative structure–activity relationships (3D–QSAR) developed at Tripos. Although the concept of the approach has been known as DYLOMMS (dynamic lattice-oriented molecular modeling system) (Wise *et al.*, 1983) for over a decade, it was not until recent years that the method became widely used after it was reborn as CoMFA in 1988 (Cramer *et al.*, 1988b, 1993). The methodology has been patented (Cramer and Wold, 1991) and the program is available as a QSAR package in SYBYL.[1]

Although CoMFA is a relatively new technique, many scientists have already had experience with this method. In addition to the inventors' publications (Cramer *et al.*, 1988b, 1993) and the SYBYL manual, extensive reviews on this subject have recently been written (Kubinyi, 1993a; Martin *et al.*, 1995). This chapter will not attempt to duplicate these reviews. The objective is to provide an introduction for the newcomers in this area.

12.1.1 Starting from classical QSAR

Most drugs exert biological activities first by binding to a target receptor molecule. It has been recognized that the binding of a molecule to a macromolecule is partly determined by the ability of the drugs to fit a cavity within the macromolecule in such a manner that both molecules are stabilized in a suitable 3D orientation to promote the observed biological activities. In other words, biological activity of a molecule is a consequence of its 3D shape, size, and geometry. Therefore, it is important to understand the nature of their binding modes in 3D shape and conformation of biological molecules in relation to the observed biological activities.

The objectives of QSAR may be described as threefold: (1) to correlate and summarize the relationships beween the biological activity and physiochemical and/or structural descriptors for better understanding of the

[1] SYBYL. Tripos Associates, St Louis, Missouri.

mode of action, (2) to optimize the structure for the best possible biological activity, and (3) to predict the biological activity of other compounds.

Classical QSAR represented by the Hansch analysis (Hansch and Fujita, 1964; Martin, 1978) has been used for decades to correlate and predict biological activities of molecules, and provided many useful insights for drug design (Hansch, 1981). Numerous successful applications are available in the literature (Martin, 1981; Hopfinger, 1985; Topliss, 1993). Then, we might ask why we need 3D–QSAR or CoMFA. One of the most often encountered problems in classical QSAR is that of missing physicochemical parameter values. When the appropriate physiochemical parameter value is not available, the compound cannot be included in the analysis. In other situations, the parameters may not be sufficient for describing the drug–receptor interactions. This is especially true with various steric descriptors. For example, the E_S steric parameters that have often been used in QSAR were derived from an intramolecular reaction rather than an intermolecular reaction, whereas drug–receptor interactions are intermolecular. This is in addition to the limitation in the experimental determination of E_S values due to the instability of the compound under experimental conditions or due to synthetic difficulty. Likewise, numerous Hammett σ constants or variations of them are essentially limited to rather simple organic functional groups. Thus, parameterization of electronic properties of various derivatives often causes a problem. We also face a similar situation with hydrophobic substituent constants, despite the fact that $\log P$ can be calculated by various methods. Classical QSAR is often confined to one or a few substitutions in a common reference structure and works best with a congeneric series. In contrast to the classical QSAR, CoMFA can include any compounds as long as they act by the same mechanism. Although organic molecules have 3D structures, classical QSAR usually does not consider the 3D structures of the molecules. Therefore, the stereochemistry or 3D structures of drugs cannot be adequately considered in classical QSAR, and scientists were often frustrated by such limitations. As both the drug and the receptor have 3D structures, it is expected that 3D–QSAR methods which are based on the 3D structures would provide better information about the drug–receptor interactions. In addition, considering 3D structures in QSAR can include more diverse compounds in the analysis. Most importantly, molecules are 3D in shape. Thus, the development of new 3D–QSAR approaches to overcome such shortcomings of the classical QSAR approach was inevitable. CoMFA is one of the new approaches.

12.1.2 3D–QSAR

Unlike the classical QSAR methods, the 3D–QSAR approaches are usually much more complex partly because more heterogeneous structures are often involved. In many cases, the orientations of the compounds with respect to their receptor interactions are not obvious even when we assume they interact with their biological target in the same mode of interaction. Therefore, the conformational flexibilities and bioactive conformations of the compounds should be carefully considered before we can perform a QSAR analysis with 3D structures.

If the binding modes of the compounds are not known, their proper orientations become the critical step in 3D–QSAR analysis. In such cases, a pharmacophore hypothesis is often used for their orientation. However, the problem can be further complicated because of the possibility of multiple binding modes even among closely related analogs (Mattos and Ringe, 1993).

Several 3D–QSAR approaches have been proposed by different scientists. Among them are molecular shape approaches (Simon *et al.*, 1977; Hopfinger, 1980; Simon *et al.*, 1980; Simon, 1993), the distance geometry approach (Crippen, 1980), the binding-site model (Crippen, 1987; Ghose and Crippen, 1990; Srivastava *et al.*, 1993), the hypothetical active-site lattice (HASL) approach (Doweyko, 1988; Wiese, 1993), the use of difference maps (Folkers *et al.*, 1993), the molecular similarity approach (Good *et al.*, 1993a,b) and CoMFA. Among these, CoMFA is the most widely used. Although CoMFA has its own limitations and disadvantages (Folkers *et al.*, 1993; Kubinyi and Abraham, 1993), it also offers many advantages over the classical QSAR (Kim, 1993d).

12.1.3 CoMFA

CoMFA is a 3D–QSAR technique employing both interactive graphics and statistical techniques for correlating shapes and properties of molecules with their biological activity. Bioactive conformations of each compound are chosen, and they are superimposed in a manner defined by the supposed mode of interaction with the target receptor. CoMFA then compares, in three dimensions, the steric and electrostatic fields calculated around the molecules with various probe groups, and extracts the important features related to the biological activity. In doing so, CoMFA tries to identify the quantitative influence of specific chemical features of molecules on their potencies. The results are then displayed in contour plots showing the important regions in three-dimensional space that are highly associated with the biological activity (Figure 12.1).

There are many aspects to be considered for a CoMFA analysis: (1) biological data, (2) selection of compounds and series design, (3) generation of 3D structure of the ligand molecules, (4) conformational analysis of each molecule, (5) establishment of the bioactive conformation of each molecule, (6) binding mode and superimposition of the molecules, (7) position of the lattice points, (8) choice of force fields and calculation of the interaction energies, (9) a statistical analysis of the data and the selection of the 3D–QSAR model, (10) display of the results in contour plots and interpretations of them,

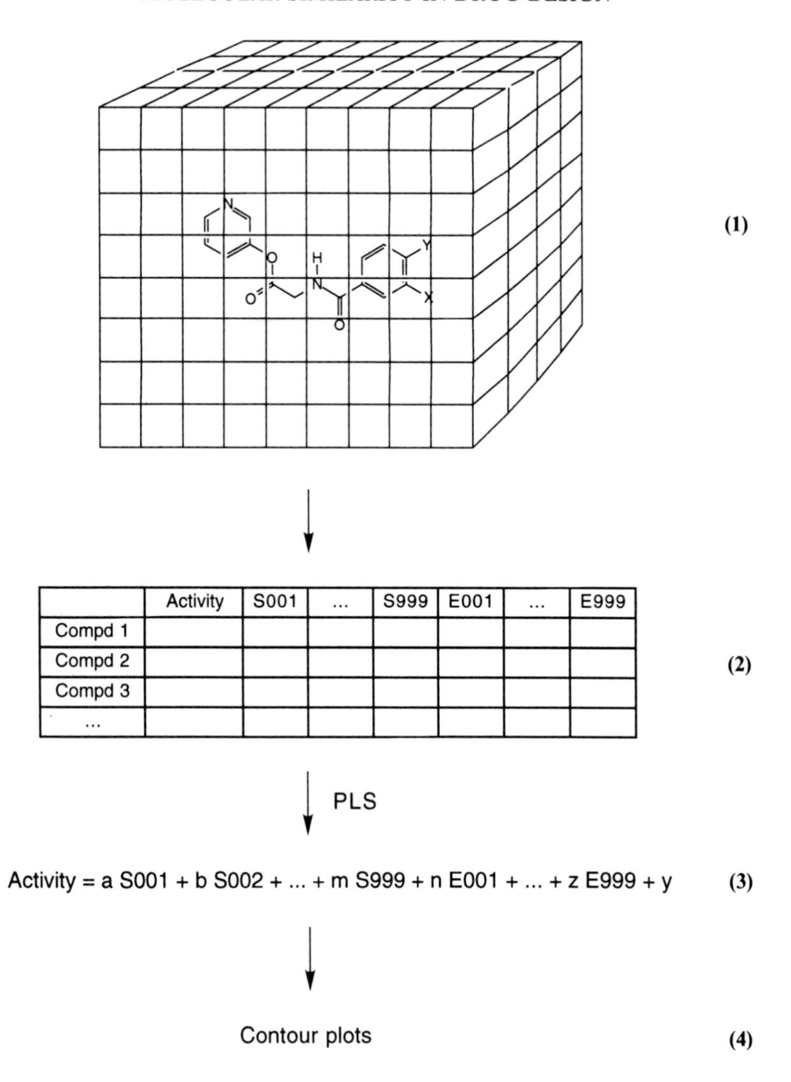

Figure 12.1 CoMFA process. (1) The selected bioactive conformation of each compound is superimposed. (2) The steric, electrostatic, and sometimes other fields are calculated with various probe groups around the molecules. (3) The important features related to the biological activity are extracted by partial least-squares statistical analysis. (4) The results are displayed in contour plots.

and (11) design and forecasting the activity of unknown compounds. In the following sections, each of these aspects will be discussed in more detail.

12.2 Biological data

Accurate biological data are essential for any QSAR technique. The biological data must be obtained for a set of ligands using uniform protocols and ideally from a single source. The biological activity ranges covered should be as large as possible, yet their mode of action should be identical. Otherwise, the data could have discrepancies that may be misleading.

Regardless of the methods of evaluation and model development, any data analysis works better when the biological data are fairly symmetrically distributed around their mean, and their precision is evenly distributed over its range of variation. An unbalanced and unrepresentative set of compounds yields a poor quality or unreliable QSAR, and the results are difficult to interpret. Since a log transform of the biological data usually removes much of the skewness in the data (Wold *et al.*, 1993), the quantitative measure of biological potency in classic QSAR as well as 3D–QSAR is normally defined as $\log(1/C)$ where C is the molar concentration of drug producing a standard response. A wide range of activity values is preferred for the analysis and should cover a range much larger than the standard deviations of the data; more than two logarithm units with an even spread of the data is preferred. Further discussion on biological data is available in the literature (Martin, 1978; Kubinyi, 1990).

Possible conformational changes of the receptor are presumed to be equivalent among all the compounds. It is important to remember that the biological activity of a drug measured *in vivo* is a composite of many interactions, including the rate of absorption, distribution, metabolism and excretion of the drug.

12.3 Selection of compounds and series design

The application of CoMFA is a process for optimizing the desired biological activity while eliminating or reducing the side-effects by structural modifications. When we modify the structure of a molecule, we simultaneously make a variety of changes. These changes not only affect the desired potency of the drug, but also absorption, disribution, metabolism, and excretion.

Among the many possible compounds that can be chosen for synthesis, how can we choose compounds for synthesis to gain the maximum amount of information possible with a minimum number of compounds? It is usually desirable to choose compounds such that their variables for hydrophobic, steric, and electrostatic properties behave independently. However, it is not easy to make such a decision by inspection of the compounds. Thus, one of the serious problems in classical QSAR is often the intercorrelation among different properties. Although it may be less of a problem in 3D–QSAR, since a wider variety of compounds are often included in the analysis, it is still important to design the series of compounds to be tested.

There are three major issues in choosing substituents for the modification of compounds: (1) minimization of colinearity, (2) maximization of variance, and (3) mapping of substituent space with the smallest number of compounds. The use of series design strategies to select compounds for QSAR studies is particularly valuable for obtaining the widest range of information and a cost-effective investment and for deriving reliable QSAR models (Wootton *et al.*, 1975; Cramer, 1980; Austel, 1982; Pleiss and Unger, 1990; Pastor and Alvarez-Builla, 1991; Kubinyi, 1993b). Techniques such as a Craig plot (Craig, 1971), the Topliss decision tree (Topliss, 1974a,b), sequential simplex method (Darvas, 1974), cluster analysis (Hansch *et al.*, 1973), Fibonacci search technique (Wilde, 1964; Bustard, 1974; Santora and Auyang, 1975; Franke, 1984), factorial design technique (Austel, 1982; Clementi *et al.*, 1993), and principal component analysis (Cramer, 1980), D-optimal design (Clementi *et al.*, 1993), or a combination of these (Lin *et al.*, 1990), provide means to design series for synthesis, although they do not solve all of the problems in drug modification. Some of these approaches are more suitable for classical QSAR studies.

In structural modification, we often encounter one of two situations. The first is the situation in which the synthetic chemist can prepare derivatives relatively quickly compared to the pharmacologist's testing procedures. The second is the situation in which pharmacological testing is faster compared to synthesis time. In the first situation, the chemist must design and make a set of compounds, submit them for testing, and await the results. In the second situation, feedback from the pharmacologist is available from each new synthesis. Depending on the situation, different approaches are used to optimize the structural modification.

In practice, it is normally not possible, for a variety of reasons, to choose compounds that can minimize the collinearity, maximize dissimilarity and orthogonality, and synthetic accessibility simultaneously, but a compromise is often made to achieve an overall optimum. The selection of well-spanned representative compounds is important not only in the design of an initial set of compounds for synthesis, but also in the selection of compounds for a training set in 3D–QSAR analysis. Several studies are devoted to this subject in CoMFA (Lin *et al.*, 1990; Norinder, 1990, 1991b; Clementi *et al.*, 1993; Cocchi and Johansson, 1993). Techniques such as principal component analysis, cluster analysis, factorial design, and D-optimal design are utilized in these studies.

12.4 Generation of three-dimensional structure of the ligand molecules

There are two aspects to be considered in the 3D structure of the ligand molecules. The first is how to accurately represent the molecular structure in three dimensions. In representing the molecular structure, two points need to be discussed: generation of the starting structure and optimization of the starting structure. The second is how to determine the bioactive conformation. These aspects will be discussed in this and the following sections.

Both X-ray crystallography and molecular modelling can provide the starting structures of the molecules. There are about 120 000 well-resolved experimentally determined crystal structures in the Cambridge Structural Database (Cambridge Crystallographic Data Centre, 1989). The advantage of using crystal structures is that some conformational information of the flexible molecule is included. The difference in the nature of the experimental and modelling approaches is of fundamental importance. Modelling can easily be performed for compounds that have not been made or cannot even exist under normal conditions. Of course, the main concern about calculation is how reliable the results of the modelled structures are. The interpretation based on molecular modelling depends on the quality of the computations. Thus, it is important to understand the concepts, the strengths and weaknesses, and the limitations of the calculations of the common computational methods associated with the molecular modelling technique.

The computational methods for the 3D-structure generation can be classified into manual, numerical, and automatic methods (Sadowski and Gasteiger, 1993). In the manual method, the user constructs a 3D structure interactively on a 3D computer graphics interface from scratch or from an existing 3D structure including those in various fragment libraries. The manually constructed 3D structure can be used as input to generate many conformations or to further refine by several numerical methods such as distance geometry, quantum or molecular mechanical methods. When the structures of relatively small numbers of compounds are desired, as in the case for 3D–QSAR studies, this approach is usually used. The automatic methods are often used for building a 3D-structure database (Pearlman, 1993). Although rapid generation of approximate 3D structures is essential for building a 3D structure database, it is usually not as critical for 3D–QSAR study.

After the initial structure is generated, the geometry has to be refined. Structure optimization procedures begin with a chemically reasonable 3D starting structure and attempt to improve the quality of the structure by minimizing its conformational energy by molecular mechanics or molecular orbital methods. Computational speed can be an important factor in structure optimization. Currently there are three commonly used theoretical calculation methods for the structure optimization: (1) molecular mechanics, (2) semiempirical molecular orbital, and (3) *ab initio* molecular orbital methods.

Molecular mechanics methods are fast and accurate for molecules within the range of the force field parameterization, and can handle very large molecules, such as enzymes. *Ab initio* methods are more time consuming, but have an accuracy comparable with experiments for those such as heat of formation and molecular geometries. The advantage of the *ab initio* approach is its generality. However, the disadvantage is its greatly increased requirements for central processing unit time and disk space. Semi-empirical methods occupy a position between the *ab initio* and molecular mechanics methods. Like molecular mechanics methods, semi-empirical methods use parameters derived from experimental data; like *ab initio* methods, they are basically quantum mechanical approaches.

The major distinction between the molecular mechanical and molecular orbital approaches is that the former does not usually consider the electrons in the molecule explicitly while the latter is concerned with the 3D distribution of electrons around the nuclei. Thus, molecular mechanics is not appropriate for studying electronic properties of the molecules. In these situations, a molecular orbital approach is better. The main difference between the semi-empirical and *ab initio* methods is that the former employs an extensive use of approximations. Thus, a missing parameter value can be a problem with both the molecular mechanics and the semi-empirical methods.

In practice, the geometry of a molecule is usually optimized by molecular mechanics methods, and its atomic charges are calculated by semi-empirical methods or less often by *ab initio* methods.

12.5 Conformational analysis of each molecule

Although some small rigid molecules have only one low-energy conformation, larger and more flexible molecules have several conformations populated at room or physiological temperature. Therefore, it is important to consider multiple conformations of all molecules included in a 3D–QSAR study. Currently, several methods are available for the generation of multiple conformations of a molecule. The choice of the methods may depend on the kinds of compounds in the study. Good reviews on this subject are available (Howard and Kollman, 1988; Burt and Greer, 1988; Leach, 1991; see also Chapter 3).

Perhaps, the most thorough conformational analysis is done by the *systematic search method*. Since greater energy is needed to change bond length or bond angles compared to a dihedral angle, the major difference between various conformations of a molecule is generally due to a dihedral angle. In the systematic search method, conformations are generated by systematically varying each of the dihedral angles in the molecule by some increment whilst keeping the bond lengths and bond angles fixed. Such systematic search is often called 'grid search'. Assuming the increment for the torsion angle variation is appropriately small, this method is an exhaustive search technique and will generate all the conformations possible. The disadvantage of this method is that the time required for systematic search increases exponentially with each additional rotatable bond, and the generation and energy evaluation of the conformations can easily become impractical. Another disadvantage of this method is that the systematic search method must deal with the ring-closure problem for cyclic structures and further limits its efficiency. Various procedures such as tree searching algorithms and build-up procedures are used to improve the method, which allows study of compounds with more than 20 dihedral angles (Howard and Kollman, 1988; Marshall, 1993).

The second conformational analysis method is the *molecular dynamics method*. Molecular dynamics is a method of studying the motions and the configurational space of the molecular system in which the time evolution or trajectory of a molecule is described by the classical Newtonian equations of motion. Thus, the molecule is described as a dynamic structure which changes with respect to time as influenced by its kinetic energy and by the interaction forces of surrounding atoms. The potential energy function and its associated force field typically include bonded interactions involving bond stretches, bond and dihedral angle bends, and non-bonded interactions including electrostatic and van der Waals terms. Procedures such as the SHAKE algorithm or the simulated annealing method have been used to improve this method.

The *Monte Carlo method* provides the third conformational analysis approach. In Monte Carlo methods, the dynamic behavior of a molecule is simulated by random changes in the structure. The energy of the changed configuration is calculated, and if the energy is lower than the previous configuration, the new configuration is accepted. If the energy is higher, an algorithm is used to determine whether the new configuration is to be accepted. Monte Carlo methods are in general not as susceptible to getting trapped in a local minimum, and can be used in the simulated annealing method. Improved Monte Carlo methods were found to be more efficient in generating conformations than molecular dynamics (Howard and Kollman, 1988).

Another frequently used conformational analysis method is the *distance geometry approach*. Distance geometry methods are often used to generate a set of 3D starting conformations. This method generates a random set of coordinates by selecting random distances within each pair of upper and lower bounds. Although the distance geometry method samples conformational space rapidly and efficiently, it does not guarantee production of all of the possible conformations. However, the time required for distance geometry calculations depends only on the total number of atoms and not on the total number of rotatable bonds in the molecule. Thus, this method can still be practical for structures that are too large for the systematic search method. Distance geometry combined with molecular mechanics usually gives superior results to molecular mechanics alone.

A complication in conformational analysis for drug molecules is that the preferred conformation of an isolated molecule does not necessarily correspond to the bioactive conformation.

12.6 Establishment of the bioactive conformation of each molecule

The bioactive conformation of a drug molecule refers to the conformation when it is bound to its target receptor. When CoMFA analysis is performed on a series of simple analogs, it may not be necessary to establish a bioactive conformation; we may simply choose the same conformation for all compounds assuming that the molecules superimpose over their common structure moiety (Kim, 1992a,b,c, 1993a). However, most compounds have various degrees of conformational freedom, and their bioactive conformations can be affected by the intrinsic forces between the atoms in the molecule as well as by the extrinsic forces between the molecule and its molecular environment. The reliability of CoMFA results depends on the establishment of the bioactive conformations. Several approaches aimed at establishing the bioactive conformation of a drug molecule have been used in the conformational analysis (Marshall and Naylor, 1990; Martin, 1991; Martin *et al.*, 1993a,b; Wermuth and Langer, 1993; Golender and Vorpagel, 1993).

Experimentally, the bioactive conformation of a drug can be obtained from structure determination of the drug–receptor complex by X-ray crystallography or NMR spectroscopy (Blaney and Hansch, 1990; Erickson and Fesik, 1992). At present, *X-ray crystallography* is the only known method for determining exactly the three-dimensional shape of macromolecules. However, there are several disadvantages in the X-ray crystallography approach (Andrews and Tintelnot, 1990):

1. The protein must be crystallized, and the crystallizing medium is usually far from the physiological conditions. This may mean that the crystalline enzyme is in an inactive conformation.
2. Data collection usually takes a long time, and this produces a time-averaged structure.
3. Distortions in the structure involving the active site may arise from the packing of the enzyme in the crystal.
4. It may not be possible to diffuse substrates or other biologically relevant molecules into the existing crystals because of either crystal instability or active-site occlusion. Although many of these disadvantages can be overcome to some extent, and a number of 3D structures of ligand-bound biomolecules determined by X-ray crystallography are increasingly available, there remains still an open question of whether the drug–receptor complex exists in a different conformation in solution from that in the crystallized form.

NMR spectroscopy is another way of obtaining 3D structural data of proteins, and is in many ways complementary to X-ray crystallography. Unlike X-ray crystallography, however, the NMR method does not require crystals of the protein and considerable variations in the solution conditions (pH, ionic strength, substrate, temperature, etc.) can be employed. Important information obtainable from NMR spectroscopy is the dynamic aspect about molecular motion (Fesik, 1991). However, as the structure determined from crystallography may not correspond to the structure in the solution, the structure obtained by NMR spectroscopy may not represent the receptor-bound conformation either.

In practice, it would be fortunate if we had any experimentally determined structural information about the active or bound conformations among the compounds under study but often experimental data are not available. In this case, the bioactive conformations have to be studied theoretically. However, there is also an advantage to the theoretical approach — the physical sample is not required.

A large number of theoretical methods are proposed to perform systematic conformational searchings. All of these methods involve some type of dimensional reduction process in order for the large number of torsion angles to be scanned. The number of conformations to be examined in the exhaustive, complete torsion-angle searches depends on the number of torsion angles to be scanned and the size of torsion-angle increments. As with structures determined by X-ray crystallography or NMR spectroscopy, there are some disadvantages of the structures obtained theoretically. The structure generated by molecular modeling usually represents the equilibrium geometry of an isolated molecule in vacuum (gas-phase simulations). In addition, the global minimum energy conformation does not necessarily correspond to the receptor-bound conformation. Therefore, generating the global minimum energy conformation does not guarantee the bioactive conformation even though a very high energy conformation is not likely to be the bioactive conformation of a potent molecule (Marshall, 1987; Mattos and Ringe, 1993).

Thus, in reality it is not easy to establish the bioactive conformation, whatever approach is used. Therefore, the bioactive conformation is often assumed to be that of the compound in the crystalline state or in solution, or that of the global minimum energy conformation of the drug molecule. However, since the bioactive conformation is not necessarily the same as that in the crystalline state, in solution, or at the energy minimum, many conformations near the global minimum energy conformation are often considered as a possible bioactive conformation. When the molecules are flexible, the number of conformations to be considered can be enormous.

In establishing the bioactive conformation, it is helpful to consider various aspects that contribute to the overall free energy of ligand–receptor interaction. If different compounds contain the same pharmacophore model or binding-site model, the intermolecular interaction energy contribution of these compounds towards the stability of the ligand–receptor complex will be roughly equivalent. However, another important contribution to the overall free energy of the interaction is the energy required to change the conformation of the ligand from its preferred orientation to its bound conformation. Since this energy contributes unfavorably to the binding of the ligand to the receptor, a compound whose low-energy conformation closely resembles the bound conformation would bind more tightly than a compound whose conformation is significantly different from the bound conformation. Likewise, a rigid compound satisfying the pharmacophore model, or binding-site model, would bind even more tightly to the receptor because its interaction would not require a decrease in entropy. Thus, a conformationally rigid molecule that binds strongly to the receptor establishes the bioactive conformation of the compound and provides useful information for the bioactive conformation of other molecules.

The *active analog approach* (Marshall *et al.*, 1979; Dammkoehler *et al.*, 1989; Marshall, 1993) is one of the well-known techniques used to establish bioactive conformations of a set of compounds. The method determines the allowed conformations of all molecules in the study by systematic search, and selects conformers that satisfy the interatomic distances in the working pharmacophore. During the searching process, the conformational energies can be taken into consideration. Its computational procedure has been improved (Dammkoehler *et al.*, 1989).

Another approach to obtain bioactive conformations is the *ensemble distance geometry approach* (Sheridan *et al.*, 1986). This method rapidly determines whether any set of conformations exists, and generates a random set of conformations that satisfy the distance constraints. An advantage of the ensemble distance geometry approach is that it handles cyclic structures without the difficulty of ring closure encountered in the systematic search method.

When a rigid compound that binds tightly to the receptor is available, the *molecular-fitting technique* can also be used to derive bioactive conformations. Taking the rigid compound as a reference structure, a set of flexible molecules can be mapped onto the rigid reference compound, or forced onto it to maximize overlaps using constrained molecular mechanics and/or dynamics. This is sometimes called the *template forcing approach.*

DISCO (Martin *et al.*, 1993a) is a computer program devised to find the bioactive conformation of molecules. In order to find the bioactive conformation of each compound, it searches superposition rules that satisfy a set of conformations for all active molecules among the conformations that were supplied to the program. The input conformations can be generated by any available conformational analysis method.

Recently, several other strategies have also been suggested. They include CATALYST of BioCad[2] and APEX-3D of Biosym (Golender and Vorpagel, 1993).

12.7 Superimposition of the molecules

Once the bioactive conformations of all the compounds are determined, the next step in CoMFA is to superimpose and align them in a grid box. The superposition of the molecules is one of the most crucial steps in CoMFA, and the results of CoMFA analyses depend on the alignment of the molecules. A complication in CoMFA is that the selection of either the bioactive conformation or the superposition may be influenced by the choice of the other. Therefore, the two aspects are sometimes considered simultaneously.

Ideally, it is best if we have some experimental evidence about the bioactive conformation and relative orientation of each compound to be studied. However, if limited information is available, we usually assume that alignment of other molecules is similar to the experimentally known molecule and simply adopt that aligment. Thus, such experimental information provides an alignment rule. However, it is discouraging to find that the conformations of bound ligands can differ greatly even among close analogs. This is in addition to the negative aspect of structural information from X-ray crystallography where the crystal structure represents a static structure and may not well represent any transient structure of drug–receptor interaction. When even limited experimental data are not available for use, theoretical approaches have to be applied. Sometimes, theoretical alignment produces a statistically better CoMFA model than the alignment based on crystallographic data (Klebe, 1993; Klebe and Abraham, 1993). (This does not mean that a better CoMFA model is a better description of the binding-site structure.)

In the theoretial consideration of molecular superposition, it is assumed that certain structural features and properties common to all active molecules are essential for the desired biological activity. This assumption often includes that any missing feature may result in a decrease or total loss in activity.

When at least one of the molecules to be aligned has a conformationally restricted structure, the superposition of molecules can be accomplished relatively easily. The superposition of other compounds with similar structures can be done by assuming the same bioactive conformation. However, the superposition of highly flexible molecules is quite difficult and time consuming. An efficient method should superimpose the molecules while considering the conformational freedom of flexible molecules. Sometimes, it may be difficult to decide the best alignment, and several sets of possible superpositions can be used in 3D–QSAR for further examination. Often the bioactive conformation is selected on the basis of how well a particular conformer can superimpose with other compounds. The important aspect in alignment may be not so much the active-site geometry, but self-consistency of alignment (Cramer *et al.*, 1993).

[2] CATALYST. BioCad Corporation, Mountain View, California 94043.

During the alignment process, it is important to consider the common, different, or union volumes of the superimposed active or inactive molecules separately, as well as together. From the union volume of the active molecules, we can establish the permitted or essential area for the active molecule. From the different volumes between all the active molecules and an inactive molecule, we can estimate the area that must be avoided for activity.

Several approaches have been used for the alignment of molecules, some of which are described below (Klebe, 1993; Itai *et al.*, 1993; Marshall, 1993).

12.7.1 Alignment based on atom overlapping

The most popular and classic method for molecular alignment is the atom overlapping which gives the best matching of the preselected atom positions. Sometimes the alignment may be based on a common template. This method, often called a *pharmacophore approach*, is powerful in detecting dissimilarity between similar molecules. A disadvantage of this method is that it requires corresponding atom pairs, and thus this method cannot be applied to molecules in which the corresponding atoms are difficult to select. Furthermore, the superimposed atom positions do not necessarily coincide with the common receptor positions for all the molecules. Therefore, other methods of overlapping are preferred especially when molecules with different structural types are involved.

Recently, the *simulated annealing method* was reported as a new method for unbiased search for an optimal molecular superposition (Papadopoulos and Dean, 1991; Barakat and Dean, 1991). Although this method is fundamentally based on the atom overlapping, it does not require explicit assignment of corresponding atoms.

12.7.2 Alignment based on receptor binding sites

Although the alignment based on the atoms or the pharmacophore can be related to similar sites in the receptor, a receptor site can interact with a variety of orientations of the same functional groups in equal affinity. Thus, in the binding-site approach, the molecules are superimposed by overlapping the receptor binding sites or the groups of the receptor that interact with ligands rather than the pharmacophore groups. Despite the complications in conformational analysis due to the increased degrees of freedom, the binding-site approach is more plausible (Marshall, 1993).

12.7.3 Alignment based on fields or pseudofields

As in the field-fit option in CoMFA, the alignment can be performed using the calculated energy fields instead of atoms or binding-site points. In other approaches, electrostatic similarity or molecular surface similarity indices are used in superposition (Kato *et al.*, 1987; Dean and Chau, 1987; Dean and Callow, 1987; Dean *et al.*, 1988; Burt *et al.*, 1990; Good *et al.*, 1990; Kearsley and Smith, 1990; Kato *et al.*, 1992; Lawrence and Davis, 1992; van Geerestein *et al.*, 1992; Good, 1992; Good *et al.*, 1993a,b; Itai *et al.*, 1993; Dean, 1993; Good and Richards, 1993; Perkins and Dean, 1993). Much of this topic is also covered in Chapters 1 and 2 of this book.

12.8 Calculation of the interaction energies

When all the compounds are superimposed, they are located in a grid box for calculating interaction energies with various probes at each lattice point. The position of the lattice points, construction of the data table and interaction energy calculations, and molecular force fields are discussed below in this respect.

12.8.1 Position of the lattice points

In order to place the lattice points around the molecules, three aspects are of concern: the size of the grid spacing, the size of the grid box, and the location of the grid box. The typical choice for grid spacing is 2 Å, and the size of the grid box is about 3–4 Å larger than the union surface of the molecules. Electrostatic interactions are long range and a larger grid box may be required. In practice, however, a similar size grid box for steric interactions appears to work fine because of the intrinsic correlation between the electrostatic energies among nearby lattice points. The grid spacing for a 1 Å size requires much more computing time and disk storage space, but usually it does not change the results. The location of the grid box sometimes significantly affects the CoMFA results including the statistics and the number of components in the final CoMFA model (Cramer *et al.*, 1988b; Greco *et al.*, 1991; Cramer *et al.*, 1993; Martin *et al.*, 1995). It is considered that the probe atom at one set of lattice points may better mimic the locations of the similar atoms in the target macromolecule. Since it is difficult to know *a priori* the best location of the grid box, the best location is often chosen after deriving initial CoMFA models at various positions. Two other strategies are suggested to minimize the potential source of instability (Cramer *et al.*, 1993). One is to rotate the superimposed molecules in such a way that they are not parallel to any of the lattice edges. The other is to replace the field value at a grid point by the average of the field values at the vertices of a cube centered on the lattice point, whose side length is two-thirds of the lattice spacing.

12.8.2 Construction of data table and interaction energy calculations

Once molecules are aligned in a grid box, the next step of CoMFA is to set up a data table as in the classical QSAR. This includes the biological activity or other properties that are to be correlated and the interaction energy fields. The interaction energies for each of the compounds are calculated at particular lattice points with one or more probes. The probe may be a small molecule such as water, or chemical fragment such as a methyl group. Each probe is placed in turn at each lattice point, and the interaction energy between the probe and the molecule is calculated using various molecular force fields. The biological activity is usually in the first column of the table and the interaction energy fields are in the remaining columns. In addition to the energy fields, other physicochemical or structural parameters are sometimes entered into the table.

12.8.3 Molecular force fields

A force field is the empirical fit to the potential energy surface. It defines the mathematical form of the equations involving the coordinates and the parameters adjusted in the empirical fit of the potential energy surface. A force field uses a combination of bond distances, bond angles, torsion angles, and interatomic distances to describe both bonds as well as van der Waals and electrostatic interactions between atoms. The force-field parameters are derived from the empirical data of a small set of molecules.

The major forces involved in ligand–receptor intermolecular interactions are electrostatic, hydrogen bonding, dispersion or van der Waals and hydrophobic (Langraft *et al.*, 1989; Cohen *et al.*, 1990; Andrews, 1993). Whereas hydrogen-bonding and electrostatic interactions primarily provide specificity, hydrophobic interactions usually provide the major driving force for binding (Fersht *et al.*, 1985; Street *et al.*, 1986; Langraft *et al.*, 1989). These interactions are often described in classic QSAR by various physicochemical parameters such as σ constants, molar refractivity (MR), E_S parameter, STERIMOL parameters, partition coefficients or hydrophobic π constants, and hydrogen-bonding parameters. The basic assumption in the original CoMFA methodology is that a suitable sampling of the steric and electrostatic fields surrounding a set of molecules might provide all the information necessary for understanding their observed biological properties (Cramer *et al.*, 1988b). Therefore, one of the issues in CoMFA is the selection of force fields. Although the steric and electrostatic fields seem to be the most universal fields, some form of a hydrophobic field appears to be quite important in CoMFA (Kim, 1991, 1993b; Kim *et al.*, 1993; Cramer *et al.*, 1993).

Steric fields. The Lennard-Jones function is often used to model the van der Waals interactions although the size and hardness of the atoms may vary somewhat between different programs. Some force fields include hydrogen-bonding information in steric interaction by using a very small radius for the hydrogen involved in a hydrogen bond (Cramer *et al.*, 1993). Other functional forms are also used (Seibel and Kollman, 1990; Wade, 1993). The steric interactions are usually calculated using a methyl or C sp^3 probe in SYBYL.

Electrostatic fields. Electrostatic interactions are usually calculated from the Coulombic potential using a charged probe that has no radius (Cramer *et al.*, 1988b; Kim and Martin, 1991). Electrostatic properties of the molecules are typically described by point charges at the center of the atoms. They are usually calculated from semi-empirical or *ab initio* molecular orbital methods, but simpler methods are also available (Gasteiger and Marsili, 1980; Sanderson, 1983; Wade, 1993). It is important that all the charges are consistent with each other and use one method to calculate them for all the molecules in a CoMFA study. In SYBYL, the electrostatic energies are usually calculated with H^+ probe alone or C sp^{3+} probe for including steric energies. The electrostatic energies can be calculated with or without considering the value of the steric field (see p. 309).

One of the parameters related to the electrostatic interaction energy calculation is the dielectric constant; the Coulombic energy between two partial atomic point charges is inversely related to the dielectric constant of the surrounding medium. The dielectric constant between two species has a shielding value of approximately unity when the interaction distance is close to the contact distance and approaches that of the bulk-solvent dielectric constant at a farther distance (Hopfinger, 1977; Wade, 1993). Since biomolecular systems are heterogeneous media, the assignment of a dielectric constant value is not simple. Commonly, a constant value between that of the solute and the solvent is used, for example 78–80 for water and 2–5 for the core of a protein. The distance-dependent dielectric function is also used. The latter is the default option in SYBYL.

Hydrogen-bonding fields. Some force fields consider hydrogen-bonding interactions to be electrostatic in nature, and others add or substitute a hydrogen-bonding term to the electrostatic term (Cramer *et al.*, 1988b, 1993; Boobbyer *et al.*, 1989; Kellogg *et al.*, 1991; Wade *et al.*, 1993; Wade and Goodford, 1993). (See also Tripos Associates' Advanced QSAR Module.[3]) The SYBYL force field belongs to the former, whereas the GRID force field belongs to the latter. When a hydrogen-bonding field is used, it is usually the GRID program and the various GRID probes such as H_2O, COO^-, NH_3^+, NH_2, CO, CH_3, etc. that are used in CoMFA.

[3] Advanced QSAR Module. Tripos Associates, St Louis, Missouri.

Hydrophobic fields. Although the steric and the electrostatic fields seem to cover most of the drug–macromolecular interactions, some form of a 'hydrophobic field' is likely to be quite important in CoMFA (Cramer *et al.*, 1988b; Kim, 1991; Kellogg *et al.*, 1991; Kim *et al.*, 1993; Kim, 1993a; Abraham and Kellogg, 1993; Martin et al., 1995). The GRID water probe and the GRID hydrogen-bonding potential were shown to describe the biological potencies of several series that had previously been correlated with the hydrophobic parameters in the traditional Hansch analysis. This approach is particularly attractive because the 'hydrophobic field' can be calculated in the same scale as the steric and the electrostatic fields, and the scaling of such data is much less of a problem than for the data composed of the mixture of energy fields and traditional parameters (Kim, 1993a; Cruciani *et al.*, 1993). Kellogg *et al.* (1991) developed an empirical hydrophobic field that can be imported into CoMFA. A number of workers have used the hydrophobic parameter (log P or π) or molecular lipophilicity potential along with the steric and the electrostatic fields in CoMFA (Audry *et al.*, 1986; Furet *et al.*, 1988). The rationale behind the use of the GRID water probe for hydrophobicity has been discussed previously (Kim, 1993a; Martin *et al.*, 1995).

12.9 Pretreatment of data

When the data table is constructed, it is analysed with statistical methods such as partial least squares (PLS) to extract important features related to the biological activity. However, before performing the data analysis, the raw data are always pretreated for reduction based on the standard deviation cut-off and energy cut-off. Electrostatic field values are sometimes adjusted depending on the location of the lattice point. The variable selection method is also sometimes applied. Likewise, scaling and centering of the data may be part of the pretreatment.

12.9.1 Reduction of the data

Much of the information in the energy field data table may be redundant due to the high correlations among different energy columns or due to little variance in the energy values among the compounds. Thus, elimination of such information is warranted. Elimination of the irrelevant data also increases the chance of finding a CoMFA model (Clark and Cramer, 1993). Additionally, the computational time for a cross-validation test will be shortened if all the energy values are not included.

Reduction by standard deviation. The energy columns with a low standard deviation do not significantly influence the CoMFA model while they require much longer computing time. Thus, it is common practice to eliminate such columns. The standard deviation cut-off value is called minimum sigma in SYBYL and the default value is 2.0.

In general, the precision of the CoMFA model gradually worsens, and the number of optimum components often varies as the standard deviation cut-off level increases (Kim, 1993b; Martin *et al.*, 1995). Usually a value between 0.05 and 2.0 may be used without affecting the results significantly. Cramer *et al.* (1993) reported that the greatest influence of the standard deviation cut-off value was on the relative contribution of the steric and electrostatic fields to the model due to the differences in the magnitudes of the two fields.

Reduction by energy cut-off. The repulsive energies in a lattice point may become very large if there is close contact between the probe and the molecule at this point. On the other hand, the attractive forces are often smaller. Thus, the energy values usually vary from −10 kcal to infinite. This skewness in energy values is usually handled by cutting off all repulsive values larger than a certain value. The result is that the cut-off value is assigned to any lattice point where the calculated field value is greater than the cut-off value. In SYBYL, the default cut-off values for steric and electrostatic energy are both 30 kcal/mol. This allows the probe atom closer to the molecule and thus includes hydrogen-bonding information (Cramer *et al.*, 1993). Statistically, the imbalance between repulsive and attractive force values may be better corrected by lowering the cut-off to 5 kcal than to 30 kcal (Wold *et al.*, 1993; Cruciani *et al.*, 1993). Thus, an energy cut-off value of 5–10 kcal seems more appropriate than 30 kcal.

Various intermolecular interactions between each ligand and the target macromolecule are affected by the position of individual molecules. Crystallographic evidence shows that the conformations, geometries, and orientations of the common core may not be identical even in structurally very close analogs (Abola *et al.*, 1985; Klebe, 1993; Mattos and Ringe, 1993; Martin *et al.*, 1995). Based on such information, it is considered that about 1 Å rms for the superposition of the pharmacophore points, whether they are by atom overlapping, binding-site overlapping, or various other similarity overlappings, may be reasonable (Martin *et al.*, 1995). Furthermore, since repulsive steric interactions increase with the 10th or 12th power of the distance between the interacting groups, a very small change in molecule alignment can give a larger change in repulsive energies. Because of these uncertainties in the superposition of the molecules, a lower cut-off value than 30 kcal is often preferred in our CoMFA studies. Since the use of a cut-off introduces an extreme discontinuity in the repulsive energy force, SYBYL uses a smoothing function to avoid such a discontinuity (Cramer *et al.*, 1993).

Reduction in electrostatic contribution. If the electrostatic fields describe an intermolecular interaction, the contribution of the lattice points that are inside the volume of any molecule in the dataset is sometimes excluded (Kim and Martin, 1991a,b; Kim, 1992c). Although this is one of the options in SYBYL, these points are included in default settings by replacing the value with the average of the accessible values (Cramer *et al.*, 1988b).

Reduction by variable selection. It is suggested that the elimination of variables should not be guided by the degree of fit or the prediction errors, but should be done to simplify the model (Wold *et al.*, 1993). When the elimination of variables is based on the fit, it may lead to partly spurious models and seriously overfit the data (Wold *et al.*, 1993). GOLPE (generating optimal linear PLS estimations) variable selection procedure (Baroni *et al.*, 1993; Clementi *et al.*, 1993), which uses a D-optimal design to preselect non-redundant variables, improves the predictive power of the CoMFA model (Cruciani *et al.*, 1993). However, opposite results are also observed in other studies using GOLPE or other similar approaches (Allen *et al*, 1992; Martin *et al.*, 1995). Nevertheless, this method may be worth further investigation.

12.9.2 Scaling of the data

The contribution of the energy fields and other descriptor variables to the extracted latent variables depends on the magnitude of one variable relative to another. Those variables with a high absolute value and/or with a large covered range are more likely to have significant influence. Furthermore, PLS is likely to succeed in finding a relationship with a good cross-validation only when the magnitude (signal) of the variance with respect to the noise in the data is high enough (see section 12.10.5) (Clark and Cramer, 1993; Cramer, 1993; Kim, 1993e).

When the energy fields are comparable and measured in the same relative scale, the variations in the original energy values are considered to be important. Thus, there is no problem in the analysis performed without scaling such data. Nevertheless, scaling is sometimes recommended. However, it is essential if the data include descriptors other than the energy fields such as HINT field values, $\log P$, MR, pK_a, etc. (Allen *et al.*, 1990; Clark *et al.*, 1990; Kellogg, *et al.*, 1991; Thomas *et al.*, 1991; Osabe *et al.*, 1992; Diana *et al.*, 1992; Altomare *et al.*, 1992; Greco *et al.*, 1992; Loughney and Schwender, 1992; McFarland, 1992; Miyashita *et al.*, 1992; Cramer *et al.*, 1993; Kim, 1993e; Davis *et al.*, 1993; Floersheim *et al.*, 1990; DePriest *et al.*, 1993; Liu and Matheson, 1994). Scaling of such descriptors may have the advantage of allowing us to rank the relative importance of individual variables.

Besides the scaling methods described below, several other methods are also suggested. They include block-adjusted non-scaling and weighting-up of the dependent variable if the data table includes more than one activity value (Wold *et al.*, 1993).

Autoscaling to unit variance. In a typical QSAR data table, each variable column has independent variables in arbitrary units. When the relative importance of the variables is not known, the usual approach to provide each variable with the same weight is to transform each variable so that it has a zero mean and a unit standard deviation. This is done by dividing each column with the column's standard deviation. Although this approach is often used in classical QSAR, it is not useful for the individual energy field in CoMFA (Wold *et al.*, 1993; Cramer *et al.*, 1993). Autoscaling can give unwanted consequences when there are many of one kind and just a few of another (Wold *et al.*, 1993; Cramer, 1993).

Block-scaling to constant group variance. Block-scaling provides each category of variables with the same weight. This can be done by dividing the initial autoscaling weights of variables in one category by the square root (or some other variation) of the number of variables in that category (Cramer *et al.*, 1993; Wold *et al.*, 1993). For the CoMFA energy fields, the overall variance in each field is calculated separately among all compounds. This is the method of the 'CoMFA standard scaling' option in SYBYL and the default option.

Block-adjusted scaling. Sometimes, variables other than energy fields are included in CoMFA analysis. In such a case, a scaling is necessary in order to give other variables a comparable weight to the total variables. Thus, the energy fields can be scaled so that their total influence on a PLS result is internally the same as that of any other variables in the CoMFA standard scaling. Alternatively, the energy field may be unscaled while other variables are scaled to have the same weight among them (Wold *et al.*, 1993).

12.9.3 Centering of the data

Centering the data is recommended for ease of interpretation and for numerical stability (Wold *et al.*, 1993). This can be done by subtracting the column averages from all data. Wold *et al.* (1993) indicated that centering of the data does not change any coefficient values or relative weights of variables, but that the number of significant components from PLS may be one less than from the data without centering.

12.10 Statistical analysis of the data and the selection of 3D–QSAR model

12.10.1 Partial least-squares (PLS) analysis

The classical multiple linear regression analysis was designed to correlate a relatively large number of compounds with few descriptor variables. It assumes that the descriptor variables are independent and contain no error.

A typical CoMFA data table usually contains hundreds or thousands of columns of interaction energy values, and the number of compounds included in the study is relatively much smaller than the number of the energy columns. Thus, a mathematical difficulty arises because a large number of descriptor variables is used to describe the biological activity of a much smaller number of compounds. For this reason, the multiple linear regression technique cannot be used directly without the danger of chance correlation.

PLS (Wold *et al.*, 1984a, 1993; Lorber *et al.*, 1987; Hoskuldsson, 1988; Bush and Nachbar, 1993; Phatak *et al.*, 1993; Cramer, 1993) was developed for such problems. Not only can it be applied to solve an equation having hundreds or thousands of variables while involving only a small number of biological data, but also it can be used with variables that are highly collinear among one another. Multiple biological activities can be simultaneously handled to derive models. In this latter case, some missing values can be handled without any problem.

PLS is an iterative procedure that produces its solutions, and summarizes the hundreds or thousands of descriptor variables as a few, orthogonal new variables called latent variables or scores. PLS generates iteratively one component at a time by maximizing the degree of commonality between all of the descriptor variables collectively and the biological data. The process stops when the requested number of components is extracted. The number of significant PLS components (latent variables) is determined by a cross-validation test.

12.10.2 Validation of CoMFA

The most important criterion for selecting a CoMFA model is how well the model can predict the activity of other compounds outside the model rather than how well the model reproduces the biological activity of the compounds included in the model. Various approaches are used for this purpose.

Cross-validation. Cross-validation (Stone, 1974; Efron and Gong, 1983; Cramer *et al.*, 1988a) is most often used for the criterion of selecting the optimum number of components for a CoMFA model. In cross-validation, the predictive ability of the model is estimated by repeatedly leaving out one (or more) compound(s) at a time until each compound is excluded exactly once. Using the reduced set of data, PLS analysis is performed to derive a model, and using the model the activity of the compound that was left out is predicted. During the cross-validation test, the sum of the squared prediction errors called the predictive residual sum of squares (PRESS), the cross-validated correlation coefficient (R^2_{cv}), and the cross-validated standard error of estimate (s_{cv}) are calculated for the model with each PLS component. The analysis continues until the PRESS no longer decreases significantly. The optimum number of components for the final 3D–QSAR model is chosen as the number of components that significantly reduces s_{cv} or increases R^2_{cv}, or that corresponds to the minimum s_{cv} or the maximum R^2_{cv}. A smaller s_{cv} and a larger R^2_{cv} indicate the model's good predictability. The R^2_{cv} value of 0.0 means that the mean square error of the model is equal to the standard deviation of the original data, and a negative R^2_{cv} value means that the mean square error of the model is greater than the standard deviation of the original data. All things being equal, a simple model (the smallest number of components) should always be selected (Principle of Parsimony, Occam's Razor). Although the leave-one-out method of cross-validation, also known as a jack-knife validation test, is often used, leave-*n*-out cross-validation is sometimes used (Crucini *et al.*, 1992, 1993; Kim, 1993b; Martin *et al.*, 1995).

The PRESS statistic is calculated as the sum of squares of the differences between the predicted and the observed values of the activity:

$$\text{PRESS} = \Sigma(Y_{obs} - Y_{pred})^2$$

The R^2_{cv} and s_{cv} are calculated by the following formulations:

$$R^2_{cv} = (\text{PRESS}_0 - \text{PRESS})/(\text{PRESS}_0)$$

$$s_{cv} = (\text{PRESS}/(n - L - 1))^{1/2} \text{ or } s_{cv} = (\text{PRESS}/n)^{1/2}$$

where PRESS_0 is the mean of the observed activity values, n is the number of compounds, and L is the number of latent variables (components) used in the fit. In the calculation of s_{cv}, two formulas are used. Both are statistically correct. However, the former penalizes the model with the same PRESS but with a larger number of components (Wold *et al.*, 1993). Both R^2_{cv} and s_{cv} are used as an estimate of the predictability of the model and are better criteria for the predictability of the CoMFA model than R^2 and s. Because the penalty for the larger number of latent variables is included only in the calculation of s_{cv}, but not in R^2_{cv}, the optimum number of components suggested by CoMFA in SYBYL implementation sometimes causes confusion. SYBYL selects the optimum component number based on R^2_{cv}, not on s_{cv}. However, it is recommended that the model with fewer components should be selected whenever the increase in R^2_{cv} is 'small' (less than 5%) (Kubinyi and Abraham, 1993). Occasionally Q^2 is used as an alternative symbol for the squared correlation coefficient from cross-validation instead of R^2_{cv}.

The cross-validated R^2_{cv} or s_{cv} are very sensitive to any large error of prediction; even one very poor prediction can yield a low R^2_{cv} or a large s_{cv} depending on the number of compounds in the data set. Thus, it is important to examine whether any one (or a few) compound caused a large error. It is worth checking the original biological data value or its alignment, or trying to understand whether the abnormal behavior of such a compound can be accounted for. In cross-validation, if a compound is unique, it is often difficult to predict its activity from others. If it cannot be accounted for, one can repeat the cross-validation test omitting the compound and compare the results with those including it. (See the discussion below on the consideration of all compounds and omission of compounds.)

Cross-validation is not infallible. Strongly grouped data with much fewer real degrees of freedom than the number of compounds may give an improper result and give an over-optimistic cross-validation (Wold *et al.*, 1993). In other situations, if all the compounds in the data are unique, cross-validation may improperly indicate a lack of correlation (Kim, 1993d). Some authors (Cramer *et al.*, 1993) believe that R^2_{cv} values in general are not very sensitive indicators of the quality of a CoMFA analysis, and attribute little significance to R^2_{cv} differences of less than 0.20.

Bootstrapping. Bootstrapping (Efron and Gong, 1983) is another technique that can be used in model selection along with the cross-validation test (Cramer *et al.*, 1988a; Mager, 1991). It is based on simulating a larger number of data sets sampled from the original data set that are of the same size as the original. The same data can be sampled more than once. The statistical analysis is performed on each of the simulating data sets. The component model with consistent results is then chosen as the final model. In addition, the mean value and the standard deviation of the statistical calculations provide the confidence intervals of the various parameters in the model. Bootstrapping requires heavy computation with relatively small possible gains compared to the cross-validation technique. Since the computational demands in 3D–QSAR are already heavy, the additional demands in computation make this technique less attractive (Wold *et al.*, 1993).

Random change of the dependent variable value. We can study the robustness of the 3D–QSAR model by comparing the CoMFA results using the original biological activity data (dependent variable) that are randomly scrambled (Mager, 1991; Clark and Cramer, 1993). We can repeatedly interchange random pairs of biological data or randomly scramble them. We can also replace the biological data with random values.

Dividing the original set into the training set and test set. If there are enough compounds available for a CoMFA study, the original data set can be divided into a training set and a test set. When a CoMFA model is derived from the training set, the model can then be used to predict the activity of the compounds in the test set as a way to validate the derived model. It is important that the training set contains a variety of compounds in order to gain maximum information. On the other hand, any unique compound may cause poor cross-validation results. Since about four or five compounds per component are suggested for PLS (Wold *et al.*, 1993), the number of compounds required for the training set depends on the number of components in the final CoMFA model. Naturally, if a large variation has been made in the structures, a larger number of compounds are required in the training set.

12.10.3 Derivation of 3D–QSAR model

After the optimum number of components are chosen from the cross-validation test, the CoMFA model is derived from all compounds and the optimum number of components. For this model, R^2 and s are calculated in the same way as R^2_{cv} and s_{cv}, except that PRESS is replaced by the sum of the squares of the differences between the calculated and the observed biological activity values.

Because it is easier to fit the observed biological data than to predict them from a CoMFA model, the R^2 of the model is always higher than R^2_{cv}, and the s is always smaller than the corresponding s_{cv}. However, the precision of the CoMFA model should never be greater than the error in their measurement of biological activity.

If multiple fields are considered to be important for biological potency, they can be examined simultaneously, or individually one after another (Martin *et al.*, 1995). The results from three different approaches in considering multiple fields were recently explored (Kim, 1994).

Chance correlation. Since the risk of chance correlation was a concern in classical QSAR, a number of different approaches were used to investigate the risk of chance correlations in CoMFA (Wakeling and Morris, 1993; Clark and Cramer, 1993). For example, PLS analyses were performed using the biological data values that were scrambled either by interchanging random pairs or by replacing them with random numbers, or using all the data filled with random numbers with varying numbers of rows and columns. The results indicated that the risk of chance correlation in PLS is very low as long as the number of columns is larger than the number of rows. Whereas classical QSAR tends to overfit the data, PLS tends to underfit it.

Collinearity. One of the serious problems often encountered in classical QSAR is collinearity among different properties. Thus, the correlation matrix among different variables is regularly examined. Similar information can be obtained from the correlation matrix of latent variables (PLS scores) from each field type (Cramer *et al.*, 1993; Kim *et al.*, 1993). A high correlation of the latent variables indicates that the information in the corresponding fields is redundant.

Number of compounds. There are two aspects in regard to the number of compounds in 3D–QSAR analysis in general. First is the number of compounds included in the analysis. Sufficient information cannot be obtained unless an appropriate number of compounds as well as a variety of compounds is included in the analysis. However, it is equally important to select proper compounds so that maximum information can be gained with a minimum number of compounds. Commonly, at least ten or more compounds are preferred. Second is the number of compounds with respect to the number of components (latent variables) in the CoMFA model. Even though the risk of chance correlation in PLS is much smaller than the classical multiple regression analysis, a similar number of compounds per component is recommended for the CoMFA model; the maximum number of components may be around four or five compounds per component or less than a quarter of the number of compounds (Wold *et al.*, 1993). In practice, the number of components in CoMFA is usually found to be smaller than five or six.

12.10.4 Outlier detection

Outliers and other inhomogeneities can easily be detected from the deviations in the calculated values of the model or by examining the corresponding residual plot. They can also be detected by inspecting the residuals from the cross-validation test.

The score (latent variable) plots can also be used to identify outliers; outliers are indicated by any points far from the others in the score plots (Wold *et al.*, 1984b, 1993; Clementi *et al.*, 1984). Outliers sometimes provide useful information.

12.10.5 Pitfalls

There are several pitfalls that require careful attention in QSAR in general (Clementi, 1984; Wooldridge, 1984; Plummer, 1990; Kubinyi, 1993b) and several potential difficulties associated with CoMFA (Cramer *et al.*, 1988b; Kim, 1992b). One particular pitfall in PLS analysis is that there is no guarantee that PLS will find all the relations contained in the data (Smith, 1988; Cramer *et al.*, 1993; Clark and Cramer, 1993). Even though the data table contains a perfectly correlated pair of columns, the probability of PLS recognizing the perfect correlation or a significant correlation decreases as an increasing number of columns in the data table is filled with random uncorrelated values.

12.10.6 Consideration of all compounds and omission of compounds

Unless part of the compounds are purposely left out as a test set, all compounds of the original data set should be used in deriving the model. If any compound is omitted, it must be stated which compounds are not included in the final equation, and the reason for omission should be described. It is possible that the compound is unique. In cross-validation, a unique compound often causes a problem (see the discusion of cross-validation above). The omission of compounds is usually a negative statement on the quality of the correlation or the biological data. At worst, such an omission can create a statistical artifact. However, outliers sometimes provide useful information and insight about the mode or mechanism of interactions. When the observed activity is much lower than that calculated, the cause might be metabolic degradation, so that a vulnerable moiety might be discovered which can sometimes be protected by chemical alterations. When the observed activity is much higher, it is possible that the key to a new class of bioactive compounds or a special functional group may have been discovered.

12.11 Display of the results in contour plots and their interpretation

Several types of fields are retrievable from a CoMFA analysis (Cramer *et al.*, 1993). The coefficient contour plots are the most often examined but sometimes PLS plots provide useful information.

12.11.1 Contour plots

The results of CoMFA are an equation showing the contribution of energy fields at each lattice point. In order to facilitate the interpretation of the results, they are also displayed as coefficient (or standard deviation times coefficient) contour plots showing the regions in space where specific molecular properties increase or decrease the potency.

Typically there are two contour levels for each type of CoMFA energy field: the positive and the negative contours. SYBYL 6.04 offers two options (by actual field value or by contribution) for the 'standard deviation times coefficient' contouring among the four methods discussed (Cramer *et al.*, 1993). The default setting is 80% and 20% by the 'contribution method' for both steric and electrostatic contour levels, where 0% corresponds to 'no contribution below this value' and 100% corresponds to 'all contributions below this value'. The contours are colored in green and yellow for positive and negative steric effects, respectively, and blue and red for positive and negative electrostatic effects, respectively. Positive steric contours show the regions where substituents increase the biological potency if occupied, and the negative steric contours show the area where substituents decrease the potency. The positive electrostatic contours indicate the regions where positive charges increase the potency, whereas the negative electrostatic contours display the regions where negative charges increase the potency. The choice of the contour levels is arbitrary, and contour levels that provide enough signal without too much noise are chosen by an iterative process. SYBYL also provides a histogram of field values that can be used as a guide for choosing the contour levels when the 'actual field value method' is used.

Although the structure of the binding site may be hypothesized from CoMFA for designing new molecules or forecasting their potency, the contours may not truly represent the details of 3D structures (Cramer *et al.*, 1993). However, there are some indications that they may reflect the active site (G. Greco, Y. Martin, M. Schiffer and L. Dutton, unpublished results; Diana *et al.*, 1992).

12.11.2 PLS plots

Two kinds of plots can be examined from PLS models: score plots and loading/weight plots. Score plots between Y-scores (biological activity) and X-scores (latent variables) from PLS models show the connection between the activity and the structures, whereas plots of X-scores show the structural similarities and dissimilarities between the compounds as well as their clustering tendencies (Wold *et al.*, 1993).

12.12 Design and forecasting the activity of unknown compounds

12.12.1 Design of new compounds

The ultimate goal of QSAR is to design a new potent molecule that can be clinically useful. Currently, new compounds may be designed visually on the basis of the contour plots of the CoMFA model derived. The potency of the designed compounds can then be forecasted by the CoMFA model. It appears that SYBYL may soon be able to offer a way of providing a chemical guide to modification of existing compounds to enhance the biological potency (Cramer *et al.*, 1993). Tripos' LEAPFROG program also has a mode that uses CoMFA information to design new compounds. In the meantime, various 3D-structure searching techniques (Martin, 1992; Bures *et al.*, 1994) may provide a number of possible candidate compounds from which to select promising compounds according to the forecasted potency by the CoMFA model. In this situation, however, the superimposability of searched compounds over the existing compounds is an important prerequisite.

12.12.2 Forecasting the activity of unknown compounds

The derived CoMFA model can be used to predict the activity of new compounds in the hope of increasing the desired biological activity. The prediction is done by transforming, scaling, and centering, if applicable, the structure descriptor values (energy fields and other variables if included) in the same way as the compounds in the model (often called the training set), and applying them to the CoMFA model.

In order to evaluate the predictability of the model, the predictive R^2 (R^2_{pred}) and s (s_{pred}) can be calculated. However, there is disagreement about what value is used for the 'mean' in the calculation (Cramer *et al.*, 1993).

In predicting the activity of new compounds, it is useful to know how similar the new compounds are to those in the training set. A diagnostic method is suggested for this purpose (Wold *et al*, 1993). If new compounds are similar to any of the compounds used in the model, the model will forecast its potency well, but if new compounds are not similar, the forecast may be poor. It is important to remember that no QSAR model can tell how the structural change in a certain region affects the activity if no change has previously been made in that region. Over-optimistic estimations of forecasting ability of a CoMFA model can result for several reasons (Martin *et al.*, 1995): the data may contain subgroups, the bioactive conformation or superposition rule may have been chosen based on a preliminary CoMFA model (Nicklaus *et al.*, 1992) or the descriptor variables for a final CoMFA model were selected based on a preliminary CoMFA model (Allen *et al.*, 1992; Baroni *et al.*, 1992). Recent literature survey revealed that the activity of 297 compounds in 25 datasets was forecasted with the average root mean square error of 0.70 logarithm unit (Martin *et al.*, 1995).

When the compounds are synthesized and biological activity determined, the observed activity values are compared with the predicted values from the model. If they do not agree well, the model is updated including new compounds, and a new prediction is made. The iterative process continues until either a satisfactory prediction can be made, no further improvement can be made, or a compound with desired potency is obtained.

12.13 Miscellaneous aspects of CoMFA

12.13.1 QSAR validation of CoMFA methodology

CoMFA has been validated by showing that it fits and correctly forecasts traditional physicochemical parameters used in classical QSAR (Kim *et al.*, 1991; Kim and Martin, 1991a; Kim, 1993b) as well as various types of biological potencies (Kim, 1991, 1992d, 1992a,b,c, 1993a,b, 1993d; Kim *et al.*, 1993). A number of CoMFA studies in the literature also supports the methodology (Norinder, 1990; Thomas *et al.*, 1991; Carroll *et al.*, 1991; DePriest *et al.*, 1991; Naerum and Jensen, 1991; Allen *et al.*, 1992; Diana *et al.*, 1992; McFarland, 1992; Kim, 1992d; Akamatsu and Fujita, 1992; El-Bermawy *et al.*, 1992; Waller and McKinney, 1992; Kim, 1993e; Davis *et al.*, 1993; DePriest *et al.*, 1993; Akamatsu *et al.*, 1993; Agarwal *et al.*, 1993; Agarwal and Taylor, 1993; Avery *et al.*, 1993; Debnath *et al.*, 1993; Martin *et al.*, 1993c; Waller *et al.*, 1993; Waller and Marshall, 1993; Wong *et al.*, 1993; Pellicciari *et al.*, 1993; Horwitz *et al.*, 1994). Since most of these work, and as a comparison between the classical QSAR (Kim, 1993d) and CoMFA is summarized in two recent publications (Kubinyi 1993a; Martin *et al.*, 1995), they are not repeated here.

12.13.2 Non-linear relationships in CoMFA

In classical QSAR studies, non-linear relationships are often observed with both *in vivo* and *in vitro* biological activity data (Hansch, 1981; Topliss, 1993). CoMFA describes non-linear dependence of biological activity on hydrophobic or steric effects. The contours from non-linear relationships show a characteristic pattern; positive and negative contours are adjacent to each other (Kim *et al.*, 1993c). The optimum hydrophobicity can be estimated in a series where the substituents vary in length (Kim, 1992a). Interesting parabolic or non-linear patterns were observed in PLS score plots from several homologous series (Kim *et al.*, 1993c). However, CoMFA has not yet been successful in describing other non-linear relationships particularly when the non-linearity depends on the same lattice point.

Non-linear PLS has been developed, but the experience with non-linear models in CoMFA is very limited. Wold reported that the non-linear models seem to overfit the data, and the benefits of them may not compensate for the complications involved in the models (Wold *et al.*, 1993).

12.13.3 Indicator variable in CoMFA

Indicator variables have long been used in classical QSAR (Kubinyi, 1990). Indicator variables have also been used in CoMFA along with the steric and electrostatic fields (Floersheim *et al.*, 1993; DePriest *et al.*, 1993; Kim, 1993e; Liu and Matheson, 1994). Although indicator variables can sometimes be used effectively, their use in CoMFA should be carefully exercised because the results can be significantly influenced by the weighting factor for the indicator variables (Kim, 1993e; DePriest *et al.*, 1993).

12.13.4 Hydrophobic effects and drug transport, distribution, and elimination

Two different requirements must be satisfied for a drug to exert its activity: transport to the target receptor site and binding to the receptor. CoMFA was developed to describe the ligand–macromolecule interactions at the receptor site. Thus, it is stated that a ligand interaction-based model such as the CoMFA method should not be used to model the effects arising from transport and distribution, but that suitably weighted $\log P$ values should be used instead (Kubinyi, 1993b). Certainly, it is easier to use $\log P$ values for the effects of drug transportation. However, the weighting of the $\log P$ column plays an important part and influences the results significantly.

The CoMFA approach has been shown to describe electronic parameters such as pK_a, and σ (Kim *et al.*, 1991; Kim and Martin, 1991a; Kim, 1993d), steric and bulk parameters such as E_S, surface area, and van der Waals volume (Kim and Martin, 1991a; Kim *et al.*, 1991; Kim, 1993d; Kim and

Martin (in preparation)) as well as lipophilic parameters such as $\log P$ and $\log k'$ (Kim, unpublished; Kim, 1993d). Therefore, it is not unreasonable to extend the use of the interaction-based model to describe the effects arising from transport or distribution. Furthermore, unlike the case with $\log P$ or HINT hydrophobic field, the use of 'hydrophobic potential' in the same scale or same force field, as in the case of using the GRID water probe, has a great advantage of avoiding the weighting process. However, the influence of molecular superposition on the results needs further investigation.

12.13.5 Limitations in CoMFA

Although CoMFA offers many advantages over the classical QSAR (Kim, 1993d) it also has several limitations and disadvantages (Folkers *et al.*, 1993). One of the most significant limitations in CoMFA is its dependence on the superposition of molecules, which is the most critical aspect in CoMFA; the classical QSAR does not have this problem. Another limitation is that CoMFA procedure includes many adjustable parameters; many of these often affect the CoMFA results significantly. Several practical problems also exist in PLS analysis (Kubinyi and Abraham, 1993; Cramer, 1993).

12.13.6 Multiple or alternate binding mode

Proteins have high specificity and rely for their action on unique interactions with ligands. In CoMFA, it is assumed that all the compounds included in the study bind to the same sites in the same mode of interaction. The same assumption also applies to any new compounds suggested from the study. It is also assumed that the target protein does not change its conformation when the ligand binds.

Although the binding conformations of many compounds that interact with the protein are found to be identical, the results from recent X-ray crystallography revealed that some compounds bind with alternative orientations in the binding site, or bind to different site points within the same binding region even among very close analogs (Meyer *et al.*, 1986; Diana *et al.*, 1992; Mattos and Ringe, 1993). Therefore, both the possible presence of multiple binding modes of ligands and of multiple binding sites on the protein and its possible consequences should be kept in mind in a CoMFA study.

12.13.7 Checklist for CoMFA publications

The following list is provided in order to help preparing a CoMFA report. Many of these are recommended in the literature (Thibaut *et al.*, 1993; Martin *et al.*, 1995). Most of these items should be described in the report if applicable.

The coordinates of the molecules should be readily available to others or deposited as supplementary material.

1. Biological data: the source and/or method of determination for biological data, their error range, and their transformation.
2. Starting geometry: the source of the starting geometries and the choice of particular stereoisomer.
3. Geometry optimization: the method used for geometry optimization and relevant conditions.
4. Conformational analysis: the method used for establishing particular bioactive conformations. The conformational energy of the chosen conformation relative to the global minimum energy conformation if available.
5. Charges: the method used in charge calculation along with the state of formal charge.
6. Superposition of molecules: the alignment method and rules.
7. Lattice position: the method used for the choice of grid box, its size, and location.
8. Interaction energy calculations: the kind and charge of the probe atom or groups, and molecular force fields used.
9. Inclusion of other descriptors: additional descriptor variables included.
10. Reduction of energy fields: the methods for the reduction of the raw data including the value of standard deviation cut-off, energy cut-off, and the electrostatic contribution for the lattice points inside a molecule but outside another molecule.
11. Data pretreatment: the scaling/weighting, centering, and/or transformation of the descriptors.
12. Variable selection: the variable selection method and detailed description.
13. Validation for model selection: the methods of cross-validation and other validation procedures used such as bootstrapping or random change of the dependent variable value. If the original data set is divided into the training set and the test set, the selection criteria of the training set.
14. Results of statistical analysis: the number of compounds included in the model, the optimum number of components (latent variables), R^2 and s values for the fitted model, the cross-validated test, and the forecasted compounds; F-statistics (optional); the observed and calculated biological activity values.
15. Contour plots: contour plots, levels, and their interpretations.
16. PLS plots: score plots and loading/weighting plots (optional).
17. Outliers: description of any outliers and possible cause.
18. Omission of compounds: description of the omitted compounds and reasons; the predicted activity values of the omitted compounds.
19. Forecasting new compounds: the forecasted activity values of new compounds and their experimentally measured activity values if available.

20. Design of new compounds: description of the method used in the design of new compounds.
21. Coordinates of molecules: coordinates of molecules, including partial atomic charges, in the superposition used in the study.

12.14 Conclusions

CoMFA is a valuable 3D–QSAR technique despite some potential difficulties with the methodology as Cramer *et al.* (1988b) have pointed out. This methodology has made significant advances in the last few years, and there is no doubt that it will continue to improve along with the advances in other related fields. As many of today's shortcomings are resolved, and various uncertainties are clarified, and the experiences of researchers grow, CoMFA will continue to play an important role in the QSAR field and in drug research.

References

Abola, E.E., Bernstein, F.C. and Koetzle, T.F. (1985) The Protein Data Bank. In *The Role of Data in Scientific Progress* (ed. P.S. Glaeser). Elsevier Science Publishers, New York.

Abraham, D.J. and Kellogg, G.E. (1993) Hydrophobic fields. In *3D QSAR in Drug Design Theory, Methods and Applications* (ed. H. Kubinyi), pp. 506–522. Escom, Leiden.

Agarwal, A. and Taylor, E.W. (1993) 3-D QSAR for intrinsic activity of 5-HT_{1A} receptor ligands by the method of comparative molecular field analysis. *Journal of Computational Chemistry*, **14**, 237–245.

Agarwal, A., Pearson, P.P., Taylor, E.W., Li, H.B., Dahlgren, T., Herslof, M., Yang, Y.H., Lambert, G., Nelson, D.L., Regan, J.W. and Martin, A.R. (1993) 3-Dimensional quantitative structure–activity relationships of 5-HT receptor binding data for tetrahydropyridinylindole derivatives—a comparison of the Hansch and CoMFA Methods. *Journal of Medicinal Chemistry*, **36**, 4006–4014.

Akamatsu, M. and Fujita, T. (1992) Quantitative analyses of hydrophobicity of di- to pentapeptides having un-ionizable side chains with substituent and structural parameters. *Journal of Pharmaceutical Sciences*, **81**, 164–174.

Akamatsu, M., Fujita, T., Ozoe, Y., Mochida, K., Nakamura, T. and Matsumura, F. (1993) Three-dimensional QSAR of insecticidal dioxatricycloalkene and its related compounds. In *Trends in QSAR and Molecular Modelling '92* (ed. C.G. Wermuth), pp. 525–526. Escom, Leiden.

Akamatsu, M., Nishimura, K., Osabe, H., Ueno, T. and Fujita, T. (1994) Quantative structure–activity studies of pyrethroids, 29. Comparative molecular-field analysis (3-dimensional) of the knockdown activity of substituted benzyl chrysanthemates and tetramethrin and related imido- and lactam-*N*-carbinyl esters. *Pesticide Biochemistry and Physiology*, **48**, 15–30.

Allen, M.S., Tan, Y.-C., Trudell, M.L., Narayanan, K., Schindler, L.R., Martin, M.J., Schultz, C., Hagen, T.J., Koehler, K.F., Codding, P.W., Skolnik, P. and Cook, J.M. (1990) Synthetic and computer-assisted analyses of the pharmacophore for the benzodiazepine receptor inverse agonist site. *Journal of Medicinal Chemistry*, **33**, 2343–2357.

Allen, M.S., LaLoggia, A.J., Corn, L.J., Martin, M.J., Constantino, G., Hagen, T.J., Koehler, K.F., Skolnik, P. and Cook, J.M. (1992) Predictive binding of β-carboline inverse agonists and antagonists via the CoMFA/GOLPE approach. *Journal of Medical Chemistry*, **35**, 4001–4010.

Altomare, C., Carrupt, P.-A., Gaillard, P., Tayar, N.E., Testa, B. and Carotti, A. (1992) Quantitative structure–metabolism relationship analyses of MAO-mediated toxication of 1-methyl-4-phenyl-1,2,3,6-tetrahydropyridine and analogues. *Chemical Research in Toxicology*, **5**, 366–375.

Andrews, P.R. (1993) Drug–receptor interactions. In *3D QSAR in Drug Design. Theory, Methods and Applications* (ed. H. Kubinyi), pp. 13–40. Escom, Leiden.

Andrews, P.R. and Tintelnot, M. (1990) Intermolecular forces and molecular binding. In *Quantitative Drug Design* (ed. C.A. Ramsden). Volume 4 of *Comprehensive Medicinal Chemistry* (eds C. Hansch, P.G. Sammes and J.B. Taylor), pp. 321–347. Pergamon Press, Oxford.

Audry, E., Dubost, J.P., Colleter, J.C. and Dallet, P. (1986) Une nouvelle approche des relations structure activité: le potentiel de lipophilie moléculaire. *European Journal of Medicinal Chemistry*, **21**, 71–72.

Austel, V. (1982) Selection of test compounds from a basic set of chemical structures. *European Journal of Medicinal Chemistry*, **17**, 339–347.

Avery, M.A., Gao, F.G., Chong, W.K.M., Mehrotra, S. and Milhous, W.K. (1993) Structure–activity relationships of the antimalarial agent artemisinin 1 synthesis and comparative molecular-field analysis of C-9 analogs of artemisinin and 10-deoxoartemisinin. *Journal of Medicinal Chemistry*, **36**, 4264–4275.

Barakat, M.T. and Dean, P.M. (1991) Molecular structure matching by simulated annealing III. The incorporation of null correspondences into the matching problem. *Journal of Computer-Aided Molecular Design*, **5**, 107–117.

Baroni, M., Clementi, S., Cruciani, G., Costantino, G., Riganelli, D. and Oberrauch, E. (1992) Predictive ability of regression-models. 2. Selection of the best predictive PLS model. *Journal of Chemometrics*, **6**, 347–356.

Baroni, M., Costantino, G., Cruciani, G., Riganelli, D., Valigi, R. and Clementi, S. (1993) Generating optimal linear PLS estimations (GOLPE): an advanced tool for handling 3D-QSAR problems. *Quantitative Structure–Activity Relationships*, **12**, 9–20.

Blaney, J.M. and Hansch, C. (1990) Application of molecular graphics to the analysis of macromolecular structures. In *Quantitative Drug Design* (ed. C.A. Ramsden). Volume 4 of *Comprehensive Medicinal Chemistry* (eds C. Hansch, P.G. Sammes and J.B. Taylor), pp. 459–496. Pergamon Press, Oxford.

Boobbyer, D.N., Goodford, P.J., McWhinnie, P.M. and Wade, R.C. (1989) New hydrogen-bond potentials for use in determining energetically favorable binding sites on moleules of known structure. *Journal of Medicinal Chemistry*, **32**, 1083–1094.

Bures, M.G., Martin, Y.C. and Willett, P. (1994) Searching techniques for databases of three-dimensional chemical structures. *Topics in Stereochemistry*, **21**, 467–511.

Burke, B.J. and Hopfinger, A.J. (1993) Advances in molecular shape analysis. In *3D QSAR in Drug Design. Theory, Methods and Applications* (ed. H. Kubinyi), pp. 276–306. Escom, Leiden.

Burt, S.K.L. and Greer, J. (1988) Search strategies for determining bioactive conformers of peptides and small molecules. *Annual Reports in Medicinal Chemistry*, **23**, 285–294.

Burt, C., Richards, W.G. and Huxley, P. (1990) The application of molecular similarity calculations. *Journal of Computational Chemistry*, **11**, 1139–1146.

Bush, B.L. and Nachbar, R.B., Jr (1993) Sample-distance partial least squares: PLS optimized for many variables, with application to CoMFA. *Journal of Computer-Aided Molecular Design*, **7**, 587–619.

Bustard, T.M. (1974) Optimization of alkyl modifications by Fibonacci search. *Journal of Medicinal Chemistry*, **17**, 777–778.

Cambridge Structural Database System User's Manual (1989) Parts 1 and 2. Cambridge Crystallographic Data Centre, Cambridge.

Carroll, F.I., Gao, Y., Rahman, M.A., Abraham, P., Parham, K., Lewin, A.H., Boja, J.W. and Kuhar, M.J. (1991) Synthesis, ligand binding, QSAR, and CoMFA study of 3β-(*p*-substituted phenyl)tropane-2β-carboxylic acid methyl esters, *Journal of Medicinal Chemistry*, **34**, 2719–2725.

Clark, M. and Cramer, R.D. III (1993) The probability of chance correlation using partial least squares (PLS). *Quantitative Structure–Activity Relationships*, **12**, 137–145.

Clark, M., Cramer, R.D. III, Jones, D.M., Patterson, D.E. and Simeroth, P.E. (1990) Comparative molecular field analysis (CoMFA). 2. Toward its use with 3D-structural databases. *Tetrahedron Computer Methodology*, **3**, 47–59.

Clementi, S. (1984) Statistics and drug design. In *Drug Design: Fact or Fantasy?* (eds G. Jolles and K.R.H. Wooldridge), pp. 73–94. Academic Press, London.

Clementi, S., Cruciani, G., Baroni, M. and Costantino, G. (1993) Series design. In *3D QSAR in Drug Design. Theory, Methods and Applications* (ed. H. Kubinyi), pp. 567–582. Escom, Leiden.

Cocchi, M. and Johansson, E. (1993) Amino acids characterization by GRID and multivariate data analysis. *Quantitative Structure–Activity Relationships*, **12**, 1–8.

Cohen, N.C., Blaney, J.M., Humblet, C., Gund, P. and Barry, D.C. (1990) Molecular modeling software and methods for medicinal chemistry. *Journal of Medicinal Chemistry*, **33**, 883–894.

Craig, P.N. (1971) Interdependence between physical parameters and selection of substituent groups for correlation studies. *Journal of Medicinal Chemistry*, **14**, 680–684.

Cramer, R.D. III (1980) BC(DEF) parameters. 1. The intrinsic dimensionality of intermolecular interactions in the liquid. *Journal of the American Chemical Society*, **102**, 1837–1849.

Cramer, R.D. III (1993) Partial least squares (PLS): its strengths and limitations. *Perspectives in Drug Discovery and Design*, **1**, 269–278.

Cramer, R.D. III and Wold, S.B. (1991) Comparative molecular field analysis (COMFA). US Patent 5,025,388.

Cramer, R.D. III, Bunce, J.D., Patterson, D.E. and Frank, I.E. (1988a) Crossvalidation, bootstrapping, and partial least squares compared with multiple regression in conventional QSAR studies. *Quantitative Structure–Activity Relationships*, **7**, 18–25.

Cramer, R.D. III, Patterson, D.E. and Bunce, J.D. (1988b) Comparative molecular field analysis (CoMFA). 1. Effect of shape on binding of steroids to carrier proteins. *Journal of the American Chemical Society*, **110**, 5959–5967.

Cramer, R.D. III, DePriest, S.A., Patterson, D.E. and Hecht, P. (1993) The developing practice of comparative molecular field analysis. In *3D QSAR in Drug Design. Theory, Methods and Applications* (ed. H. Kubinyi), pp. 443–485. Escom, Leiden.

Crippen, G.M. (1980) Quantitative structure–activity relationships by distance geometry: systematic analysis of dihydrofolate reductase inhibitors. *Journal of Medicinal Chemistry*, **23**, 599–606.

Crippen, G.M. (1987) Voronoi binding site models. *Journal of Computational Chemistry*, **8**, 943–954.

Cruciani, G., Baroni, M., Clementi, S., Costantino, G., Riganelli, D. and Skagerberg, B. (1992) Prediction ability of regression models. Part I. The SDEP parameter. *Journal of Chemometrics*, **6**, 335–346.

Cruciani, B., Clementi, S. and Baroni, M. (1993) Variable selection in PLS analysis. In *3D QSAR in Drug Design. Theory, Methods and Applications* (ed. H. Kubinyi), pp. 551–564. Escom, Leiden.

Dammkoehler, R.A., Karasek, S.F., Shands, E.F.B. and Marshall, G.R. (1989) Constrained search of conformation hyperspace. *Journal of Computer-Aided Molecular Design*, **3**, 3–21.

Darvas, F. (1974) Application of the sequential simplex method in designing drug analogs. *Journal of Medicinal Chemistry*, **17**, 799–804.

Davis, A.M., Gensmantel, N.P. and Marriott, D.P. (1993) Use of the GRID program in the 3-D QSAR analysis of a series of calcium channel agonists. In *Trends in QSAR and Molecular Modelling '92* (ed. C.G. Wermuth), pp. 156–518. Escom, Leiden.

Dean, P.M. (1993) Molecular similarity. In *3D QSAR in Drug Design* (ed. H. Kubinyi), pp. 150–172. Escom, Leiden.

Dean, P.M. and Callow, P. (1987) Molecular recognition: identification of local minima for matching in rotational 3-space by cluster analysis. *Journal of Molecular Graphics*, **5**, 159–164.

Dean, P.M. and Chau, P.-L. (1987) Molecular recognition: optimized searching through rotational 3-space for pattern matches on molecular surfaces. *Journal of Molecular Graphics*, **5**, 152–158.

Dean, P.M., Callow, P. and Chau, P.-L. (1988) Molecular recognition: blind-searching for regions of strong structural match on the surfaces of two dissimilar molecules. *Journal of Molecular Graphics*, **6**, 28–34, 38.

Debnath, A.K., Hansch, C., Kim, K.H. and Martin, Y.C. (1993) Mechanistic interpretation of the genotoxicity of nitrofurans as antibacterial agents using quantitative structure–activity relationships (QSAR) and comparative molecular field analysis (CoMFA). *Journal of Medicinal Chemistry*, **36**, 1007–1016.

DePriest, S.A., Shands, E.F.B., Dammkoehler, R.A. and Marshall, G.R. (1991) 3D-QSAR: further studies on inhibitors of angiotensin-converting enzyme. In *QSAR: Rational Approaches to the Design of Bioactive Compounds* (eds C. Silipo and A. Vittoria), pp. 405–414. Elsevier Science Publishers, Amsterdam.

DePriest, S.A., Meyer, D., Naylor, C.B. and Marshall, G.R. (1993) 3D-QSAR of angiotensin-converting enzyme and thermolysin inhibitors — a comparison of CoMFA models based on deduced and experimentally determined active-site geometries. *Journal of the American Chemical Society*, **115**, 5372–5384.

Diana, G.D., Kowalczyk, P., Treasurywala, A.M., Oglesby, R.C., Pevear, D.C. and Dutko, F.J. (1992) CoMFA analysis of the interactions of antipicornavirus compounds in the binding pocket of human rhinovirus-14. *Journal of Medicinal Chemistry*, **35**, 1002–1008.

Doweyko, A.M. (1988) The hypothetical active site lattice. An approach to modelling active sites from data on inhibitor molecules. *Journal of Medicinal Chemistry*, **31**, 1396–1406.

Efron, B. and Gong, G. (1983) A leisurely look at the bootstrap, the jackknife, and cross-validation. *The American Statistician*, **37**, 36–48.

El-Bermawy, M., Lotter, H. and Glennon, R.A. (1992) Comparative molecular field analysis of the binding of arylpiperazines at 5-HT_{1A} serotonin receptors. *Medicinal Chemistry Research*, **2**, 290–297.

Erickson, J.W. and Fesik, S.W. (1992) Macromolecular x-ray crystallography and NMR as tools for structure-based drug design. *Annual Report in Medicinal Chemistry*, **27**, 271–289.

Fersht, A.R., Shi, J., Knill-Jones, J., Lowe, D.M., Wilkinson, A.J., Blow, D.M., Brick, P., Carter, P., Waye, M.M.Y. and Winter, G. (1985) Hydrogen bonding and biological specificity analysed by protein engineering. *Nature*, **314**, 235–238.

Fesik, S.W. (1991) NMR studies of molecular complexes as a tool in drug design. *Journal of Medicinal Chemistry*, **34**, 2937–2945.

Floersheim, P., Nozulak, J. and Weber, H.P. (1993) Experience with comparative molecular field analysis. In *Trends in QSAR and Molecular Modelling '92* (ed. C.G. Wermuth), pp. 227–232. Escom, Leiden.

Folkers, G., Merz, A. and Rognan, D. (1993) CoMFA: scope and limitations. In *3D QSAR in Drug Design. Theory, Methods and Applications* (ed. H. Kubinyi), pp. 583–618. Escom, Leiden.

Franke, R. (1984) *Theoretical Drug Design Methods.* Elsevier, New York.

Furet, P., Sele, A. and Cohen, N.C. (1988) 3D molecular lipophilicity potential profiles: a new tool in molecular modeling. *Journal of Molecular Graphics*, **6**, 182–189.

Gasteiger, J. and Marsili, M. (1980) Iterative partial equalization of orbital electronegativity — a rapid access to atomic charges. *Tetrahedron*, **36**, 3219–3288.

Ghose, A.K. and Crippen, G.M. (1990) Modeling the benzodiazepine receptor binding site by the general three-dimensional structure-directed quantitative structure–activity relationship method REMOTEDISC. *Molecular Pharmacology*, **37**, 725–734.

Golender, V.E. and Vorpagel, E.R. (1993) Computer-assisted pharmacophore identification. In *3D QSAR in Drug Design. Theory, Methods and Applications* (ed. H. Kubinyi), pp. 137–149. Escom, Leiden.

Good, A.C. (1992) The calculation of molecular similarity: alternative formulas, data manipulation and graphical display. *Journal of Molecular Graphics*, **10**, 144–151.

Good, A.C. and Richards, W.G. (1993) Rapid evaluation of shape similarity using Gaussian functions. *Journal of Chemical Information and Computer Sciences*, **33**, 112–116.

Good, A.C., Hodgkin, E.E. and Richards, W.G. (1992) Utilization of Gaussian functions for the rapid evaluation of molecular similarity. *Journal of Chemical Information and Computer Sciences*, **32**, 188–191.

Good, A.C., So, S.S. and Richards, W.G. (1993a) Structure–activity relationships from molecular similarity-matrices. *Journal of Medicinal Chemistry*, **36**, 433–438.

Good, A.C., Peterson, S.J. and Richards, W.G. (1993b) QSARS from similarity matrices — technique validation and application in the comparison of different similarity evaluation methods. *Journal of Medicinal Chemistry*, **36**, 2929–2937.

Greco, G., Novellino, E., Silipo, C. and Vittoria, A. (1991) Comparative molecular field analysis on a set of muscarinic agents. *Quantitative Structure–Activity Relationships*, **10**, 289–299.

Greco, G., Novellino, E., Silipo, C. and Vittoria, A. (1992) Study of benzodiazepines receptor sites using a combined QSAR–CoMFA approach. *Quantitative Structure–Activity Relationships*, **11**, 461–477.

Hansch, C. (1981) The physicochemical approach to drug design and discovery (QSAR). *Drug Development and Research*, **1**, 267–309.

Hansch, C. and Fujita, T. (1964) Rho sigma pi analysis. A method for the correlation of biological activity and chemical structure. *Journal of the American Chemical Society*, **86**, 1616–1626.

Hansch, C., Unger, S.H. and Forsythe, A.B. (1973) Strategy in drug design. Cluster analysis as an aid in the selection of substituents. *Journal of Medicinal Chemistry*, **16**, 1217–1222.

Hopfinger, A.J. (1977) *Intermolecular Interactions and Biomolecular Organization*, pp. 320–324. John Wiley, New York.

Hopfinger, A.J. (1980) A QSAR investigation of dihydrofolate reductase inhibition by Baker triazines based upon molecular shape analysis. *Journal of the American Chemical Society*, **102**, 7196–7206.

Hopfinger, A.J. (1985) Computer-assisted drug design. *Journal of Medicinal Chemistry*, **28**, 1133–1139.

Horwitz, J.P., Massova, I., Wiese, T.E., Besler, B.H. and Corbett, T.H. (1994) Comparative molecular field analysis of the antitumor activity of 9H-Thioxanthen-9-one derivatives against pancreatic ductal carcinoma 03. *Journal of Medicinal Chemistry*, **37**, 781–786.

Hoskuldsson, A. (1988) PLS regression methods. *Journal of Chemometrics*, **2**, 211–228.

Howard, A.E. and Kollman, P.A. (1988) An analysis of current methodologies for conformational searching of complex molecules. *Journal of Medicinal Chemistry*, **31**, 1669–1675.

Itai, A., Tomioka, N., Yamada, M., Inoue, A. and Kato, Y. (1993) Molecular superposition for rational drug design. In *3D QSAR in Drug Design. Theory, Methods and Applications* (ed. H. Kubinyi), pp. 200–225. Escom, Leiden.

Kato, I., Itai, A. and Iitaka, Y. (1987) A novel method for superimposing molecules and mapping. *Tetrahedron*, **43**, 5229–5234.

Kato, Y., Inoue, A., Yamada, M., Tomioka, N. and Itai, A. (1992) Automatic superposition of drug molecules based on their common receptor site. *Journal of Computer-Aided Molecular Design*, **6**, 475–486.

Kearsley, S.K. and Smith, G.M. (1990) An alternative method for the alignment of molecular structures: maximizing electrostatic and steric overlap. *Tetrahedron Computer Methodology*, **3**, 615–633.

Kellogg, G.E., Semus, S.F. and Abraham, D.J. (1991) HINT: a new method of empirical hydrophobic field calculations for CoMFA. *Journal of Computer-Aided Molecular Design*, **5**, 545–552.

Klebe, G. (1993) Structural alignment of molecules. In *3D QSAR in Drug Design. Theory, Methods and Applications* (ed. H. Kubinyi), pp. 173–199. Escom, Leiden.

Kim, K.H. (1991) A novel method of describing hydrophobic effects directly from 3D structures in 3D-quantitative structure–activity relationships study. *Medicinal Chemistry Research*, **1**, 259–264.

Kim, K.H. (1992a) 3D-quantitative structure–activity relationships: nonlinear dependence described directly from 3D structures using comparative molecular field analysis (CoMFA). *Quantitative Structure–Activity Relationships*, **11**, 309–317.

Kim, K.H. (1992b) 3D-quantitative structure–activity relationships: investigation of steric effects with descriptors directly from 3D structures using a comparative molecular field analysis (CoMFA) approach. *Quantitative Structure–Activity Relationships*, **11**, 453–460.

Kim, K.H. (1992c) 3D-quantitative structure–activity relationships: description of electronic effects directly from 3D structures using a GRID-comparative molecular field analysis (CoMFA) approach. *Quantitative Structure–Activity Relationships*, **11**, 127–134.

Kim, K.H. (1992d) Description of nonlinear dependence directly from 3D structures in 3D-quantitative structure–activity relationships. *Medicinal Chemistry Research*, **2**, 22–27.

Kim, K.H. (1993a) 3D-quantitative structure–activity relationships — describing hydrophobic interactions directly from 3D structures using a comparative molecular-field analysis (CoMFA) approach. *Quantitative Structure–Activity Relationships*, **12**, 232–238.

Kim, K.H. (1993b) Use of the hydrogen-bond potential function in comparative molecular field analysis (CoMFA): an extension of CoMFA. In *3D QSAR in Drug Design. Theory, Methods and Applications* (ed. H. Kubinyi), pp. 245–251. Escom, Leiden.

Kim, K.H. (1993c) Nonlinear dependence in comparative molecular field analysis (CoMFA). *Journal of Computer-Aided Molecular Design*, **7**, 71–82.

Kim, K.H. (1993d) Comparison of classical and 3D QSAR. In *3D QSAR in Drug Design. Theory, Methods and Applications* (ed. H. Kubinyi), pp. 619–642. Escom, Leiden.

Kim, K.H. (1993e) Use of indicator variable in comparative molecular field analysis. *Medicinal Chemistry Research*, **3**, 257–267.

Kim, K.H. (1994) Separation of electronic, hydrophobic, and steric effects in 3D-quantitative structure–activity relationships with descriptors directly from 3D structures using a comparative molecular field analysis (CoMFA) approach. *Research Trends*. In press.

Kim, K.H. and Martin, Y.C. (1991a) Direct prediction of dissociation constants (pK_a's) of clonidine-like imidazolines, 2-substituted imidazoles, and 1-methyl-2-substituted-imidazoles from 3D structures using a comparative molecular field analysis (CoMFA) approach. *Journal of Medicinal Chemistry*, **34**, 2056–2060.

Kim, K.H. and Martin, Y.C. (1991b) Direct prediction of linear free energy substituent effects from 3D structures using comparative molecular field analysis. 1. Electronic effects of substituted benzoic acids. *Journal of Organic Chemistry*, **56**, 2723–2729.

Kim, K.H. and Martin, Y.C. (1991c) Evaluation of electrostatic and steric descriptors for 3D-QSAR: the H^+ and CH_3 probes using comparative molecular field analysis (CoMFA) and the modified partial least squares method. In *QSAR: Rational Approaches to the Design of Bioactive Compounds* (eds C. Silipo and A. Vittoria), pp. 151–154. Elsevier Science Publishers, Amsterdam.

Kim, K.H. and Martin, Y.C. (1995). In preparation.

Kim, K.H., Greco, G., Novellino, E., Silipo, C. and Vittoria, A. (1992) Use of the hydrogen bond potential function in a comparative molecular field analysis (CoMFA) on a set of benzodiazepines. *Journal of Computer-Aided Molecular Design*, **7**, 263–280.

Klebe, G. and Abraham, U. (1993) On the prediction of binding properties of drug molecules by comparative molecular field analysis. *Journal of Medicinal Chemistry*, **36**, 70–80.

Kubinyi, H. (1990) The Free-Wilson method and its relationship to the extrathermodynamic approach. In *Quantitative Drug Design* (ed. C.A. Ramsden). Volume 4 of *Comprehensive Medicinal Chemistry* (eds C. Hansch, P.G. Sammes and J.B. Taylor), pp. 589–643. Pergamon Press, Oxford.

Kubinyi, H. (ed.) (1993a) *3D QSAR in Drug Design. Theory, Methods and Applications.* Escom, Leiden.

Kubinyi, H. (1993b) *QSAR: Hansch Analysis and Related Approaches*, pp. 109–114. VCH, Weinheim.

Kubinyi, H. and Abraham, U. (1993) Practical problems in PLS analyses. In *3D QSAR in Drug Design. Theory, Methods and Applications* (ed. H. Kubinyi), pp. 717–728. Esco, Leiden.

Langraft, B., Cohen, F.E., Smith, K.A., Gadski, R. and Ciardelli, T.L. (1989) Structural significance of the *C*-terminal amphiphilic helix of interleukin-2. *Journal of Biological Chemistry*, **264**, 816–822.

Lawrence, M.C. and Davis, P.C. (1992) CLIX: a search algorithm for finding novel ligands capable of binding proteins of known three-dimensional structure. *Proteins: Structure Function and Genetics*, **12**, 31–41.

Leach, A.R. (1991) A survey of methods for searching the conformational space of small and medium sized molecules. In *Reviews of Computational Chemistry*, Vol. II (eds K.B. Lipkowitz and D.B. Boyd), pp. 1–55. VCH Publishers, New York.

Lin, C.T., Pavlik, P.A. and Martin, Y.C. (1990) Use of molecular fields to compare series of potentially bioactive molecules designed by scientists or by computer. *Tetrahedron Computer Methodology*, **3(6C)**, 723–738.

Liu, R. and Matheson, L.E. (1994) Comparative molecular field analysis combined with physicochemical parameters for prediction of polydimethylsiloxane membrane flux in Isopropanol. *Pharmaceutical Research*, **11**, 257–266.

Lorber, A., Wangen, L.E. and Kowalski, B.R. (1987) A theoretical foundation for the PLS algorithm. *Journal of Chemometrics*, **1**, 19–31.

Loughney, D.A. and Schwender, C.F. (1992) A comparison of progestin and androgen receptor binding using the CoMFA technique. *Journal of Computer-Aided Molecular Design*, **6**, 569–581.

Mager, P.P. (1991) Non least-squares jackknife regression in drug design. *Drug Design Delivery*, **7**, 119–129.

Marshall, G.R. (1987) Computer-aided drug design. *Annual Reviews in Pharmacology and Toxicology*, **27**, 193–213.

Marshall, G.R. (1993) Binding-site modeling of unknown receptors. In *3D QSAR in Drug Design. Theory, Methods and Applications* (ed. H. Kubinyi), pp. 80–116. Escom, Leiden.

Marshall, G.R. and Naylor, C.B. (1990) Use of molecular graphics for structural analysis of small molecules. In *Quantitative Drug Design* (ed. C.A. Ramsden). Volume 4 of *Comprehensive Medicinal Chemistry* (eds C. Hansch, P.G. Sammes and J.B. Taylor), pp. 431–458. Pergamon Press, Oxford.

Marshall, G.R., Barry, C.D., Bosshard, H.E., Dammkoehler, R.A. and Dunn, D.A. (1979) The conformation parameter in drug design: the active analog approach. In *Computer-Assisted Drug Design* (eds E.C. Olson and R.E. Christoffersen), pp. 205–226. American Chemical Society, Washington, DC.

Marshall, G.R., Mayer, D., Naylor, C.B., Hodgkin, E.E. and Cramer, R.D. III (1989) Mechanism-based analysis of enzyme inhibitors of amide bond hydrolysis. In *QSAR: Quantitative*

Structure–Activity Relationships in Drug Design (ed. J.L. Fauchere), pp. 287–295. Alan R. Liss, New York.

Martin, Y.C. (1978) *Quantitative Drug Design*, p. 425. Dekker, New York.

Martin, Y.C. (1981) A practitioner's perspective on the role of quantitative structure–activity analysis in medicinal chemistry. *Journal of Medicinal Chemistry*, **24**, 229–237.

Martin, Y.C. (1991) Computer-assisted rational drug design. *Methods in Enzymology*, **203**, 587–613.

Martin, Y.C. (1992) 3D database searching in drug design. *Journal of Medicinal Chemistry*, **35**, 2145–2154.

Martin, Y.C., Bures, M.G., Danaher, E.A., DeLazzer, J., Lico, I. and Pavlik, P.A. (1993a) A fast new approach to pharmacophore mapping and its application to dopaminergic and benzodiazepine agonists. *Journal of Computer-Aided Molecular Design*, **7**, 83–102.

Martin, Y.C., Bures, M.G., Danaher, E.A. and DeLazzer, J. (1993b) New strategies that improve the efficiency of the 3D design of bioactive molecules. In *Trends in QSAR and Molecular Modelling '92* (ed. C.G. Wermuth), pp. 20–26. Escom, Leiden.

Martin, Y.C., Lin, C.T. and Wu, J. (1993c) Application of CoMFA to the design and structural optimization of D_1 dopaminergic agonists. In *3D QSAR in Drug Design. Theory, Methods and Applications* (ed. H. Kubinyi), pp. 643–660. Escom, Leiden.

Martin, Y.C., Kim, K.H. and Lin, C.T. (1995) *Comparative Molecular Field Analysis: CoMFA* (ed. M. Charton). In press.

Mattos, C. and Ringe, D. (1993) Multiple binding modes. In *3D QSAR in Drug Design. Theory, Methods and Applications* (ed. H. Kubinyi), pp. 226–254. Escom, Leiden.

McFarland, J.W. (1992) Comparative molecular field analysis of anticoccidial triazines. *Journal of Medicinal Chemistry*, **35**, 2543–2550.

Meyer, Jr, E.F., Radhakrishnan, R., Cole, G.M. and Presta, L.G. (1986) Structure of the product complex of acetyl-Ala-Pro-Ala with porcine pancreatic elastase at 1.65 Å resolution. *Journal of Molecular Biology*, **189**, 533–539.

Miyashita, Y., Ohsako, H., Takayama, C. and Sasaki, S.-I. (1992) Multivariate structure–activity relationships analysis of fungicidal and herbicidal thiolcarbamates using partial least squares method. *Quantitative Structure–Activity Relationships*, **11**, 17–22.

Naerum, L. and Jensen, J.S. (1991) 3D-QSAR of a number of new non-NMDA receptor antagonists. In *QSAR: Rational Approaches to the Design of Bioactive Compounds* (eds C. Silipo and A. Vittoria), pp. 489–492. Elsevier Science Publishers, Amsterdam.

Nicklaus, M.C., Milne, G.W.A. and Burke, Jr, T.R. (1992) QSAR of conformationally flexible molecules: comparative molecular field analysis of protein–tyrosine kinase inhibitors. *Journal of Computer-Aided Molecular Design*, **6**, 487–504.

Norinder, U. (1990) Experimental design based 3-D QSAR analysis of steroid–protein interactions: applications to human CBG complexes. *Journal of Computer-Aided Molecular Design*, **4**, 381–389.

Norinder, U. (1991a) 3D QSAR analysis of steroid–protein interactions — the use of difference maps. *Journal of Computer-Aided Molecular Design*, **5**, 419–426.

Norinder, U. (1991b) Theoretical amino acid descriptors. Application to bradykinin potentiating peptides. *Peptides*, **12**, 1223–1227.

Osabe, H., Morishima, Y., Goto, Y. and Fujita, T. (1992) Quantitative structure–activity relationships of light-dependent herbicidal 4-Pyridone-3-Carboxanilides III. 3-D (comparative molecular field) analysis including light-dependent diphenyl ether herbicides. *Pesticide Science*, **35**, 187–200.

Papadopoulos, M.C. and Dean, P.M. (1991) Molecular structure matching by simulated annealing. 4. Classification of atom correspondences in sets of dissimilar molecules. *Journal of Computer-Aided Molecular Design*, **5**, 119–133.

Pastor, M. and Alvarez-Builla, J. (1991) The EDISFAR programs. Rational drug series design. *Quantitative Structure–Activity Relationships*, **10**, 350–358.

Pearlman, R.S. (1993) 3D molecular structures: generation and use in 3D searching. In *3D QSAR in Drug Design. Theory, Methods and Applications* (ed. H. Kubinyi), pp. 41–79. Escom, Leiden.

Pellicciari, R., Natalini, B., Costantino, G., Garzon, A., Luneia, R., Mahmoud, M.R., Marinozzi, M., Roberti, M., Rosato, G.C. and Shiba, S.A. (1993) Heterocyclic modulators of the NMDA receptor. *Il Farmaco*, **48**, 151–157.

Perkins, T.D.J. and Dean, P.M. (1993) An exploration of a novel strategy for superposing several flexible molecules. *Journal of Computer-Aided Molecular Design*, **7**, 155–172.

Phatak, A., Reilly, P.M. and Penlidis, A. (1993) An approach to interval estimation in partial least squares regression. *Analytica Chemica Acta*, **277**, 495–501.

Pleiss, M.A. and Unger, S.H. (1990) The design of test series and the significance of QSAR relationship. In *Quantitative Drug Design* (ed. C.A. Ramsden). Volume 4 of *Comprehensive Medicinal Chemistry* (eds C. Hansch, P.G. Sammes and J.B. Taylor), pp. 561–587. Pergamon Press, Oxford.

Plummer, E.L. (1990) The application of quantitative design strategies in pesticide discovery. In *Reviews in Computational Chemistry*, Vol. I (eds K.B. Lipkowitz and D.B. Boyd), pp. 119–168. VCH Publishers, New York.

Sadowski, J. and Gasteiger, J. (1993) From atoms and bonds to three-dimensional atomic coordinates: automatic model builders. *Chemical Reviews*, **93**, 2567–2581.

Santora, N.J. and Auyang, K. (1975) Non-computer approach to structure–activity study. An expanded Fibonacci search applied to structurally diverse types of compounds. *Journal of Medicinal Chemistry*, **18**, 959–963.

Saunderson, R.T. (1983) Electronegativity and bond energy. *Journal of the American Chemical Society*, **105**, 2259–2261.

Seibel, G.L. and Kollman, P.A. (1990) Molecular mechanics and the modeling of drug structures. In *Quantitative Drug Design* (ed. C.A. Ramsden). Volume 4 of *Comprehensive Medicinal Chemistry* (eds C. Hansch, P.G. Sammes and J.B. Taylor), pp. 125–138. Pergamon Press, Oxford.

Sheridan, R.P., Nilakantan, R., Dixon, J.S. and Venkataraghavan, R. (1986) The ensemble approach to distance geometry: application to the nicotinic pharmacophore. *Journal of Medicinal Chemistry*, **29**, 899–906.

Simon, Z. (1993) MTD and hyperstructure approaches. In *3D QSAR in Drug Design. Theory, Methods and Applications* (ed. H. Kubinyi), pp. 307–319. Escom, Leiden.

Simon, Z., Badileuscu, I. and Racovitan, T. (1977) Mapping of dihydrofolate-reductase receptor site by correlation with minimal topological (steric) differences. *Journal of Theoretical Biology*, 485–495.

Simon, Z., Dragomir, N., Plauchithiu, M.G., Holban, S., Glatt, H. and Kerek, F. (1980) Receptor site mapping for cardiotoxic aglycones by the minimal steric difference method. *European Journal of Medicinal Chemistry*, **15**, 521–527.

Smith, H.J. (1988) *Introduction to the Principles of Drug Design*, pp. 240–264. Wright, London.

Srivastava, S., Richardson, W.W., Bradley, M.P. and Crippen, G.M. (1993) Three-dimensional receptor modelling using distance geometry and Voronoi polyhedra. In *3D QSAR in Drug Design. Theory, Methods and Applications* (ed. H. Kubinyi), pp. 409–430. Escom, Leiden.

Stone, M. (1974) Cross-validatory choice and assessment of statistical predictions. *Journal of the Royal Statistical Society, B*, **36**, 111–133.

Street, I.P., Armstrong, C.R. and Withers, S.G. (1986) Hydrogen bonding and specificity. Fluorodeoxy sugars as probes of hydrogen bonding in the glycogen phosphorylase–glucose complex. *Biochemistry*, **25**, 6021–6027.

Thibaut, U., Folkers, G., Klebe, G., Kubinyi, H., Merz, A. and Rognan, D. (1993) Recommendations for CoMFA studies and 3D QSAR publications. In *3D QSAR in Drug Design. Theory, Methods and Applications* (ed. H. Kubinyi), pp. 711–716. Escom, Leiden.

Thomas, B.F., Compton, D.R., Martin, B.R. and Semus, S.F. (1991) Modeling the cannabinoid receptor — a 3-dimensional quantitative structure–activity analysis. *Molecular Pharmacology*, **40**, 656–665.

Topliss, J.G. (1993) Some observations on classical QSAR. *Perspectives in Drug Discovery and Design*, **1**, 253–268.

Topliss, J.G. (1974a) Utilization of operational schemes for analog synthesis in drug design. *Journal of Medicinal Chemistry*, **17**, 799–804.

Topliss, J.G. (1974b) A manual method for applying the Hansch approach to drug design. *Journal of Medicinal Chemistry*, **20**, 463–469.

van Geerestein, V.J., Perry, N.C., Grootenhuis, P.D.J. and Haasnoot, C.A.G. (1990) 3D shape fitting and 3D DB searching by SPERM. *Tetrahedron Computer Methodology*, **3**, 595–613.

Wade, R.C. (1993) Molecular interaction fields. In *3D QSAR in Drug Design. Theory, Methods and Applications* (ed. H. Kubinyi), pp. 486–506. Escom, Leiden.

Wade, R.C. and Goodford, P.J. (1993) Further development of hydrogen bond functions for use in determining energetically favorable binding sites on molecules of known structure. 2.

Ligand probe groups with the ability to form more than two hydrogen bonds. *Journal of Medicinal Chemistry*, **36**, 148–156.

Wade, R.C., Clark, K.J. and Goodford, P.J. (1993) Further development of hydrogen bond functions for use in determining energetically favorable binding sites on molecules of known structure. 1. Ligand probe groups with the ability to form two hydrogen bonds. *Journal of Medicinal Chemistry*, **36**, 140–147.

Wakeling, I.N. and Morris, J.J. (1993) A test of significance for partial least squares regression. *Journal of Chemometrics*, **7**, 291–304.

Waller, C.L. and McKinney, J.D. (1992) Comparative molecular field analysis of polyhalogenated dibenzo-*p*-dioxins, dibenzofurans, and biphenyls. *Journal of Medicinal Chemistry*, **35**, 3660–3666.

Waller, C.L. and Marshall, G.R. (1993) 3-Dimensional quantitative structure–activity relationship of angiotensin-converting enzyme and thermolysin inhibitors. 2. A comparison of CoMFA models incorporating molecular-orbital fields and desolvation free-energies based on active-analog and complementary-receptor field alignment rules. *Journal of Medicinal Chemistry*, **36**, 2390–2403.

Waller, C.L., Oprea, T.I., Giolitti, A. and Marshall, G.R. (1993) 3-Dimensional QSAR of human-immunodeficiency-virus-1 protease inhibitors. 1.1 Å CoMFA study employing experimentally determined alignment rules. *Journal of Medicinal Chemistry*, **36**, 4152–4160.

Wermuth, C.-G. and Langer, T. (1993) Pharmacophore identification. In *3D QSAR in Drug Design. Theory, Methods and Applications* (ed. H. Kubinyi), pp. 117–136. Escom, Leiden.

Wiese, M. (1993) The hypothetical active-site lattice. In *3D QSAR in Drug Design. Theory, Methods and Applications* (ed. H. Kubinyi), pp. 431–442. Escom, Leiden.

Wilde, D.J. (1964) *Optimum Seeking Methods.* Prentice-Hall, Englewood Cliffs, New Jersey.

Wise, M., Cramer, R.D., Smith, D. and Exman, I. (1983) Progress in three-dimensional drug design: the use of realtime colour graphics and computer postulation of bioactive molecules in DYLOMMS. In *Quantitative Approaches to Drug Design* (*Proceedings of the 4th European Symposium on Chemical Structure–Biological Activity: Quantitative Approaches*) (ed. J.C. Dearden), pp. 145–146. Elsevier, Amsterdam.

Wold, S., Ruhe, A., Wold, H. and Dunn, W.J., III (1984a) The collinearity problem in linear regression. The partial least squares (PLS) approach to generalized inverses. *SIAM Journal of Scientific and Statistical Computing*, **5**, 735–743.

Wold, S., Dunn, W.J., III and Hellberg, S. (1984b) Pattern recognition as a tool for drug design. In *Drug Design: Fact or Fantasy?* (eds G. Jolles and K.R.H. Wooldridge), pp. 95–117. Academic Press, London.

Wold, S., Johansson, E. and Cocchi, M. (1993) PLS—partial least-squares projections to latent structures. In *3D QSAR in Drug Design. Theory, Methods and Applications* (ed. H. Kubinyi), pp. 523–550. Escom, Leiden.

Wong, G., Kroehler, K.F., Skolnick, P., Gu, Z.Q., Ananthan, S., Schonholzer, P., Hunkeler, W., Zhang, W.J. and Cook, J.M. (1993) Synthesis and computer-assisted analysis of the structural requirements for selective, high-affinity ligand-binding to diazepam-insensitive benzodiazepine receptors. *Journal of Medicinal Chemistry*, **36**, 1820–1830.

Wooldridge, K.R.H. (1984) The virtues of present strategies for drug discovery. In *Drug Design: Fact or Fantasy?* (eds G. Jolles and K.R.H. Wooldridge), pp. 209–216. Academic Press, London.

Wootton, R., Cranfield, R., Sheppey, G.C. and Goodford, P.J. (1975) Physicochemical–activity relationships in practice. 2. Rational selection of benzenoid substituents. *Journal of Medicinal Chemistry*, **18**, 607–613.

Index